Holographic Staring Radar

Related titles on radar:

Advances in Bistatic Radar Willis and Griffiths
Airborne Early Warning System Concepts, 3rd Edition Long
Bistatic Radar, 2nd Edition Willis
Design of Multi-Frequency CW Radars Jankiraman
Digital Techniques for Wideband Receivers, 2nd Edition Tsui
Electronic Warfare Pocket Guide Adamy
Foliage Penetration Radar: Detection and characterisation of objects under trees Davis
Fundamentals of Ground Radar for ATC Engineers and Technicians Bouwman
Fundamentals of Systems Engineering and Defense Systems Applications Jeffrey
Introduction to Electronic Warfare Modeling and Simulation Adamy
Introduction to Electronic Defense Systems Neri
Introduction to Sensors for Ranging and Imaging Brooker
Microwave Passive Direction Finding Lipsky
Microwave Receivers with Electronic Warfare Applications Tsui
Phased-Array Radar Design: Application of radar fundamentals Jeffrey
Pocket Radar Guide: Key facts, equations, and data Curry
Principles of Modern Radar, Volume 1: Basic principles Richards, Scheer and Holm
Principles of Modern Radar, Volume 2: Advanced techniques Melvin and Scheer
Principles of Modern Radar, Volume 3: Applications Scheer and Melvin
Principles of Waveform Diversity and Design Wicks *et al.*
Pulse Doppler Radar Alabaster
Radar Cross Section Measurements Knott
Radar Cross Section, 2nd Edition Knott *et al.*
Radar Design Principles: Signal processing and the environment, 2nd Edition Nathanson *et al.*
Radar Detection DiFranco and Rubin
Radar Essentials: A concise handbook for radar design and performance Curry
Radar Foundations for Imaging and Advanced Concepts Sullivan
Radar Principles for the Non-Specialist, 3rd Edition Toomay and Hannan
Test and Evaluation of Aircraft Avionics and Weapons Systems McShea
Understanding Radar Systems Kingsley and Quegan
Understanding Synthetic Aperture Radar Images Oliver and Quegan
Radar and Electronic Warfare Principles for the Non-specialist, 4th Edition Hannen
Inverse Synthetic Aperture Radar Imaging: Principles, algorithms and applications Chen and Martorella
Stimson's Introduction to Airborne Radar, 3rd Edition Baker, Griffiths and Adamy
Test and Evaluation of Avionics and Weapon Sytems, 2nd Edition McShea
Angle-of-Arrival Estimation Using Radar Interferometry: Methods and applications Holder
Biologically-Inspired Radar and Sonar: Lessons from Nature Balleri, Griffiths and Baker
The Impact of Cognition on Radar Technology Farina, De Maio and Haykin
Novel Radar Techniques and Applications, Volume 1: Real Aperture Array Radar, Imaging Radar, and Passive and Multistatic Radar Klemm, Nickel, Gierull, Lombardo, Griffiths and Koch
Novel Radar Techniques and Applications, Volume 2: Waveform Diversity and Cognitive Radar, and Target Tracking and Data Fusion Klemm, Nickel, Gierull, Lombardo, Griffiths and Koch
Radar and Communication Spectrum Sharing Blunt and Perrins
Systems Engineering for Ethical Autonomous Systems Gillespie
Shadowing Function from Randomly Rough Surfaces: Derivation and applications Bourlier and Li
Photo for Radar Networks and Electronic Warfare Systems Bogoni, Laghezza and Ghelfi
Multidimensional Radar Imaging Martorella
Radar Waveform Design Based on Optimization Theory Cui, De Maio, Farina and Li
Micro-Doppler Radar and its Applications Fioranelli, Griffiths, Ritchie and Balleri
Maritime Surveillance with Synthetic Aperture Radar Di Martino and Antonio Iodice
Electronic Scanned Array Design Williams
Advanced Sparsity-Driven Models and Methods for Radar Applications Gang Li
Deep Neural Network Design for Radar Applications Gurbuz

Holographic Staring Radar

Gordon Oswald and Chris Baker

The Institution of Engineering and Technology

Published by SciTech Publishing, an imprint of The Institution of Engineering and Technology, London, United Kingdom

The Institution of Engineering and Technology is registered as a Charity in England & Wales (no. 211014) and Scotland (no. SC038698).

First published 2021

The Institution of Engineering and Technology
Michael Faraday House
Six Hills Way, Stevenage
Herts, SG1 2AY, United Kingdom

www.theiet.org

British Library Cataloguing in Publication Data
A catalogue record for this product is available from the British Library

ISBN 978-1-78561-389-0 (hardback)
ISBN 978-1-78561-390-6 (PDF)

Typeset in India by Exeter Premedia Services Private Limited
Printed in the UK by CPI Group (UK) Ltd, Croydon

Contents

About the authors

Gordon Oswald was the chief science officer of Aveillant Ltd, UK. He studied physics at Oxford, then radioglaciology at Cambridge, applying radar in Arctic and Antarctic geophysics. He joined Cambridge Consultants (Arctic remote sensing; European and US anti-air weapons evaluation; automotive collision avoidance); founded Aveillant (aviation and wind turbine reconciliation; UAV threat classification). He has served as a research professor at the University of Maine, chairman of the British Association of Remote Sensing Companies, and organising secretary of the Cambridge Society for the Application of Research.

Chris Baker holds the chair in Intelligent sensors systems at the University of Birmingham and prior to this was the chief technology officer of Aveillant Ltd., UK. He has also held positions at The Ohio State University, Australian National University (ANU), University College London and UK government research laboratories. He is the author of over three hundred publications, including "Stimsons introduction to airborne radar". He is a fellow of the IET and the IEEE and holds a number of visiting positions at leading universities.

Acknowledgements:

By **Gordon Oswald:**

First, my wife, Catherine, who has endured much absence, many ambitions, many projects and many places; she has loved and stood with me through a mix of triumphs and inevitable surprises, and has let me believe it's all worth while.

Stan Evans, who addicted me to radar and showed me how far you can get with an intuitive grasp plus respect for the numbers (even beneath an ice sheet).

Ian Aitchison, my physics tutor, who made me aware that 'all these equations really only work for two bodies at a time'.

By **Chris Baker:**

I'd like to acknowledge the love and support of my wife Janet, who has journeyed far and wide with me throughout my career in radar and has always attended to the multitude of things that need attention, which can be lacking from an overworked author like this one.

Foreword: Directions of travel for radar air surveillance

The development of radar has been one of the most successful direct applications of physics ever attempted, and then implemented and applied at large scale.

Radar continues to provide the key surveillance component for both air defence and air traffic control. As we drive past them, rotating radar antennas at airfields and the Pyramid at Fylingdales are familiar sights as they search their airspace.

Certain watchwords of radar engineering have underpinned many of the developments of the past 80 years, and remain potential avenues for improvement. For example; 'Narrow beams are good', 'Fast detection is good', 'Agility is good', and 'Clutter is bad'.

All of these statements of merit are true. Radar has become excellent at looking into airspace and delivering information in answer to repeated interrogation 'Is there anything there… or there… or there?'. Modern radar is getting better at asking the same question, but with the extensions that 'anything' can be 'anything like this… or like this… ', and 'there' can be anywhere, not just the next azimuth value. The next question is 'Is that the whole picture?'.

To find the answer we look to the basis of radar in theory, in combination with today's available technology. The underlying principles for all of these developments have been and are the laws of physics, and they provide full support for current radar designs. The laws also show that each of these statements is really a design choice for the time, under prevailing technical limits, rather than their necessary, sufficient and permanent consequence. As direct consequences of the laws, they would necessarily and properly limit any development in other directions, but as design choices they leave us the freedom to ask more fundamental questions. The purpose of this book is to return to the physical laws to ask those questions again.

The most fundamental question for designers is 'What is needed foreseeably from air surveillance?'. As airspace becomes more congested, its occupants more varied, and the human need to discriminate between them more critical, a good answer is 'Current and continuing intelligence on significant occupants of airspace'. Not only must presence and positions of radar targets be reported, but also a measure of their significance: 'Are they Radar Targets of Interest'?

Once 'significance' enters the list of requirements, the basis for design choices must change, and each statement of merit needs to be re-examined, now on the basis that the fundamental good demanded from radar is not just the presence of, but relevant information about, all targets.

What is the 'information', and where does it come from? Earlier forms of radar yield 'information' consisting of their pointing direction, a time delay and the fact of detection, and all of these can be seen as artefacts of the radar. On this basis, the involvement of the target is as a passive scatterer of the incident EM field; the radar appears to initiate and select the information.

While this perception underlies radar development, it is easy to understand why the focus is on agile detection of targets.

In truth the radar originates the incident field and its look directions, but not the target information. That information is created by the target, irreversibly scattering unique and precise amplitude and phase sequences from an almost uniform incident field. It is encoded in the sequence and spatial distribution of those scattered signals.

These forms of radar then decode information for that target only when the beam overlaps with the target, when received phases align across the receiving aperture, and when the pulse arrival time coincides with a particular threshold operation, and the flow of information ceases when the beam dwell time ends.

Narrow beams and short dwell times still leave open the question of whether the information obtained by radar in this way is the whole available picture. We start with the target; a unique, pervasive and continuing sequence of encoded scatters must contain more precise, relevant information about the target and its motion, only provided that the target is irradiated and the scatter can be decoded. The relevant laws do not limit solutions of Maxwell's equations to calculating the presence and position of an object. Its continuing motion, its shape, its orientation and even its materials create the signals; specifically, the sequence of complex field amplitudes scattered and received in any direction in the course of observation.

The Electromagnetic Uniqueness Theorem states that the whole, coherent picture is available – hence the term 'Holographic'. Whether the scatters can be decoded depends first on whether the sensitivity budget of the radar is adequate, but also on whether the encode-decode sequence continues through time and is received by the radar. If it is, then the target and its behaviour can be measured. If not, because narrow transmission beams and fastest detection are pursued in the design, then further relevant measurements will be difficult.

The initial statements of merit are true, but under what conditions are they applicable?

1a. Narrow beams are good since they imply accuracy in discriminating between signals received from different directions.
1b. Narrow transmit beams are undesirable to the extent that each target is then illuminated only for short periods at extended intervals, interrupting the information flow (as with short FFTs in measuring Doppler shifts).
2a. Fast detection is good in that it can occur before the beam moves away and there is a short delay before reports can occur.
2b. Fast detection is undesirable in cutting short observations that can yield more accurate information, such as in high-resolution Fourier analysis of Doppler shifts, and speed at this scale is unnecessary when the beam does not move.
3a. Agility appears good if finite radar resources must be shared between different radar functions.

3b. Agile observation is undesirable when observation under one function is curtailed to meet the requirements of another. Resource management implies prioritisation between information, functions and users, while staring allows the search to occur in computer memory, not in sequential access to space.

4a. Clutter is bad because it can yield unwanted paints or reports and confuse the surveillance picture.

4b. Clutter, static or moving, is a feature of airspace, and information is the best way to avoid reporting it.

This book aims to explore whether these divergences between the initial 'statements of merit' and aspects of the staring mode of operation are in fact consequences of the governing laws, in which case modelling under those laws will reveal disadvantages in performance, or whether, as the alternative, continual observation of radar targets makes available a larger information set that naturally supports performance in discrimination and classification.

The first part of the book introduces various forms of staring radar, which include the earliest and simplest forms of electromagnetic surveillance and its users. The next two chapters summarise what are the physical laws under which all radar operates, and what are the requirements that these systems need or will need to meet to fulfil a range of applications. We are then able to be specific about the technology needed to implement staring radar.

Two configuration models are introduced that will be used to quantify what is to be expected in operation, how this expanded information flow may be assembled, and its significance identified, while recognising that more information means more processing, which requires appropriate and affordable numerical resources.

The available computing technology has been a determining factor for holographic staring radar sensors to become practicable. These radars deploy hundreds or thousands of beams, and only once certain forms of parallel, floating-point numerical processing have become available has it been credible to transfer the burden of beamforming from a hardware-intensive reflector or passive array to a fully-digital, multi-look receiving function. That technology arrived a number of years ago, and industrial investigation of HSR is now fully practical. In fact, better information depends on more, fast computing, based on the expanded information flow, and the cost is minimal on the scale of radar systems.

All divergences in design approach may introduce vulnerabilities, and a number have been identified for staring radar. One objection arises from the increased computing burden itself, but fails to recognise that information is the primary output requirement, and it originates with the target, not with the radar. Computing capacity is the fundamental enabler in communicating from the target to the radar user. The book moves on to investigate more significant arising objections and the possible sources of resilience, which respectively will present obstacles to progress in this direction or can translate into additional sources of information.

These are followed by descriptions of opportunities that, using these models, arise from the flexible target dwell times, the usefulness of multilook surveillance, potential multi-functionality, spectrum efficiency and resilience to interference.

There have been concerns among Government defence research agencies that, although developers perceive avenues for significant improvements in agile forms of radar, actual advances in implementation, deployment and performance have been slowing over the past two decades. If this is true, one reason may be not that the efforts have been insufficient, but that the hurdles have been mis-identified. It seems possible that efforts have been directed too much towards agile, adaptive beams, waveforms and algorithms, and too little towards allowing the input signal information stream to expand to the point where the flow is pervasive and continuous.

In conclusion, the book will end with an assessment of these risks and opportunities, capabilities and costs. 80 years on from Chain Home, the quantitative information content of continuing target-scattered signals can now be evaluated, as the data source for coherent radar target analysis.

Chapter 1
Introduction to holographic staring radar

Radar, the use of radio signals non-cooperatively to determine the presence and trajectory of air and marine craft, grew out of an urgent need for long-range threat warning. It has since become a customary feature of airports, ships, military aircraft and ground operations, and more recently of cars on public roads. During World War II, radar both exploited and drove the development of radio systems and electronics, playing a major role in several theatres of war. Its earliest forms were ground-based, massive, static structures that were soon to be replaced by more compact systems capable of both search and tracking functions.

This text begins with history, then recognises the applications and the constraining physical laws, proposes two concepts that can be tested against various specific surveillance requirements, assesses the possible benefits and vulnerabilities of each and discusses its possible extension in terms of surveillance capacity.

In Chapter 2, we include an outline of current users and uses of radar, its challenges and their solutions as provided in traditional radar configurations. Chapter 3 aims to describe the actual constraints imposed by the laws of physics and their consequences with respect to scanning and staring radar, and identifies some potential for growth in radar capabilities. Chapter 4 outlines applications for staring radar, and the requirements that will arise and must be met by a staring radar solution. Chapter 5 sets out conceptual configurations for staring surveillance radar and outlines their possible gains and vulnerabilities. Chapter 6 describes aspects of resolution cell discovery, and the measurements that can be made from which target reports can be delivered. Chapter 7 provides more detail about the vulnerabilities of staring radar and possible methods for recovery of performance under those conditions. Chapter 8 discusses the availability in a staring sensor of historical but continuing coherent signal data and target information, and the uses to which they can be put. Chapter 9 deals with aspects of multipath propagation that apply particularly to staring radar. Chapter 10 explores aspects of coherent surveillance networking, the necessary occupation of the radio spectrum by staring radar and opportunities for improved spectral efficiency. Finally, Chapter 11 summarises the findings of earlier chapters and concludes on the qualification of holographic staring radar as a guide for development of ground-based radar surveillance.

In summary, this book aims to bring out the significant characteristics of staring radars, promised for over a decade but not yet realised, as a major contributor either to the civil or the military surveillance sectors. The potential benefits of that contribution, the necessary engineering capabilities and product opportunities that arise from an optimal matching of technique with requirements (and not vice versa) are described, within the constraints imposed by time, money and the laws of physics. Working in the context of the radar paradigm of the past 70 years, dominated by beams and increasingly agile scanning, this book aims to provide a reassessment of the balance between the ideals of agile spatial searching and of persistent, ubiquitous and coherent signal acquisition and analysis.

1.1 Early history

Starting in 1936, a form of radar was developed and deployed in Britain that contributed to the outcome of the 1940 Battle of Britain and reduced the effectiveness of the extensive bombing of British cities until 1943. It had no moving parts and explored a wide 'Field of Regard' (FoR) within its surroundings using a fixed, broad-beamed transmitter and a receiver of signals arriving from any part of the FoR. It was adjustable or configurable to measure directions of arrival. This was the Chain Home air defence radar system; further examples are found in certain over-the-horizon radars, including Cobra Mist. More recent developments in such 'staring radars' are the subject of this book.

1.2 Distinct forms of radar

Later in the 1940s during World War II, with growing requirements for airborne surveillance and with access to microwave frequencies at high power, a form of radar was developed to deploy a narrow transmit/receive beam that could be swept sequentially around a chosen FoR. This 'beam scanning radar' (BSR) form has several advantages in its potential for simple and rapid detection, positioning and display, while minimising the volume of data needing real-time analysis. Radar has continued since then to develop primarily as a scanning, narrowly directional search tool. Such radar instruments, based on parabolic reflecting antennas or passive analogue arrays, require a minimum of high-performance radio-frequency electronics and minimum-capacity computing equipment to deal with the single beam and the scanning mechanism. For active arrays to emulate and then exceed the performance of reflectors and passive arrays, especially at high frequencies, it has the implication that large numbers of components, which may be expensive, are needed for agile beamforming, and fast computing must be provided, leading initially to very high costs. For many years the prospect of using such arrays to form and process ubiquitous beams simultaneously was forbidding. The paradigm of sequentially scanning one or a few beams has remained effective as the basis for radar development over many decades, meeting performance requirements in terms of the radar operating characteristic (ROC), or the probability of detection (P_d) versus the probability of false alarm (P_{fa}) (or the False Alarm Rate (FAR)), and has persisted into the twenty-first century.

If the only advantage of extending the available dwell time on target was to recover the loss of gain in a broad-beamed transmitter, and especially at the cost of slower detection during which the target may move, turn, or be obscured, then a staring approach would be ruled out. Why then is staring radar worthy of further consideration?

In the present day the requirements of high P_d with low FAR are, increasingly, not sufficient to support effective surveillance. There are an ever-increasing number of detectable objects in the airspace, only some of which are of interest. Successful discrimination between different types of target and clutter is then not advanced by focusing solely on the ROC, since it may work as well for clutter as for targets.

The reason for further considering a staring approach is to show whether it offers additional information to support target discrimination. We therefore ask what information do the laws of physics allow beyond the tracking and reporting of detectable targets?

1.3 Physical constraints and complexities

The constraining and enabling laws of physics are Maxwell's equations (MEs), the Electromagnetic Uniqueness Theorem (EUNIT), Huygens' Principle (HP) and the Reciprocity Theorem (RT). Here the key constraint is the speed of light arising from Maxwell's equations, the enabler is the EUNIT and the HP and RT provide means for modelling the relationships between targets, signals and the radar.

The uniqueness theorem in electrostatics and its derivative for dynamic systems, the EUNIT, provide that for EM fields observed over an aperture and over time, and scattered from within a 3-dimensional Volume of Regard (VoR), if there is a solution of Maxwell's equations meeting the sensitivity and resolution requirements then that is the only analytical solution.

The EUNIT only applies if targets in the VoR are interrogated persistently and in compliance with the Nyquist criterion. Measurements made intermittently are non-compliant under the EUNIT. They introduce many alternate physically possible solutions that must be distinguished statistically, and the resulting performance will be inferior to the persistently acquired EUNIT result.

An aspect of radar function that may have restrained development within narrow bounds is a perception that information used and processed and reported by radar is created at the point of detection of a signal, and that further analysis can only benefit from repeated, equivalent detections. The 'enemy' in detection is seen as the unpredictable nature of thermal noise, which enforces a high threshold of detection.

In fact, the information used by the radar is created at the point of scattering by the target; it arrives at the receiver as provided by the MEs, and may be determined according to the EUNIT. Where the reception is dominated by noise, the information is not destroyed but merely hidden, waiting to be discovered in association with other measurements (as in the application of a Fourier transform to a sequence of noisy observations).

Noise within linear receiving electronics may obscure signal information but does not destroy it. Even when the signal is obscured by noise, it exists, and will not be destroyed until the radar process becomes non-linear. Noise is not the enemy – its steady and symmetric distribution provides an excellent reference, rather than a source of uncertainty; the longer the dwell time the more precisely its distribution is known, and the greater the potential to locate information underlying it.

A challenge is presented by targets whose trajectory involves non-linear or accelerating motion during the extended dwell period. A cursory assessment might be that where the target deviates from a constant radial speed, the usual procedures for coherent integration will be degraded and that sensitivity will be compromised, casting doubt on the staring radar concept.

To assess the potential of staring radar we will explore the requirements for resolution and coherence of extractable target information. We first recognise the use of Fourier analysis to extract underlying sinusoids; the most basic form of coherence. The concept of coherence then needs expansion. Monochromaticity itself is not a necessity, but the sequence of received complex amplitudes must be consistent with realistic target motion; e.g., a polynomial motion sequence representing a ballistically moving, inertial target. This is what the EUNIT supports.

More complex than the case where signal is hidden within noise or beneath clutter is the case where a single target interacts with the radar via more than one propagation path. This is the realm of 'multipath', and represents a significant complication for the EUNIT and for radar – and potentially more so for staring radar. We shall address the issue of multipath in detail in Chapters 7 and 9, in which that complication is recognised, and the measures necessary to manage its effect are described. The lack of a single, unique path does not falsify the EUNIT or render it inapplicable, but may require significant additional information and processing to obtain the solution.

1.4 What is radar for, how has it developed and what is its potential?

The term 'RADAR' (from RAdio Direction And Ranging) here refers primarily to airspace surveillance, and that is the field in which the term itself was first applied, replacing the earlier term Radio Direction Finding (RDF). However, in developing these techniques there is much to learn from other areas of application of non-cooperative interrogation by radio. In the authors' experience, the study of surface imaging from the air, airborne and surface-borne ice sheet research, surface movement and automotive collision warning, automotive occupant tracking, missile trajectory measurements and marine security have all contributed to perceptions of radar's range of usefulness.

In this book we shall deal primarily with the field of air surveillance; mostly to detect and track aircraft, mostly in connection with a facility such as an airport or a service such as the National Air Traffic Service (NATS), and mostly based on the ground. Surveillance radar may be deployed in ships, in aircraft and in spacecraft, but here the focus is ground-based. In any of these cases, radar can give rise to considerable strategic or tactical, military or commercial benefits. It therefore attracts the interest of several communities, and our intent is to explore a direction of growth for those engaged in choosing, using, procuring, specifying, building, designing or developing these devices; those engaged in defence planning, transport planning, spectrum management and planning; and those engaged in technical and military education, air traffic control and ATC training, etc.

The function of surveillance radar is to interrogate a volume of airspace at known radio frequencies, to receive signals at each point in a receiving array observing the same volume, to interpret them and determine whether they indicate the presence of objects of interest. Objects of interest are to be found among (a) scattering objects within the VoR that are, (b) moving smoothly, or that (c) behave dynamically or scatter in a way that meets the interest of the user.

Radar technology and development has become a systematic progression in which the barriers to improvement perceived by users are addressed in sequence by engineers, and in a way in which each incremental step advances the existing art. Radical changes may be researched, but may be more difficult or expensive, or seen as too risky to accommodate.

Notwithstanding the difficulty, we consider whether a different approach might be taken in exploiting the interaction of radio waves and materials, allowing a greater proportion of the total information about objects, materials, targets and clutter to be acquired and exploited, with the potential to discriminate precisely between target types, trajectories, events and intentions.

1.4.1 Antecedents for a surveillance radar approach

In Britain, at the time prior to the onset of World War II when the unmet need for effective air surveillance and early warning of attack was most critical, radar was a late arrival among developments based on acoustics and infra-red sensing. Against the established flow of research and development, radar became an essential enabling component of the ultimately successful air defence operation that followed. Indeed, while the technical advances that were needed and made were substantial, the advocates of radar achieved what might be seen as an even more improbable success: the re-orientation of the Air Ministry's procurement programme in time for it to support the establishment of the coastal Chain Home air defence radar system when it was needed.

In the course of meeting the surveillance requirement either for civil or military air operations; to seek, detect and track airborne threats or assets, qualitative changes have occurred in the situation facing the user. Early air defence surveillance requirements related to the approach of enemy aircraft in the region near the front lines in World War I, when detection ranges of a few miles would be useful, and approaches based on detection of the sound of aircraft engines or on the heat of their exhausts appeared appropriate as the ground on which to develop air defence in the late 1930s. Infra-red was the preferred approach of Churchill's scientific advisor, Frederick Lindemann, and acoustic lenses were built at several locations in preparation for German airborne threats, but both were challenged by the need for detection and tracking at longer ranges than a few miles. Early demonstrations of RDF, led by Robert Watson-Watt, showed that although a radio 'death ray' could not be built, target detection could be achieved at much greater ranges by the transmission, scattering and reception of radio waves. These resulted in the rapid development and successful deployment of Chain Home, a large-scale staring air defence surveillance system that became operational in 1939.

1.4.2 The sequential-scanning radar approach

Following the introduction and successful use of Chain Home, the design practices and physical construct it used (a static, staring, HF radio design) have been almost completely superseded. Excellent reasons, many associated with the need for airborne surveillance, led to a focus on combining small mechanical size with narrow search beams. The highest possible radio frequencies were needed, with a sequentially scanning search pattern. This had the implicit benefit of limiting processing requirements to 'one beam at a time'.

The concept of a 'beam' has become a defining characteristic of radar design. First the beam is bounded and pointed, and the match with reception is built in; the receiver determines the presence of a target, and its position is decided or updated while the next beam is formed in the scan. A paradigm has developed under which the transmitted beam and the receiver's VoR are both circularly scanned and as closely matched as possible, and a vast body of successful research, development and surveillance practice has taken place on this basis.

A metaphor for the distinction between scanning and staring can be found in nature. There are predators, including the owl and the echo-locating bat, whose surveillance is directed, persistent and acquisitive, and there are prey whose eyes stare widely and defensively into its surroundings. However, we know of no surviving species that relies for its defence on a sequentially scanning, momentarily focused surveillance method analogous to scanning radar.

In this book we will often refer to the practices of scanning radar, both to acknowledge how great success has been achieved in that way, but also to point out where performance is limited, where staring may offer advantages, and how its own limitations may be successfully overcome.

1.4.3 Staring by comparison

Within the scanning paradigm no target can be under persistent surveillance, but it is its evolution with time that separates a target of interest from objects of distraction. In the presence of complex and unstable clutter, moving ground and sea clutter, aerial intrusion, unmanned drones, aviation incidents, etc., surveillance requirements are moving away from mere detection. A focus on target analysis and discrimination over time must call into question the intermittent interrogation of targets that is inherent in scanning radar and limits the precision of Doppler or image information.

It is only relatively recently that the term 'staring' or 'ubiquitous' radar has come into more common usage as part of the overall radar landscape. Indeed, ubiquitous radar is defined according to the IEEE standard radar definitions (2017).

> *A radar that looks everywhere all the time and which performs multiple functions simultaneously instead of sequentially is a ubiquitous radar. The concept is characterized as having a broad beamwidth (quasi-isotropic) transmitting antenna and a staring array of multiple narrower receiving beams covering the same solid angle. The output of each beam has a receiver and processor to provide various radar functions simultaneously and independently in each beam.*

Even here, the association between 'each beam' and 'a receiver and processor' is superseded by a receiving system in which each receiver contributes to many if not all beams. A possible, more complete definition for staring radar that is based on its function rather than its configuration is:

> *'A persistent radar that transmits a known electromagnetic sensing field within and throughout a volume and receives scattered returns from all directions within the volume simultaneously'. The returns provide a solution to Maxwell's Equations and under the Electromagnetic Uniqueness Theorem in relation to the scattering charges and conductors within the volume. Actual target positions, motions and behaviour are translated directly into signal information without the constraining effects of spatial sampling as practised in scanning radar, and signal information is translated into numerical measurements of targets and their trajectories.*

This leads to the configuration of a broad-beamed and persistent transmission, with reception over an extended antenna array, able to transform measurements of complex signal information directly to target positions, motions and more detailed characteristics. It is further possible to benefit to an effectively unlimited extent from the continuity provided by persistent, coherent interrogation, and to acquire precise, complex, time-varying spectral information related to any target.

We will use the term Holographic Staring Radar (HSR) to refer to surveillance sensors that exploit these characteristics.

This definition is necessarily broad but captures the essence of a staring radar as one that has broad illumination and the ability to search, with high 3-dimensional resolution, a volume matched to the total volume illuminated. It differs from a staring but directionally agile single-target tracking radar, where there are single, coincident transmit and receive beams that are trained to follow a moving target. Single-target trackers and staring radar are not unrelated, but the static nature of staring illumination coupled with the use of multiple simultaneous direction finding 'beams' distinguish the two forms. More recently, advanced electronically scanned array radar systems have been developed that inherently have great flexibility such that they can, in principle, mimic staring and many other types of radar. More typically, they can either scan a beam or series of beams and also implement directed, single transmit and receive beams for special applications such as tracking or weather modes.

Such electronically scanned arrays are capable of performing multiple functions. However, this means that decisions have to be made adaptively about the nature of its resources and the time for which they should be applied to any given function. This is referred to as 'resource management' and, along with their very high costs, is a notoriously challenging aspect of this form of radar. In a sense, the available processing power is turned inward, managing the radar, rather than outward, learning about targets.

To emphasise this point, the contrast is between systems designed to achieve a threshold functionality (P_d vs FAR) using specific, simply and quickly detectable signal features, and systems designed to find and extract the maximum target information whenever it may arise in the available signals, thereby allowing targets to self-classify.

This dependence of radar functionality on signal information and processing capacity can be traced in hindsight. In the case of Chain Home, targets were first detected in the range/time dimension, without any indication of direction, by observation of an oscilloscope. As the signals persisted, their direction was identified by manually turning a goniometer to direct a receiver null to a position at which the target of interest was suppressed, while targets at other directions remained on the oscilloscope screen.

For the purposes of this book we shall consider primarily ground-based surveillance applications, and leave some of the most demanding aspects of radar to later studies including space, weight and power constraints and the requirements of mobile, airborne and spaceborne radar. However future electronics and processing systems may be expected to allow these restrictions to be overcome, and for this approach to become applicable to a wider range of surveillance requirements and functions.

1.5 Historical background

It can be argued that the history of 'staring radar' is as old as the history of radar itself. Christian Hulsmeyer from Germany is usually credited as the inventor of radar. His original 'Telemobiloskop', first demonstrated on May 18, 1904, had fixed illumination and receive beams. An extract from Hulsmeyer's patent is shown in Figure 1.1. Any target entering the illuminated zone triggered a warning alarm such that evasive action could be taken to avoid collisions. Despite his system proving itself in numerous tests and demonstrations, commercial success did not follow. Neither the maritime industry nor the German navy were sufficiently convinced of the role of this device, and Hulsmeyer was forced to apply his considerable talents elsewhere.

It was not until over thirty years later that the concept of radar raised its head again with the experiment conducted near Daventry in England on February 26, 1935, as events, if not policies, were converging towards World War II. Having shown that an electromagnetic 'death ray' as a defensive weapon was not achievable, Arnold Wilkins and Robert Watson-Watt demonstrated the electromagnetic observation of aircraft at distance, and subsequently that this should be preferred both to acoustic and infra-red methods of remote, early detection of airborne threats. Once the value of the role to be played by radar could be seen, this led quickly (in the time scale of military R&D) to the 'Chain Home' series of systems that provided air defence for British shores. Chain Home was a system that quite literally provided radio 'floodlight illumination' with echoes detected via a series of direction-measuring crossed dipoles connected to a low-noise, high-gain receiver. Figure 1.2 shows part of the Chain Home radar system. Chain Home owed its architecture to the Daventry test, and to the need to minimise the innovative steps required in transitioning from an exploratory demonstration to a deployed system tasked with a critical role in enabling the air defence of Great Britain. The system proved capable of measuring the direction and range of aircraft. Together with an effective communication and plotting service, without which its technical capabilities would have been neutralised, Chain Home provided the RAF with sufficient warning and positional accuracy to engage the attacking force, and has been credited with a major role in determining the outcome of the Battle of Britain, and possibly of World War II. Its success led to a secure base for radar research in Britain, supporting later breakthroughs such as the cavity magnetron, needed to enable physically smaller and higher frequency airborne radar. The success of these compact sensors led to the adoption of the radar architecture subsequently, and with few exceptions used by both military and civil air services.

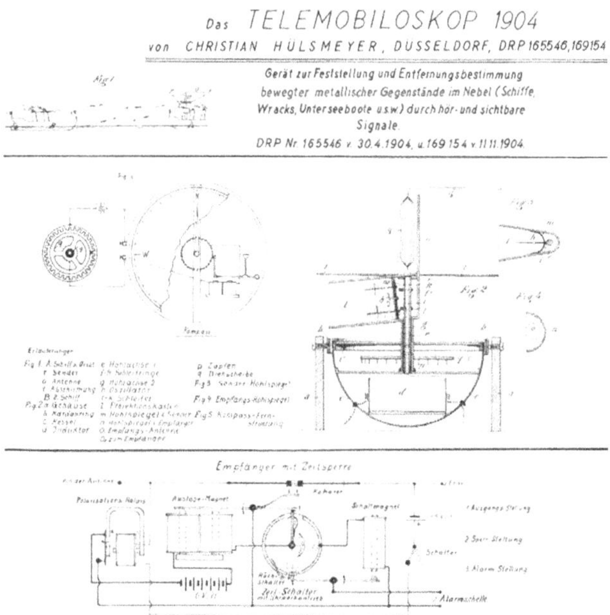

Figure 1.1 Hulsmeyer's telemobiloskop

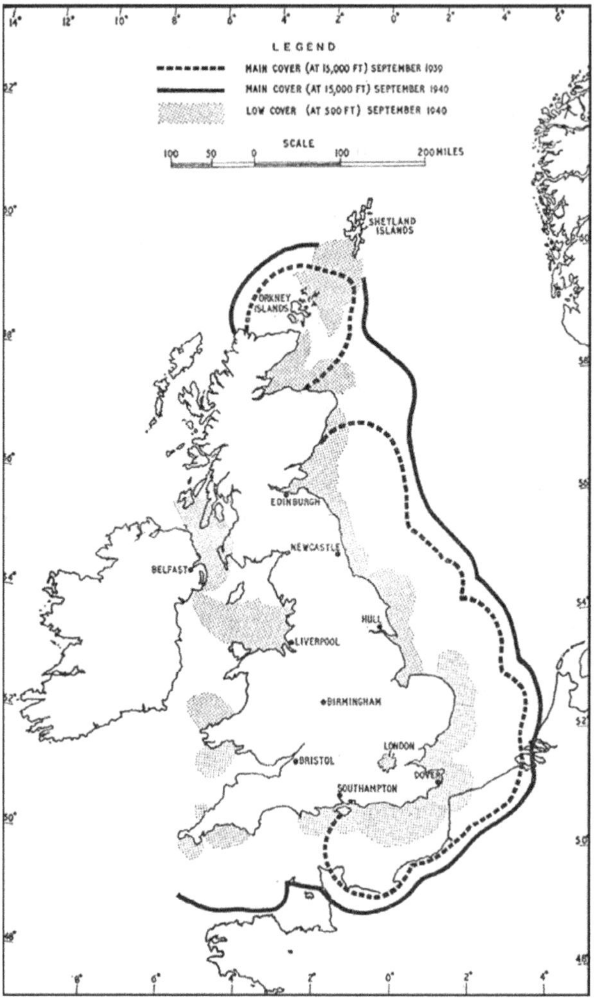

Figure 1.2 Coverage and outline of chain home station

A combination of developments fuelled by a desire to have more compact systems, and particularly the need for airborne sensors able to detect submarines, and others requiring all-round surveillance, led both to the need for higher-frequency sensors and mechanically scanned antennas, used in both transmission and reception. The development of the transmit/receive or T/R switch was also a key enabling technology. Subsequently, the simple geometry and scanning process of these radar systems made them the overwhelming design of choice. So much so that, with a few notable exceptions such as imaging radars, mechanically scanned radar systems have become the norm and are a familiar part of everyday life. Figure 1.3 shows a typical marine radar installation.

Figure 1.3 *Maritime surveillance radar enabling the voyage data recording (VDR) system*

Scanning radar finds numerous applications in civil and military, aerial and maritime use. In Figure 1.3, a luxury cruiser is equipped with a wide range of radio, acoustic and optical navigation aids.

More recently, while the process of sequential scanning has been retained, radar developments have moved in the direction of electronic scanning rather than mechanical. Figure 1.4 shows the PAVE PAWS phased array radar (PAR) that replaced the mechanically scanned BMEWS radar at Fylingdales in North Yorkshire in the 1990s.

The staring architecture of Chain Home, fundamentally different from the scanning approach, has never been further exploited.

Figure 1.4 *The fylingdales PAVE PAWS missile defence radar (creative commons)*

In addition to the scanning or staring approaches to radar, further alternative approaches are found between 'monostatic', in which transmitter and receiver are co-located, 'bistatic' and 'multistatic' radars, in which transmitters and receivers are geographically separated, and also 'passive' radars in which transmissions from other radio stations are detected and used as the VoR-illuminating wave.

There have been attempts to explore bistatic radar geometries, where the transmitter and receiver are placed in quite separate locations; however, the more complicated geometries associated with bistatic geometries and the need for synchronisation across large distances have inhibited their development. Bistatic radar is beginning to show signs of coming into more common usage, especially with the increasing interest in passive radar. In fact, passive radar leads directly to a natural extension of bistatic radar into multistatic or networked systems as multiple transmitters and/or receivers are exploited. These are more complex still but are beginning to emerge from research laboratories with increasing take-up by industrial companies. Passive radars are also, usually, examples of staring radars. However, a full description of passive radar lies outside the scope of this text and the interested reader is referred to [1].

As with the bistatic concept it is only relatively recently that the notion of 'traditional' scanning radar has started to be challenged. Although the basic concept of a staring radar has this long history, it was not until the 1980s that concepts began to emerge that were described in open publications. Three early and significant reports come from Thomson-CSF (now Thales) and ONERA in France, and refer to a concept called 'RIAS' [2–4]. RIAS is a mnemonic that means 'Radar a Impulsion et Antenne Synthetique' (or Synthetic pulse and antenna radar); RIAS has two concentric arrays one for transmit and one for receive, as shown in Figure 1.5. The transmit array simultaneously radiates a set of orthogonal waveforms. As each transmitter uses a separate waveform this can be thought of as an early form of multiple input,

Figure 1.5 *The RIAS concept as implemented by Thomson CSF (now Thales) in 1990*

multiple output or MIMO radar. The system has total hemispheric instantaneous coverage by the virtue of multiple digital beam forming on receive so as to achieve permanent search in all directions and continuous tracking of the detected target. Experimental results using RIAS were described by Luce *et al.* [5].

RIAS operated in the VHF band, chosen as a counter to stealth technology and, most likely, to reduce the number of required elements and, therefore, system cost and complexity. Indeed, the authors, with some foresight, point out that the price to be paid for such a digital staring array is in the requirement for high-computational capacity. This is still true today but it is a price that has dropped significantly, year on year. In 'Phased Array Radars in France: Present and Future' in a review article [6] Colin makes specific reference to RIAS indicating the importance of this program and its contribution to the general development of digital beam-forming. More recently, the ideas behind RIAS have been taken up by Jianqi *et al.* [7], from China. They describe a system very similar in concept to RIAS but also emphasise the MIMO nature of the hardware configuration. They call their system SIAR (Synthetic impulse and aperture radar). This work has culminated in a book [8], which describes in some detail the basic concept along with long-time integration strategies that follow from a staring radar concept and extensions to operation in other frequency bands.

Second, in the mid- to late 1980s, Wirth, in Germany, published a series of papers on a concept termed 'Floodlight radar' [9, 10]. Here the main motivation was to reduce the vulnerability of radar to interception by electronic warfare equipment, especially those associated with anti-radiation missiles (ARMs). A system was implemented, known as the Omni-directional Low Probability of Intercept (OLPI) radar. The system used a continuous-wave transmission, operated in S-band and had a transmit power of 10W. It used sector illumination covering a region that was 120° in azimuth and 20° in elevation via a column of eight vertical dipoles. The broad illumination combined with a CW waveform minimises the emitted power density and correspondingly minimises the likelihood of intercept. The receive array (shown in Figure 1.6) consisted of a planar array of 64 columns, each containing eight dipoles making 512 elements in total. Multiple beams were formed in analogue using a 64-port Butler matrix. Each formed beam was then digitised prior to further processing. Overall, the staring nature of the concept coupled with digital beam forming opened up many new processing possibilities such as long-term coherent integration for fine Doppler resolution as well as the application of techniques such as 'sequential processing' for improved target detection. Again, this was part of a broader research effort centred on phased arrays, much of which was brought together in a book authored by Wirth and published in 2001 [11].

A third and very significant effort was led by Merrill Skolnik of the Naval Research Laboratory (NRL) in the USA. Skolnik's ideas were first introduced in 1998 [12] and later in 1999 [13] as 'a far term concept for a radar that could look everywhere, all the time'. He termed this 'Ubiquitous radar', and a comprehensive account of the concept and its advantages and disadvantages can be found in an NRL report authored by Skolnik [14] and summarised in a 2003 conference paper [15]. In the NRL report Skolnik describes ubiquitous radar as inherently having fixed low

Figure 1.6 The OLPI receive array

gain or quasi omni-directional transmission, digital beam forming on receive and a computational capacity to perform multiple signal processing functions in parallel. Skolnik also recognises that a staring radar leads to long coherent integration times, enhanced Doppler filtering and the overall ability to achieve radar capabilities otherwise not possible.

Also in 2003, Rabideau and Parker explored a variety of ubiquitous radar concepts within a MIMO radar framework as applied to ground surveillance applications [16]. Alter *et al.* [17] implemented a version of ubiquitous radar that operated in L-band and had a peak power of 1kW. It employed a transmit array of half wavelength space emitters arranged in 19 rows and 11 columns making 209 elements in all. This gave coverage of a 90° sector in azimuth and 15° in elevation. The receive array consisted of 1 590 elements arranged as 53 rows and 30 columns. This gave receive beamwidths of 2.7° in elevation and 4.7° in azimuth. Multiple beams could be formed digitally to cover the whole of the illuminated volume. The analysis of results highlights numerous advantages of the ubiquitous radar concept, without necessarily exploring all possibilities. It also points out that there can be a need to compensate for range and Doppler walk during a coherent processing interval. As we shall see in Chapter 7, very high radial speeds and highly dynamic trajectories may threaten the staring radar, since range and Doppler may vary in the course of the long coherent integration times on which the staring approach relies. However, under these conditions, and provided that sampling rules are observed, we shall see that it is possible to maintain performance through processing that may enhance detection as well as improving clutter rejection.

Subsequently, the ubiquitous radar concept has been extended to use with conformal arrays [18] employing a conical structure. Their concept can be operated with both narrow and broad beams on transmit and can mimic both electronic scanning and staring. In [18] there is a discussion and computation of the requirements for additional gain as the system becomes closer and closer to the ubiquitous case. This extra gain is achieved by long coherent processing times making the range capability of traditional radar and staring radar the same. Harman and Hume [19] have reported a staring radar concept that uses an architecture consisting of a number of fixed transmit beams with corresponding digital beamforming such that all round surveillance is possible. This has been embodied in the ALARM and Obsidian radar systems operating in C and X bands, respectively.

Overall, Skolnik's ubiquitous radar concept is very much the foundation upon which the descriptions in this book build. The advent of highly capable digitisation technology and subsequent signal processing has allowed Skolnik's ideas to be implemented and to be extended to a new level such that all of the advantages of staring can be fully exploited.

However, despite successful experiments with various forms of ubiquitous radar, and across the radar community, development efforts remain focused elsewhere; either on cost reduction of radars directed towards the civil aviation field, or for the military, on multi-function radar on the basis of ever more agile spatial scanning, array reconfiguration and time-varying waveforms.

Sequential spatial sampling reduces the number of resolution elements that need to be simultaneously processed and, therefore, the overall processing burden. However, it destroys the simultaneity and continuity of illumination of the whole FoR, and the coherence in time and space of the observed fields. It makes dwell times dependent on the required return time, the angular resolution and the number of parallel beams, and inevitably focuses effort on critical resource management under immovable time limitations. A scanning radar may deploy more than one beam; however, the process remains dominated by a scan sequence in which the resources of the radar are focused in one or a small number of directions at a time and require intensive management.

A staring radar searching the FoR coherently and processing beams in parallel removes the constraint on dwell time and provides the possibility of continuous signal analysis, at the cost of multiplying the processing burden. However, the constraints that arise because many ranges and directions need to be interrogated, many pulses filtered and Doppler bins computed and many targets detected and tracked simultaneously do not challenge the laws of physics. They can be addressed by high-performance computing; a commodity that technologies such as Field Programmable Gate Arrays (FPGAs), Graphics Processing Units (GPUs) and other parallel processors can supply cost-effectively.

The speed of light is finite, noise is proportional to bandwidth, the scale of resolution is proportional to wavelength and inversely to aperture and bandwidth, and sensitivity is proportional to power and array size. These laws of physics are not elastic. However, the precision of target information grows with extended dwell times, and the capacity of available processors continues to expand, without a perceptible limit. We conjecture that to meet future surveillance requirements, growth in radar capability must be met by designing to exploit the growth in processing capacity.

In radar technology development, these interactions of requirements and constraints are complex. Requirements and their prioritisation evolve with time, as do available materials, techniques and mathematical and analytical tools, and so changes in design practice are a persistent necessity. However, also necessarily, changes in design approaches introduce uncertainty and cost in already-expensive development programs. They also threaten suppliers' ability to justify their investments in evolutionary change.

Despite the need for stability in methods and in procurement, in this book we set out a case for change in design practices in surveillance radar by reassessing the physical constraints. The case is that for many surveillance applications a staring paradigm offers greater capabilities in target and clutter discrimination than a process of sequential scanning, and greater flexibility in responding to changes in requirements. The opportunity arises in general because the laws of physics themselves allow for broader exploitation of information encoded in EM waves, in pursuit of defence, security and safety. It arises now because of the rapid increase both in affordable computing capacity (an increase that does not arise from military requirements but from the explosion of digital processing and communications that has become the basis of much of human commercial and recreational activity), and because clear discrimination between different targets and target types is increasingly important.

In this book we aim to refocus attention on the requirements for more radical improvements in radar performance, and on opportunities that the laws of physics may underpin as well as the constraints that they impose on present radar designs.

The book presents answers to the question: 'With advances in high-capacity computing, does Holographic Staring Radar represent a new departure and a unique set of capabilities, or is it simply a more complex subset of the evolving family of radar sensors?'

References

[1] Griffiths H.D., Baker C.J. *An Introduction to Passive Radar*. Norwood, MA: Artech House; 2017.

[2] Dorey J., Blanchard Y., Christophe F. *Le project 'RIAS': Une approche nouvelle des surveillance aerienne*. Colloque international sure la radar, ONERA TP No. 1984-20; Versailles; 1984. pp. 15–20.

[3] Dorey J., Garnier G. 'The RIAS pulsed synthetic-antenna radar'. *L'Onde electrique*. 1989;**69**:36–44.

[4] Dorey J., Garnier G., et Auvray G. 'RIAS radar a impulsion et antenne synthetique'. *Collique International sur le Radar*. Paris; 1989. pp. 556–62.

[5] Luce A-S., Molina H., Muller D., Thirard V. 'Experimental results on RIAS digital beam-forming radar'. *Proceedings of the International Conference on Radar – RADAR 92*; 1992. pp. 74–7.

[6] Colin J-M. 'Phased array radars in France: present and future'. *Proceedings of the IEEE International Symposium on Phased Array Systems and Technology*; 1996. pp. 458–62.

[7] Jianqi W., Ruilong H., Kai J. 'Researches of a new kind of advanced metric wave radar'. *Proceedings of the IEEE International Radar Conference*; 1999. pp. 194–7.

[8] Baixiao C., Jianqi W. *Synthetic Impulse and Aperture Radar (Siar): a Novel Multi-Frequency MIMO Radar*. Hoboken: Wiley; 2001.

[9] Wirth W.D. 'Omni-directional low probability of intercept radar'. *Proceedings of the International Conference on Radar*; 1989. pp. 25–30.

[10] Wirth W.D. 'Long term coherent integration for a floodlight radar'. *Proceedings of IEEE International Radar Conference*; 1995. pp. 698–703.

[11] Wirth W.D. *Radar Techniques Using Array Antennas*. 2nd edn. London: IET Radar, Sonar and Navigation Series; 2013.

[12] Skolnik M. 'A different system approach for future air surveillance radars'. *The Digest of the US National Fire Control Symposium*. San Diego; 1998.

[13] Skolnik M. 'Improvements for air surveillance radar'. *Proceedings of the IEEE radar conference*; 1999. pp. 18–21.

[14] Skolnik M. 'Systems aspects of digital beam forming ubiquitous radar'. NRL/MR/5007—02-8625. Washington, DC: U.S. Naval Research Laboratory; 2002.

[15] Skolnik M. 'Attributes of the ubiquitous phased array radar'. *Proceedings of the IEEE International Conference on Phased Array Systems and Technology*; 2003. pp. 101–6.

[16] Rabideau D.J., Parker P. 'Ubiquitous MIMO multifunction digital array radar'. *Proceedings of the 37th Asilomar Conference on Signals, Systems and Computers*. 2003;**1**:1057–64.

[17] Alter J.J., White R.M., Kretschmer F.F., Olin I.D., Temes C.L. 'Ubiquitous radar: an implementation concept'. *Proceedings of the IEEE Radar Conference*; 2004. pp. 65–70.

[18] Carta P., Galati G., Piracci E.G., Madia F., Ronconi R. 'Implementation of the ubiquitous radar concept with a conformal array'. *Proceedings of the European Radar Conference*; 2015. pp. 441–4.

[19] Harman S.A., Hume A.L. 'Applications of staring radar'. *Proceedings of the International Radar Conference*; 2015. pp. 0270–3.

Chapter 2

Users and uses of surveillance radar

In this chapter the roles and requirements for surveillance radar are briefly reviewed in order to provide the background against which applications particularly suited to a ground-based staring radar configuration can be introduced. Most modern surveillance radar systems carry out multiple tasks such as detection, tracking and, sometimes, target classification. Examples include civil and military Air Traffic Management (ATM), maritime surveillance and military air-defence radar.

Users of surveillance radar include civil aviation operators and regulators, military transport authorities, air defence organisations, marine surveillance authorities and ship operators, weather forecasters and researchers and others. In most cases the requirements are set on the basis of detecting, as often as possible, small air targets as far away as possible, as reliably as possible and positioned as accurately as possible, and travelling and accelerating neither too fast nor too slow.

These broad requirements occur at different levels depending on whether the user serves a specific airport, or as part of a multi-radar ATM system, or monitoring over the horizon air activity, or at a forward military operating base. Over the past several decades these requirements have remained sufficiently stable that a well-understood, evolving range of technologies has achieved a high level of success. However, airspace is becoming more densely occupied and the need to discriminate between different air vehicle types has grown, adding requirements not necessarily catered for by radar sensors of the types routinely deployed.

2.1 Requirements for surveillance

The IEEE defines surveillance radar as: '*A radar used to detect, locate, and track targets over a large volume of space*' and primary surveillance radar as: '*A radar system in which the return signals are the echoes obtained by reflection from the target*'.

The users of radar surveillance have historically been primarily concerned with managing air traffic and detecting aerial threats. As the occupation of airspace increases, their need for more accurate, more timely and more detailed information about their targets of interest also grows.

Examples of primary surveillance radars are the well-known and long-established Air Surveillance Radars (ASR) used to monitor air traffic close to an airport and Air

Figure 2.1 USAF E-3 SENTRY system: AWACS

Route Surveillance Radars (ARSR) that monitor air traffic on-route to and from airports. Together these provide the backbone of modern ATM systems. There are also advanced and highly capable military surveillance radar systems that comprise ground-based, air-based and even space-based components that collectively provide comprehensive early warning of threats and enemy activity. Perhaps the best-known of these is the US E-3 SENTRY Airborne Early Warning and Control System, better known as AWACS (Figure 2.1). AWACS is able to detect both air and surface targets. There are many types of surveillance radar, the UK SENTINEL system is an example of a long-range air-to-ground surveillance radar able to provide real-time, high resolution images of the ground, as well as information regarding moving targets on the ground.

Here, we concentrate on surveillance of the airspace, although much of what follows may be applied equally to surveillance of the ground (either from ground-based systems or from air or space-based platforms). The IEEE definition for surveillance is followed but in its broadest sense taking the meaning of '*a large volume of space*' to be a volume of space of any size and for this '*space*' to be either two dimensional or three dimensional so that both air and surface applications can be considered.

Quantitative requirements for both military and civil and air and surface surveillance functions share many aspects including small targets at long range, at both low and high speeds, and the reporting of target trajectories within seconds.

What is not provided within the IEEE definition is any requirement for definition or classification of 'targets', their dynamic behaviour or intent. Radars developed according to the historic evolutionary approach have found a number of challenges in recent years, and we shall explore the potential of HSR to approach a full capability in these respects.

Note also that the IEEE recognises secondary surveillance radar and defines this as *A cooperative target identification system such as the military identification friend or foe (IFF) Mark XII or the civil air traffic control radar beacon system (ATCRBS) in which an interrogator transmits a coded signal that asks for a reply.*

The transponder on the vehicle or platform queried answers with a coded reply. As a cooperative system it has, in many ways, more in common with communications than radar and, therefore, is not considered within the scope of this book other than to note its importance in civil and military ATM.

2.2 Air surveillance

As surveillance radar covers a very wide range of applications, there is a correspondingly wide range of radar types that can be conceived. Here, to help simplify matters, this is broken down into air and ground surveillance so that more emphasis can be placed on the essentials that dictate the requirements and subsequent radar sensor system design.

Consider, first, surveillance of airspace taking long-range detection, location and tracking as a starting point. Long-range air surveillance using radar has been in existence since the Second World War and in many cases has not changed a great deal in terms of basic design since the introduction of narrow-beamed, scanning radars.

The key requirement of a surveillance radar is to detect targets and measure their range, bearing and other attributes. This is the case regardless of target type, be it aircraft, ship, land vehicle, pedestrian, land mass, precipitation, ocean, etc. Different targets return echoes of differing and varying intensities, some of which are scattered back in the direction of the transmitter where, in a monostatic radar, they are collected by the receiving antenna and radar front end. Targets may have to be detected to ranges of the order of 400 km over a full azimuth extent of 360^0. The more accurate the location of targets, the more uses to which the data can be put. However, positional accuracy is ultimately a function of the radar antenna specification, the transmission bandwidth and the ratio of target signal strength to receiver noise strength. Detections that occur in the same or closely related successive locations may be associated with one another and a track can then be formed and declared as such. The output from tracked information is a time sequence of refined estimates of the location of a target. The number of targets that can be detected, located and tracked is ultimately a function of the specification and design of the radar, the target types and behaviour and the effects of the environment such as clutter and extreme weather.

As surveillance radar designs have progressed the most significant changes have been in:

1. system architecture,
2. the ability to digitise and apply digital signal processing and
3. the speed of processing,

Most civil air traffic control, weather and related surveillance radars have continued to use mechanically scanned parabolic dish antennas. Figure 2.2 shows an example of a US NEXRAD weather radar housed inside an electrically transparent '*radome*' (which provides resilience to weather). The radar is dual polarised so that

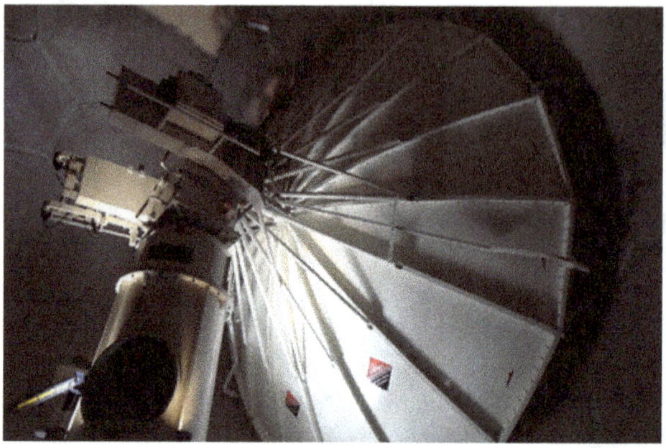

Figure 2.2 One of the US NEXRAD parabolic dish radar antennas inside a protective radome. Note, the antenna is dual polarised.

improved information can be extracted regarding the prevailing weather conditions, in particular rainfall.

A fundamental design parameter of any radar system and a major determinant of its performance is the beamwidth of the radar. The effective beamwidth of a radar system is given, approximately, by the ratio of the wavelength to the dimensions of the antenna, e.g.:

$$\theta_B = \frac{\lambda}{d}\ \text{(radians)} \tag{2.1}$$

where

θ_B= beamwidth (radians),
λ= wavelength (m) and
d= antenna dimension (m).

The beam of a radar system will have the same width in the elevation (or vertical) dimension and in the azimuth (or horizontal) dimension only if the antenna is circular or square. In many surveillance cases, it may be desirable to have unequal beamwidths and, hence, the size of the antenna will be different in the vertical and horizontal axes. For example, an antenna of height 2 m and width 4 m, operating with a 25 cm wavelength (L-band) will have beamwidths of approximately 0.125 radians (6.6 degrees) and 0.0625 radians (3.3 degrees), respectively. For the array antennas predominantly considered here, the array elements are typically spaced at half wavelength intervals to avoid grating lobes. This means equation (2.1) can be expressed as a simple numerical ratio – the inverse of the number of elements in the relevant dimension. For the example above, the vertical and horizontal beamwidths would again be 1/8 and 1/16, or 0.125 and 0.0625 radians, respectively.

Broadly, the beamwidth for surveillance radar systems is a trade-off between detection sensitivity, the number of targets in a single resolution cell, angular accuracy, clutter and interference rejection and cost. At the time of writing the vast majority of surveillance radars and certainly almost all of those carrying out ATM

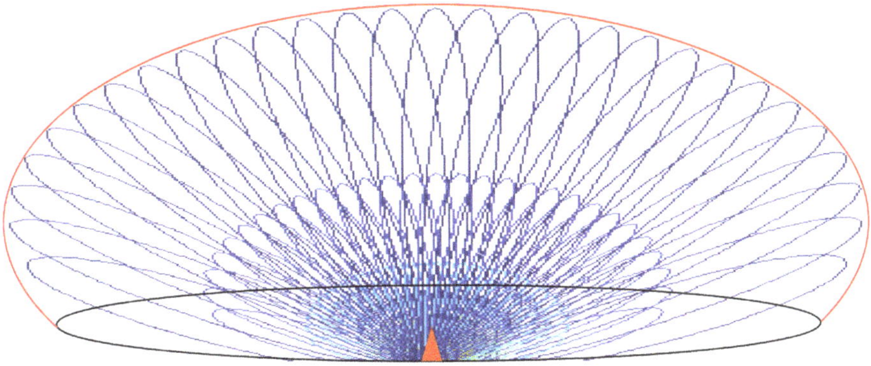

Figure 2.3 Schematic illustration of staring radar irradiation and reception. The transmit beam is outlined in red; multiple receive beams are shown in blue; the surface outline of cover is in black.

functions use mechanically scanned parabolic dish antennas and both commercial and leisure marine applications are mechanically scanned linear array antennas.

Staring radar systems typically employ separate transmit and receive antennas for their different and complementary functions. The transmit antenna is designed to fill an entire 'field of regard', i.e., the surveillance volume in a single or in multiple beams. The receive antenna may be a planar array that enables higher gain on reception, forming multiple, narrower beamwidths that fill the whole of the surveillance volume, collectively matching the transmit beam or beams. This is shown schematically for a single transmit beam case in Figure 2.3. As a consequence, the concept of scanning beams becomes redundant and the ability to exert digital control over receive beam forming opens up new signal processing options and strategies for performance improvements.

Some of these differences in processing and performance can be revealed by recalling how a mechanically scanned antenna functions within a surveillance radar system. Mechanically scanned radar systems rotate at a constant rate mapping out the sky in the two dimensions of range and azimuth (or bearing). Typical rotation rates are between 4 and 12 seconds, although there are examples of both faster and slower rates.

The scan period, i.e., the time to complete one full rotation of the antenna, is denoted by T (typically 4 seconds for civil ATM). As a rotating beam of azimuth width, θ_B, sweeps past a target, the time for which that target is illuminated by the beam, the dwell time, τ, is given by:

$$\tau = \frac{\theta_B . T}{2\pi} \text{ seconds} \tag{2.2}$$

Using the parameters introduced above for a 2 × 4 m antenna, the time that a target is illuminated is 4 × 0.0625 / 6.28 = 0.039 seconds T seconds. A typical L-band ATM surveillance radar may have a pulse repetition frequency (PRF) of the order of 450 Hz. The number of pulses emitted and received from a target, as the beam sweeps past, is given by the product of the dwell time and the PRF. For the case considered here, this is approximately 18 echoes. An L-band Air Route Surveillance radar (ARSR) might also rotate at a rate of once every 4 seconds, which means only 18 echoes are collected by the receiving antenna for each scan of the FoR. Note also that the target is only illuminated very intermittently; in this case just 1% of the scan interval. Summation of 18 pulses gives a maximum integration gain of 12 dB to maintain sensitivity. Overall, though, this places great emphasis on having sufficient sensitivity, mainly delivered through a physically large antenna, very high transmitter powers and long pulse lengths such that adequate detection can be achieved, even from a single transmit pulse. Using appropriately selected design parameters, these radars have been able to map out aircraft activity in the skies sufficiently reliably and accurately to enable safe navigation of air traffic.

Low PRFs are used to avoid 'second time around' echoes or ambiguities when operating at long ranges. The maximum PRF that avoids such ambiguities is given by:

$$PRF_{max} = c/2R_{max} \tag{2.3}$$

where C = velocity of light (m/s) and R_{max} = maximum unambiguous range (m).

Thus, a PRF of 450 Hz leads to a maximum unambiguous range of 333 km. Multiple, higher PRFs can be used to increase integration times but have to be combined and correlated to avoid range ambiguities. The PRF values have to be carefully chosen but with appropriate design can be very effective.

In stark contrast, for a staring radar, integration times are, in principle, unlimited. However, as we will highlight in Chapter 3, there are valid reasons why, while pulse repetition times have minimum limits to avoid ambiguity, integration times may have upper limits to maintain signal coherence. Here a nominal but realistic value of 2 seconds is used as an upper limit on integration time. Thus, with an unambiguous PRF of 1 kHz, this would provide 2 000 echoes to be integrated, and, assuming perfect integration efficiency, would give a gain of 33 dB, 21 dB more than for the mechanically scanned case. This allows a staring radar to exercise design freedoms not available to a scanned radar. One obvious gain trade-off that can be exploited in a staring radar is to use a smaller transmit antenna of lower gain. This has a wider beamwidth that can cover the whole of the surveillance volume as required to support the staring radar concept. A high gain receive array antenna can be used to ensure adequate system sensitivity and sufficiently high angular accuracy. Equally, there are trade-offs in terms of transmit power that can also be exploited. Overall, there are additional design freedoms that occur using a staring concept, although, as with any radar, a sufficient signal to noise ratio has to be achieved to capture the smallest RCS target at the longest desired range. We will return to this in more detail in Chapter 3 and beyond.

There are variants of 2-D mechanically scanned radar systems, usually found in military systems, that use multiple beams stacked in a vertical direction to achieve a necessary 3-D capability. Figure 2.4 shows an example of such a system where a parabolic dish antenna is fed by a vertical stack of illuminators arranged so that a series of overlapping beams are formed in the vertical plane. Each feed will have

Figure 2.4 *A 3-D mechanically scanned radar (courtesy CSIR)*

a sufficiently different transmission frequency so that information can be recovered from each beam without ambiguity.

More recently, and largely for military applications, electronically scanned antennas have been developed and are starting to become the norm. Typically, these use combined '*transmit-receive*' modules which feed small elemental radiators such as dipole antennas. A number of these are combined together, usually in a planar fashion, to form a *'phased array'* used for both transmission and reception. On receive, elements are combined and collectively fed to an analogue beam-former and digitised prior to signal processing. Alternatively, a group of elements can be combined to form a '*sub-array*'. The output of the sub-array is digitised and limited beamforming and control can be achieved digitally. The digitisation of every element has been seen as prohibitively expensive and, hence, has been the preserve of research labs. However, the additional fidelity and control through element level digitisation has been proven and represents a key future direction for this form of electronic scanning. Note, as a planar array face is used the beamwidth of the antenna is a function of the pointing angle increasing as scan angle increases. This typically restricts beam pointing to be less than plus and minus 60^0 and sometimes closer to plus and minus 45^0.

Electronically scanned arrays are able to change their pointing direction effectively instantaneously. This means that dwell times on targets can be selectable and made longer, allowing greater integration times and improved sensitivity (or the power and/or aperture can be reduced). However, this also implies that the radar has to be able to control its own beam pointing so that it divides its time appropriately between differing targets within its field of view, as well as between different functions such as classification and tracking. The radar has to do this to take advantage of its near real-time ability to re-point the radar beam, impossible for a human operator. This aspect of operation is known as 'resource management' and is a notoriously challenging research problem.

Unless achieved in a near optimal fashion it means that full performance potential is not fully realised and the expense of electronic scanning becomes difficult to justify. If a near optimal solution is in place electronic scanning may represent an effective way of detecting, locating and tracking targets, but as we shall see the benefits of persistent signal coherence for target classification are lost.

Electronic scanning also enables ‘on-the-fly’ control of the antenna beam pattern for military users, with potential for forward operating bases. This allows nulls to be generated in the direction of sources of interference or deliberate jamming, thus allowing detection performance to be maintained elsewhere. The closer the array is to full element-level digitisation, the more degrees of freedom there are for finer control of beam nulls and, thus, for coping with more and more sources of interference. Of course, the price to pay is an ever-increasing complexity and cost of the system and has led to some concepts being unaffordable.

Much of the development effort of air surveillance has also been aimed at improved extraction of echo signals from a background of clutter and noise, provision of more information to the operator, improvement of displays and increased automation. Other developments have responded to operational requirements for radars to operate in increasingly congested and hostile electromagnetic environments. For example, small auxiliary antennas have been used to enable mechanically scanned systems to form nulls to counter sources of interference.

Tracking is the process whereby multiple detections over time are associated as being from the same target. An estimate of location, in 2-D or 3-D, is made for each detection and these are subsequently filtered to generate a refined estimate of the measured location for each scan and dwell. For surveillance radar a method known as Track-While-Scan (TWS) is often used and is able to track multiple targets simultaneously. This can be achieved using a conventional air surveillance radar with a mechanically rotating antenna. Target tracking is accomplished by linking detections from one dwell to the next and so on over a fixed number of dwells. If there are sufficient detections linked by location and time then a track can be declared and maintained until the target is no longer detected, whereupon it is deleted from a track register. TWS can be in two or three dimensions depending on the capability of the surveillance radar.

There are also tracking radars used by the military that continuously follow a single target in angle (usually both azimuth and elevation) and in range in order to determine trajectory and to predict the future position. The output of a single-target tracking radar is location and is produced almost continuously. Sophisticated signal processing is employed to estimate target size or specific characteristics necessary before a decision is made to launch an attack. These radars are able to perform in this way because they dedicate their resources to a single target which is automatically followed by the antenna (mechanically or electronically). In other words, they ‘stare’ at the target continuously but ignore all other airspace in the surveillance volume. Hence, the disadvantage of this mode of operation is that only one target can be tracked at a time. This can be partly overcome with electronic scanning, which can control dwell times to allow a pre-determined level of tracking to be achieved but as the number of targets grows, the radar has a greater and greater challenge to adequately illuminate the targets.

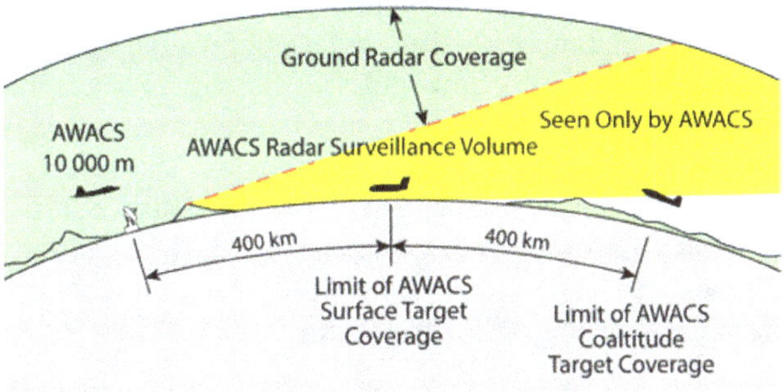

Figure 2.5 *From an altitude of 9 km, AWACS can detect sea and low-altitude targets out to 400 km and higher altitude targets out to beyond 800 km*

In addition, improvements in system architecture together with improvements in digital signal processing have enhanced the amount and quality of information that can be obtained from surveillance radar systems for both civil and military users. Further, and again exploiting digital technology, the form of display presented to an operator and the integration of the radar output with other data sources (such as secondary radar) has also greatly improved. Digitisation has not only allowed systems to carry out many more tasks including target tracking but also, particularly in the case of military systems, has enabled some classes of targets to be distinguished.

Airspace surveillance from the air is largely carried out by the military. The aforementioned US AWACS system is one such example, although other countries have developed similar capabilities. AWACS is used here as an example of how airborne airspace surveillance can be carried out.

At the heart of the AWACS system are the AN/APY-1 and AN/APY-2 radars housed in a large circular radome carried above the fuselage of a Boing E-3 Sentry aircraft, as shown in Figure 2.1. The aircraft is based on the Boeing 707 airframe and flies at an altitude of around 9 km. The radar is able to detect low altitude and even sea surface targets out to a range of around 400 km. It is also able to detect higher altitude targets to ranges of the order of 800 km and beyond. Figure 2.5 shows schematically the air and surface coverage zones in which AWACS can provide detection, location and tracking.

Naturally, by placing the radar sensor at a high altitude the horizon automatically gets extended and longer ground detection ranges become possible. This provides effective early warning of enemy activity so that appropriate reaction can be put in place in a timely fashion. Figure 2.5 also shows how a ground-based surveillance radar can be 'blind' to some targets due to the effects of mountainous terrain that places the target in a shadow region (i.e., there is not the necessary line-of-sight to the target) where such targets are revealed to an air carried radar.

In many ways staring air surveillance radar has been a natural extension of traditional scanning radar. In staring radar, echoes are measured continuously, up to the rate of the PRF, across all elements comprising the receive array antenna.

This gives rise to many new options for optimising processing still further, enabling the extraction of more and more detailed target information. Perhaps the simplest enhancement, yet one that is very powerful, is the ability to select the update rate. As has been seen, mechanically scanned systems update every dwell, which means there are several seconds or more between target reports. This is in contrast to modern cooperative sensor systems that also provide air surveillance such as Automatic Dependent Surveillance Broadcast (ADSB) systems, which provide an update every second. Staring radar can achieve this very simply whereas it is something not possible with mechanical scanning ASRs or ARSRs.

2.3 Ground surveillance

Radar systems that survey the ground can take many forms from systems carried in air and spacecraft that generate high resolution, map-like images to those at a fixed ground site that detect and may track targets moving on or close to the ground. Here, we briefly examine both types of systems as they provide data complementary to one another. Figure 2.6 depicts an ESA radar imaging system that is part of the Copernicus mission.

Air and spaceborne imaging radars have been the subject of intense research and development over the past seventy years. Modern, advanced systems can generate images with resolutions as high as 30 cm in 2-D. The ability to gather such detailed information regardless of weather or time of day has made imaging radars an invaluable remote sensing tool for both military and civil applications. Indeed, space-based systems are able to map much of the surface of our planet with an update rate of just a few days. Typically, the highest resolutions are reserved for military applications where the requirement is more localised and the ensuing very high data rates can

Figure 2.6 ESA radar imaging system within the Copernicus mission

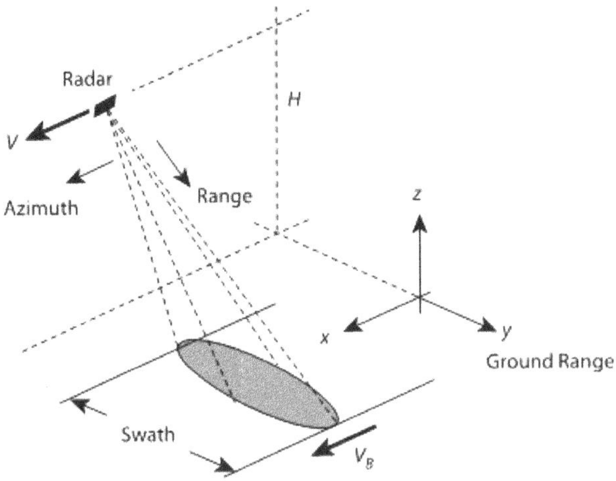

Figure 2.7 Sideways imaging synthetic aperture radar (SAR)

be accepted. However, as digital technology continues to progress, this is changing almost year on year.

Imaging radars use a sideways geometry whereby the radar is flown in an air or space craft with the antenna oriented to view the ground in a direction perpendicular to the flight of the aircraft. This is achieved using synthetic aperture radar (SAR) imaging. The sideways looking geometry is shown schematically in Figure 2.7.

Long apertures can be synthesised leading to narrow effective beamwidths that enable fine along-track resolution. The length of synthetic aperture is limited by the duration for which a target is illuminated [1] and yields along-track resolution ultimately limited to half the length of the real aperture. The real aperture size is, in part, dictated by sensitivity requirements and, hence, resolutions are limited to a fraction of the airframe length for long range systems. Wide bandwidths enable high range (across-track) resolution and these can be chosen to match the along-track resolution.

To overcome the resolution limitations of side looking SAR, the 'spotlight' mode of operation can be employed. In spotlight mode, a zone of interest is illuminated continuously by steering a beam constantly towards the target, as shown schematically in Figure 2.8. This allows much longer apertures to be synthesised and, hence, much finer along-track resolutions can be achieved.

In spotlight mode very wide bandwidths, as high as 4 GHz, can also be used leading to very fine range resolutions of the order of a few centimetres. Figure 2.9 shows an example SAR image from the Karlsruhe area in Germany that has been processed to a resolution of the order of 10 cm. The image shows very fine detail with individual parts of buildings, trees, parked cars, etc. all clearly visible.

Spotlight SAR imaging is an example of a radar concept that uses the advantages of staring, in this case, to achieve very fine image resolution. However, rather like the example of single target tracking, this comes at the expense of the size of the surveillance area on the ground. Nevertheless, there are design freedoms that

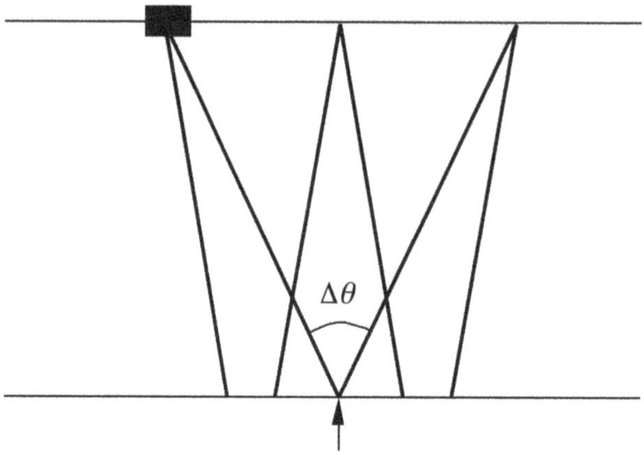

Figure 2.8 Spotlight mode SAR

enable the size of surveillance area and ultimate achievable resolution to be traded with one another [2].

Ground-based, surface surveillance radars come in multiple forms. There are marine radars that provide early warning of nearby craft and aid safe navigation of the seas. Currently, these are all mechanically scanned and use a very mature level of technology. There are also military radars that provide detections of moving objects, such as personnel, that are used to provide intruder alerts. There are example systems that use mechanical scanning [3] and others that use forms of electronic scanning [4].

Figure 2.9 SAR image of the Karlsruhe area in Germany. The image was produced from the PAMIR airborne radar systems designed and operated by the Fraunhofer Institute for High Frequency Physics and Radar techniques (FHR).

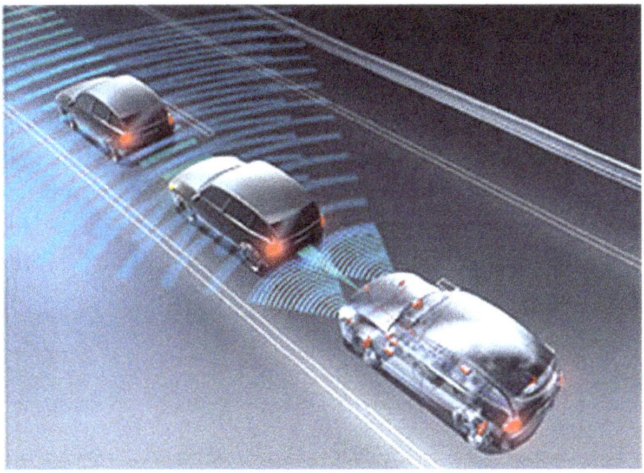

Figure 2.10 Vehicular radars using a staring mode of operation

Civil marine radars rely on short pulses and high range resolution to separate targets from clutter. They use a magnetron transmitter and are non-coherent. Naval systems are coherent and can therefore use Doppler shifts to differentiate moving targets from a stationary clutter background. Doppler resolution is directly proportional to integration time. Consequently, a scanning radar, whether electronic or mechanical, will have a limited dwell time and, hence, similarly limited Doppler resolution. Moving to a staring concept removes the restriction of limited dwell time and finer Doppler resolutions can be achieved. This not only enables moving targets to be separated from stationary clutter but also allows improved sensitivity and the potential for Doppler-based classification.

Perhaps the greatest volume use of staring radars is now in automotive applications, for private safety. The new generation of automotive radar systems not only aid navigation and collision avoidance but are also an integral part of autonomous driving. These systems have to detect all possible obstacles and ideally classify them. This implies a requirement for both fine range resolution and fine Doppler resolution. With careful waveform design and signal conditioning this can be achieved. Indeed, the current group of advanced automotive radars [5, 6] use a combination of staring and MIMO [7]. Wide bandwidths of up to 3 GHz achieve fine range resolution, MIMO helps improve angular resolution and long dwells provide fine Doppler resolution, as illustrated in Figure 2.10.

2.4 The range of uses of HSR

Overall, surveillance radar systems often have requirements which indicate that a staring mode of operation is a strong candidate for best performance, especially where fine Doppler resolution is a prime driver.

We have seen that many radar requirements arise from very long-range air surveillance systems, required to view targets to several hundred kilometres, through to relatively short-range automotive radars that only need to detect targets no further than 200 m.

These fields of application and the arising requirements indeed justify an exploration of the technical and physical context, of the possible vulnerabilities, and of the affordability of staring radar solutions. Chapter 3 will provide this context, in terms of the well-established physical laws that underpin and determine the constraints under which radar engineering design must work. Chapter 7 will discuss possible vulnerabilities that arise and the costs associated with these more information-intensive radar systems.

References

[1] Griffiths H.D., Baker C.J., Adamy D. *Stimson's Introduction to Airborne Radar*. 3rd edn. SciTech; 2014.

[2] Belcher D.P., Baker C.J. 'High resolution processing of hybrid strip-map/spotlight mode SAR'. *IEE Proceedings – Radar, Sonar and Navigation*. 1996;**143**(6):366–74.

[3] MSTAR. Available from https://en.wikipedia.org/wiki/MSTAR [Accessed 2020].

[4] Blighter Surveillance Systems. Available from https://www.blighter.com [Accessed 2020].

[5] Pfeffer C., Feger R., Wagner C., Stelzer A. 'FMCW MIMO radar system for frequency-division multiple TX-beamforming'. *IEEE Transactions on Microwave Theory and Techniques*. 2013;**61**(12):4262–74.

[6] Feng R., Uysal F., Yarovoy A. 'Target localization using MIMO-monopulse: applications on a 79GHz FMCW automotive radar'. *Proceedings of 15th European Radar Conference*; EuRad, Madrid; September 2018.

[7] Li J., Stoica P. *MIMO Radar Signal Processing*. Hoboken, New Jersey, USA: Wiley; 2009.

Chapter 3
Physics of holographic staring radar

3.1 Targets and information

Surveillance radar is required to make physical observations of airborne targets within a field of regard and deliver information derived from them.

Aspects of its historical development and the functions it has been designed to perform have been introduced. Our intent here is to define the electromagnetic processes underlying radar's surveillance function, linking those to its capabilities and limitations in delivering necessary information about such objects. An excellent introduction to electromagnetics is given in Feynman, 1977 [1].

We shall explore whether that delivery is achievable with a holographic staring radar (HSR). In that case, we will form a link between radar performance features and the quantity, the precision and the timeliness of delivered information. Some aspects may enhance its value to the user beyond the radar operating characteristic (ROC) performance criteria commonly used.

If successful, this will offer a broader landscape of opportunities for target analysis, classification and prioritisation. We will progress to a more detailed description of the applications and requirements in Chapter 4, and of concepts of staring radar in Chapter 5, followed by a description of the capture of target return signals in Chapter 6.

All forms of radar operate within the constraints of physics, but with differing emphases depending on the architecture being considered. In some cases the speed of measurement is key, while in others detailed target discrimination is the highest priority. The radar range equation (RRE) has been seen as the defining rule for meeting radar design requirements, but it meets that need only in terms of signal amplitude, not of phase progressions and potentially of target information. To cover all cases the physical influences on radar functionality include, in addition to the energy balance between signals and noise; physical laws including Maxwell's Equations, the Electromagnetic Uniqueness Theorem (EUNIT), Huygens' Principle, sampling theory, the Reciprocity Theorem and, most fundamentally, Newton's laws governing the dynamics of solid bodies in flight.

3.1.1 *Physics and signal-encoded information*

There are different ways to link the flow of information quantitatively with physical processes including thermodynamics (order, entropy, reversibility) and electromagnetics (signals, noise and linearity). References are to be found in Feynman, 1977 [2], Feynman, 1977 [3], ITT, 1977 [4]. However, the link remains fuzzy because information is difficult to define without reference to its application. For surveillance radar the requirements are detection, dynamic positioning and low rates of false reports, including mis-classification. Detection and accuracy are well-understood, but classification is challenging. It is intimately linked with false reporting, and whereas detection and positioning involve small sets of variables, information for target classification can be more complex.

A useful link is through the Second Law of Thermodynamics (Feynman, 1977 [5]). Under the Second Law, inefficiency in a thermodynamic process (related to its irreversibility) leads to the growth of entropy (a measure of disorder). By analogy, non-linearity in signal processing makes it irreversible and leads to loss of signal information.

Non-linear processing yields cross-products; jitter yields multiplicative phase noise and a single-bit detection threshold operation, intended to fix signal-derived data prior to track filtering, inevitably introduces non-linearity. It is an essential part of the argument of this book that to optimise information content is essential, and process linearity is the starting point.

The information we seek is created by an unknown event; in this case the scattering of a signal. The scattering itself is irreversible, but it embodies and disseminates information about its source (the target and the incident, ordered radar transmission). In a linear process of reception the signal with its information content may be obscured by the addition of noise; however, it is not erased and can be extracted by association within longer signal sequences. The information is encoded within the complex amplitude sequence forming the signal and can be linearly processed (the algorithm) to make its presence detectable against the (known) noise distribution. Note that it is not the algorithm that creates the information. The algorithm exists to select and separate forms of signal information that represent targets of interest.

A radar sensor should therefore be designed to maintain linearity in processing as far as is possible and affordable, before deriving qualified target reports. Early detection tends to act in the reverse sense by discarding information that does not qualify in amplitude at an early stage of this process.

3.1.2 Detection with a scanning beam

A beam scanning radar (BSR) interrogates its Volume of Regard (VoR) via a sequence of narrowly directional beams and monitors the receipt of responses scattered from targets within each. It could be said in this case that the information is set up in the radar with direction and timing, and the single point of search is a numeric threshold and range value.

The power $\mathbf{P_r}$ received from a target is given by the RRE in a version where a single antenna with gain $\mathbf{G_a}$ is used both for transmission and reception:

$$\mathbf{P_r = P_t \cdot G_a^2 \cdot \sigma \cdot (\lambda^2/4\pi)/(4\pi R^2)^2} \tag{3.1}$$

where σ is the radar cross-section of the target and **R** is the range, $\mathbf{P_t}$ is the transmitted power and λ is the radar wavelength.

To detect a target the amplitude received at each range and direction is compared with a threshold calculated from noise statistics at other positions or times. The threshold test generates a 'yes' or 'no', after which the target information is selected and stored, and performance is simply the truth or falsity of the yes or the no. To achieve a 'yes' result, the received amplitude $\mathbf{V_r}$ must exceed a chosen value $\mathbf{V_{thresh}}$ related to the noise power:

$$\mathbf{V_r} > \mathbf{V_{thresh}} \propto \mathrm{sqrt}(\mathbf{B_n}.\mathbf{kT}.\mathbf{NF}) \qquad (3.2)$$

where noise bandwidth = $\mathbf{B_n}$, **k** is Boltzmann's constant, **T** is the absolute temperature, and **NF** is the receiver noise factor.

This approach is effective and fast when pointing in the right direction and in the disturbing presence only of noise, but in scanning it is slow to return to that direction, it can be confused by clutter, and surveillance performance is limited to P(d) and P(fa) as used to define the ROC.

This description outlines the basic function of the radar. However the non-linear yes/no function is wasteful of information beyond detection and positioning. The complex scattering event by which the target modulates the incident signal, in whose scattered form any available target-specific information is encoded, is reduced to a single bit of information at the detection threshold, to which are added the known radar range and direction to form a report or to support tracking.

The VoR may contain not only targets of interest in an empty volume of space, but also objects that may pass the threshold test but do not qualify as targets of interest. Discrimination between target types is then necessary but is made more difficult by the use of narrow beam scanning. Limited additional information about the nature of the target is available with MTI radar, and some may be recoverable by recourse to expensive and time-consuming waveform or scan adaptation.

With respect to reported information, for a BSR the number of items of information available per target per scan is 4 – detection, range, azimuth and the presence of non-zero Doppler.

The approach for HSR is to continue the time series of information encoded in target-scattered signals, to process the signals linearly and to decode the information more completely and precisely. To support this approach signals are discussed and presented in linear form rather than a logarithmic form expressed in decibels.

3.1.3 Holographic staring radar and analytic solutions

Staring radar can be recognised as a progression through various concepts already mentioned – Telemobiloscop, Chain Home, Ubiquitous Radar, RIAS, OLPI and others, and towards a more general proposition about effective radar surveillance. Our purpose is not to show that the staring form of radar is a better means of detecting and tracking aircraft, but to consider whether it better exploits the opportunities and constraints represented by the physical laws.

We recognise and aim to exploit the EUNIT, with Maxwell's equations, Huygens' Principle and the behaviour of solid bodies under Newton's laws.

A simple statement of the EUNIT is that 'Providing boundary conditions for Maxwell's equations uniquely fixes a solution for those equations' (Smith, 1997 [6]).

With the power budget and the receiving aperture, the targets and physical objects within the VoR are the boundary conditions, defined by their positions, shapes and conductive or dielectric character. According to the EUNIT, accurate, fully sampled measurements of the EM waves arriving at the receiving aperture provide a unique representation of the presence and motion of those objects. The design parameters of the radar, its position and orientation, the target position and scattering pattern, through Maxwell's equations, define the extent and precision to which signals received can be used to measure uniquely the boundary conditions of interest – the targets.

The EUNIT uniquely links the observed fields to the content of the VoR. It provides that for EM fields observed and processed coherently over a receiving aperture, and scattered from within its VoR, if an analytic solution of Maxwell's equations can be found that (a) represents feasible dynamics of bodies (conductors or dielectrics), in flight (ballistic or powered), and (b) meets the sensitivity criteria under the resolution and power budget constraints of the radar, then it is the only solution. This provides a basis for solving both for trajectories and target characteristics on an analytical rather than a statistical basis.

We note that the EUNIT only applies if targets in the VoR are interrogated continually and in effective compliance with the Nyquist criterion. Measurements made under Nyquist criteria but intermittently do not meet these conditions. Each is made in isolation and relative incoherence, and the intervening gaps introduce many alternate physically possible trajectory solutions that must be distinguished statistically, but none at all in terms of target characteristics.

To fully exploit the EUNIT over the entire VoR is feasible with up to date array technology, and the reduced size and the cost of computing make it practicable to process the data fast enough for reporting. The large volume of continually coherent signal data generated is the principal benefit offered by a staring radar, and the design intent should no longer be to minimise the computing burden, but to supply adequate computing capacity to decode continuing target-return signal information accurately.

3.1.4 Extending time on target

The broad irradiation used in a staring radar implies reduced gain on transmission. To recover the necessary signal to noise ratio for detection, staring allows coherent integration of successive pulses, by which every target is irradiated.

Coherent integration by Fourier analysis also allows the HSR to resolve target returns or components at different Doppler shifts or modulation sidebands. These may be associated with different speeds, dynamics, rotors or strain movements, and in each case offer information to support target discrimination. Incoherent gain is also potentially useful over very long dwell times.

Where the objective is not merely to detect but to analyse the target and its context (i.e., to collect and assess information about its location, its motion and its dynamic behaviour, and also about those of objects in the environment) on a continuous basis, extended dwell directly enables the measurement of scattering characteristics as a function of time, supporting target discrimination.

The RRE is given below in a version that permits different values of antenna gain on transmission and reception and includes the effects of processing gain in the receiver:

$$\mathbf{P_r = P_t \cdot G_t \cdot G_r \cdot G_{co} \cdot G_{nc}} \cdot \sigma \cdot (\lambda^2/4\pi)/(4\pi \mathbf{R}^2)^2 \quad (3.3)$$

where $\mathbf{G_t}$ is the gain of the transmit antenna, $\mathbf{G_r}$ is the gain of the receiver, $\mathbf{G_{co}}$ is the coherent gain, and $\mathbf{G_{nc}}$ is the incoherent gain, either of which can increase with dwell time on target and support the potential of staring radar to meet the ROC conditions.

Looking further into this approach, Fourier analysis allows the HSR to distinguish and measure complex amplitudes at many closely spaced values of frequency offset (from the carrier), and this complex spectrum is the first step in decoding information encoded in the complex amplitude progression of scattered signals. As for scanning radar the HSR also finds range and azimuth, and in the forms envisaged it measures elevation as well. However, the most significant difference in the information it delivers is its measurement in the complex frequency domain. To recover from the low gain of the staring transmitter, the staring radar needs additional coherent integration gain yielding a factor typically over 100. For future examples, for omnidirectional transmission and the same receiving aperture this factor (less the process gain available in the scanning case) can be achieved by a Fourier transform combining 2048 time domain points. At a PRF near 1 kHz this implies a CPI of 2 seconds. We shall discuss the implications for the reporting interval in Chapters 5 and 6. Shorter or longer CPIs can be accommodated, in parallel if necessary, or selectively at different ranges and using different PRIs.

The HSR Doppler spectrum for a single range/azimuth/elevation resolution cell, based on a fast Fourier transform, may contain 2^N ($N \approx 8 - 14$) complex Doppler values, any of which, or any combination, indicates both the presence and the radial speed but also modulation effects of the scattering target or targets during the CPI. Taking $N = 11$, information derivable from this spectrum (while maintaining an undisturbed noise distribution as a reference) may be up to 100 items (spectrum indices, line amplitudes, line widths and symmetries, etc.), delivered per cell per CPI. This can be compared with the BSR value of 4 items (Section 3.1.2) per cell per scan. The anticipated benefit of staring is therefore a quantitative gain in target information. The decision whether to report the detection of a target should be taken at the output of such analysis, in the context of the whole VoR environment, and according to the definition of targets of interest.

3.1.5 Modelling a scattering target

Targets can be modelled theoretically in terms of their decomposition under Huygens' Principle into point or spherical scattering nodes, on a quarter-wavelength

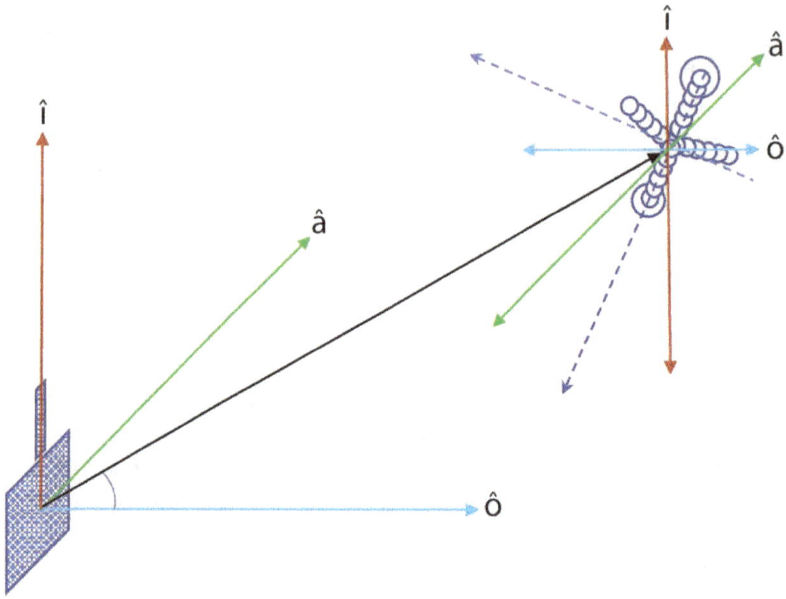

Figure 3.1 Geometry for a Huygens model of an aircraft and staring radar

scale, whose coherent superposition yields a unique complex reflection coefficient as a function of the range and 'look' direction in the target frame of reference. Of course for multistatic radar the cross-section varies with both look directions.

The radar cross-section of any real (non-spherical) target is then linked to the size and shape of the airframe and also depends strongly on its alignment relative to the radar.

To model a target with any kind of correspondence to reality requires a description of the scattering process not at the level of total scattered power, but at the level at which wavelets scattered by all of the spherical nodes are summed as complex sinusoids, at the radar look direction, with the target range and orientation, and with appropriate reference to wave and target polarisation.

To do this, in Section 3.2 we use Maxwell's equations and Huygens' principle, in the context of the EUNIT. The key steps are to calculate (1) the incident EM field at the target, (2) the scattered field at the radar and (3) the received signal at a resolution cell containing the target.

These models can then be used to explore the signal-encoded effects of shapes, trajectories and target features such as propellers, rotors, nacelles and airframe dynamics, and examples are illustrated in subsequent chapters.

Figure 3.1 illustrates the geometry of modelling in the field of view of a surface-based transmitter and receiving aperture.

A model of this kind is used with the linear field equations of EM propagation to construct signals consistent with the position, shape and motion of the target, when irradiated by the transmitting array and observed at the aperture filled with the array illustrated.

These signals can then, in principle, be used with appropriate time domain and frequency domain processes to explore the effectiveness and the burden in computing terms of finding solutions in terms of target trajectories and characteristics, and the aperture and dwell time of the radar. There certainly exist conditions under which the burden will be too great for any specific design – the question is 'can we show that capacity will be practical within the performance requirements?'

3.2 Physics fundamentals

3.2.1 Maxwell's equations

A comprehensive description of the propagation of electromagnetic waves is given by Maxwell's equations, relating electric and magnetic field components to the presence and motion of charges and the properties of the medium of propagation. These equations are the source for $\mathbf{V_c}$, the value of the speed of light, the strength of electric and magnetic forces and their time derivatives.

A formative reference for the equations and relations of classical electricity and magnetism is available in [7].

Maxwell's equations give spatial variables in implicit dimensions x, y and z, written as the derivatives $\nabla.$ and $\nabla\mathbf{x}$ of vectors $\mathbf{D}$ the displacement current in a dielectric medium, $\mathbf{B}$ the magnetic flux density, $\mathbf{F_e}$ the electric field strength and $\mathbf{F_h}$ the magnetic field strength, in terms of the vector current density $\mathbf{J}$, and scalars ρ_V, the spatial charge density and t, the time variable:

1. $\nabla.\mathbf{D} = \rho_v$
2. $\nabla.\mathbf{B} = \mathbf{0}$
3. $\nabla \times F_e = -\partial\mathbf{B} / \partial\mathbf{t}$
4. $\nabla \times F_h = \partial\mathbf{D} / \partial\mathbf{J}.$

The generation and the reception of electromagnetic waves (EMWs) at the radar are separately and uniquely determined by these equations and by the engineering design of transducers in the form of electronics, conductors and dielectric materials. However, to use these waves and signals to interrogate unknown targets requires an understanding of the interaction of EMWs with those targets. The theoretical basis for the interaction is derived from Fresnel/Fraunhofer/Kirchhoff/Huygens diffraction and scattering theory.

EMWs cannot be adequately described by a scalar equation that focuses only on the power budget. EM fields are vector quantities that, although they may simplify to scalar values after processing, can only be properly modelled as vector functions.

The equations above are in their differential forms, written in terms of the fields at a point, and the scalar charge and vector current densities. The space is defined in terms of unit vectors $\hat{\mathbf{I}}$ (up/down), $\hat{\mathbf{a}}$ (left/right) and $\hat{\mathbf{o}}$ (front/back), and is illustrated in Figure 3.1.

To determine the radiated and scattered fields for a radar, operating at a chosen frequency $\mathbf{F_c}$ and angular frequency $\omega_c = 2\pi \mathbf{F_c}$ the relevant current density $\mathbf{J_L}$ is that at the transmitting antenna, which conducts vector current elements, each of amplitude $\mathbf{I_L}$ and directed length $\partial\mathbf{L}.\hat{\mathbf{i}}$, where $\hat{\mathbf{i}}$ is a unit vector in the direction of current flow in the transmitter. The current density $\mathbf{J_L}$ is the surface derivative of $\mathbf{I_L}$ in amps per square metre normal to a surface element $\mathbf{S}$ including the transmitter conductors. $\mathbf{S}$ is perpendicular to $\mathbf{J_L}$ and $\hat{\mathbf{i}}$. In a regular, linear array operating at ω_c, $\partial\mathbf{S}$ is taken to be one half wavelength $(\lambda_c/2)$ squared. For HSR a transmitter will contain a number $(\mathbf{L_T})$ of vertically spaced half-wave elements to create a narrowed vertical beam, yielding a vertical aperture $\mathbf{a_L} = \mathbf{L_T}.\partial\mathbf{L}.\hat{\mathbf{i}}$.

The element conduction current is:

$\mathbf{I_L} = \int_S \mathbf{J_L}.\partial\mathbf{S}$, where $\mathbf{S}$ is the surface of integration.

The field emitted into free space by a transmitting array of $\mathbf{L_T}$ elements is illustrated by the red arrows in Figure 3.2.

Once the conduction current $\mathbf{I_L}$, the number of elements $\mathbf{L_T}$ the angular frequency ω_c, the element transmission gain $\mathbf{G_{tL}}(\varepsilon)$, the target azimuth and elevation angles α

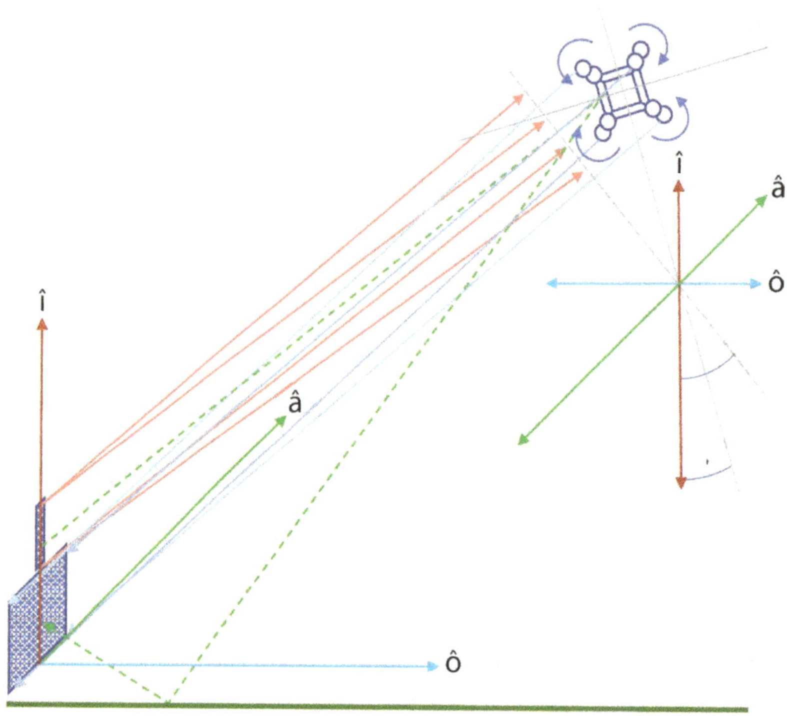

Figure 3.2 Huygens model for a drone air target, illustrating the planar incident wave, scattered signals and the potential for multipath propagation

and ε and its range $\mathbf{R_L}$ to each element are defined, the vector exciting field at the target can be determined.

The electric field strength at range $\mathbf{R}(>> \lambda_c)$ is given by:

$$\mathbf{F_e}(\varepsilon) = \partial/\partial \mathbf{t}\ ((\mu_0/4\pi \mathbf{R})\ .\ \textstyle\sum_{\mathbf{L}} (\mathbf{G_{TL}}(\varepsilon)\ .\ \mathbf{I_L}\ .\ \exp(\mathbf{j}\omega_c\ .\ (\mathbf{t} - \mathbf{R_L}/\mathbf{V_c}))) \tag{3.4}$$

where μ_0 is the permeability of free space, and the sum $\sum_L$ is over contributions from transmitting elements $1{:}\mathbf{L_T}$.

To compute field values for every point in the VoR is complex, since it extends from the induction volumes very close to transmitter conductors (for $\mathbf{R} < \lambda_c$) and also includes regions best described in terms of Fresnel diffraction, close to transmitting apertures (for $\mathbf{R} \approx\leq \mathbf{a_L}$). The results of interest for radar are those described in terms of free space propagation and Fraunhofer diffraction of sinusoids at a known frequency $\mathbf{F_c}$, where the size of the VoR and the distances to scattering objects are much greater than both the wavelength and the relevant apertures, and waves incident on scatterers or receiving elements are effectively plane waves.

For locations of interest in the VoR, also for Fraunhofer propagation:

$$\mathbf{R} \gg (\mathbf{L_T}.\partial\mathbf{L})^2/(\lambda_c/2) \tag{3.5}$$

The transmitted field can then be described in terms of the unit vectors, a direction of polarisation and the Poynting vector at any point.

3.2.1.1 Incident field

Taking the case of linear vertical polarisation, for these planar wave-fronts the incident electric field $\mathbf{F_{einc}}$ at a target at range $\mathbf{R}$, elevation ε and azimuth α is given by:

$$\begin{aligned}\mathbf{F_{einc}}(\varepsilon, \alpha, \mathbf{R}, \mathbf{t}) &= (\mathbf{Z_0}.\mathbf{I_L}.\mathbf{a_L}.\mathbf{cos}(\varepsilon)/\mathbf{R}).\exp(\mathbf{j}\omega_c(\mathbf{t} - \mathbf{R}/\mathbf{V_c})) \\ &.\mathrm{sqrt}(\mathbf{G_t}(\varepsilon)).(\hat{\imath}.\cos(\varepsilon) - \sin(\varepsilon).(\hat{a}.\cos(\alpha) + \hat{o}.\sin(\alpha))\end{aligned} \tag{3.6}$$

$\mathbf{F_{einc}}$ is linearly polarised normal to the elevation direction ε, and in the plane containing the azimuth direction α, and the current vector $\mathbf{I_T}$. The azimuth direction, α, is measured in the plane containing unit vectors $\hat{\mathbf{a}}$ and $\hat{\mathbf{o}}$. ε and α define the direction of the Poynting vector of transmission at each point in the VoR. The radiated field strength in the boresight direction, where the retarded potentials are in phase, has a value proportional to $\mathbf{L_T} \cdot \mathbf{J_L}$. Field strengths are reduced at higher elevations in the beam pattern by interference between the elements with complex gains $G_k(\varepsilon)$, and the beam pattern is formed by summing over the elements $\mathbf{L}$. Here $\mathbf{V_c}$ is the speed of light and $\mathbf{Z_0}$ is the impedance of free space.

The purpose of setting out the vector description of fields is to bring out the stages of propagation and computation at which vector superpositions, polarisation, multiple reflections and Doppler interference can be compared and prioritised. These quantities can only be treated linearly for a surveillance system where irradiation and observation are effectively continual, and in this way radar performance can be modelled to include scattering, multipath and polarisation effects.

3.2.1.2 Scattered field

Having defined the value of $\mathbf{F_{einc}}$ over space and time for the transmitted field it is possible to derive the scattered field for a spherical conductive scatterer of cross section $\boldsymbol{\sigma}$, in the direction and range of the radar at $(\mathbf{R}, -\boldsymbol{\varepsilon}, -\boldsymbol{\alpha})$.

For a target centred at $\mathbf{R}$ and oriented at azimuth α_t and made up of $\mathbf{n}$ spherical scatterers each at radius ρ_n and offset α_n from the target centroid and pointing direction, the projected range differential is

$$\Delta r_n = \rho_n \cos(\alpha_t + \alpha_n - \alpha), \tag{3.7}$$

the planar field at the radar is:

$$\begin{aligned}\mathbf{F_{esca}}(\boldsymbol{\varepsilon}, \boldsymbol{\alpha}, \mathbf{R}, \mathbf{t}) &= \textstyle\sum_n(-\mathrm{sqrt}(\sigma_{n}.\mathrm{F_{einc}}/\mathrm{R}).\exp(\mathbf{j}\omega_c(\mathbf{t} - \mathbf{2}\Delta\mathbf{r_n})/\mathbf{V_c}))\ldots \\ &.(\hat{\mathbf{I}}.\cos(\varepsilon) - \sin(-\varepsilon).(\boldsymbol{\alpha}.\cos(\alpha) + \hat{\mathbf{o}}.\sin(-\alpha)))\end{aligned} \tag{3.8}$$

3.2.1.3 Received signal

This is received at receiver element $\mathbf{m}$ as a signal voltage:

$$\begin{aligned}\mathbf{V_{rec}}(\boldsymbol{\varepsilon}, \boldsymbol{\alpha}, \mathbf{R}, \mathbf{t}, \mathbf{m}) &= \mathbf{F_{esca}}(\boldsymbol{\varepsilon}, \boldsymbol{\alpha}, \mathbf{R}, \mathbf{t}).\mathrm{sqrt}(\mathbf{G_m}(\boldsymbol{\varepsilon}, \boldsymbol{\alpha})) \\ &.\exp(\mathbf{j}\omega\mathbf{0}(\mathbf{t} - \Delta\mathbf{r_m}(\boldsymbol{\varepsilon}, \boldsymbol{\alpha})/\mathbf{V_c}))\end{aligned} \tag{3.9}$$

Provided that the scattered field is observed along the reverse direction of the Poynting vector at the target, then the scalar product of the unit Poynting vectors is unity, and those terms can be eliminated. Note that for a bi- or multistatic system this will no longer be true.

Combining Equations (3.6), (3.8) and (3.9)

$$\begin{aligned}\mathbf{V_{rec}}(\boldsymbol{\varepsilon}, \boldsymbol{\alpha}, \mathbf{R}, \mathbf{t}, \mathbf{m}) &= \textstyle\sum_n(\mathrm{sqrt}(\sigma_n).\mathbf{Z_0}.\mathbf{I_L}.\mathbf{a_L}.\cos(\varepsilon)/\mathbf{R}.\exp(\mathbf{j}\omega_c(\mathbf{t} - \mathbf{R_m}/\mathbf{V_c}))/\mathbf{R}\ldots. \\ &.\exp(\mathbf{j}\omega_c(\mathbf{t} - (\mathbf{2}\Delta\mathbf{R_n})/\mathbf{V_c})).\mathrm{sqrt}(\mathbf{G_m}(\boldsymbol{\varepsilon}, \boldsymbol{\alpha})).\exp(\mathbf{j}\omega_c(\mathbf{t} - \Delta\mathbf{R_m}(\boldsymbol{\varepsilon}, \boldsymbol{\alpha})/\mathbf{V_c})) \\ &= \textstyle\sum_n(\mathrm{sqrt}(\sigma_n).\mathrm{sqrt}(\mathbf{G_m}(\boldsymbol{\varepsilon}, \boldsymbol{\alpha})).(\mathbf{Z_0}.\mathbf{I_L}.\mathbf{a_L}.\cos(\boldsymbol{\varepsilon})/\mathbf{R^2})\ldots \\ &.\exp(\mathbf{j}\omega_c(\mathbf{2t} - \mathbf{2}(\mathbf{R_n} + \Delta\mathbf{R_n}) + \Delta\mathbf{R_m}(\boldsymbol{\varepsilon}, \boldsymbol{\alpha}))/\mathbf{V_c}))\end{aligned} \tag{3.10}$$

The model consisting of node cross-sections σ_n, node positions ρ_n, α_n at pointing direction $\boldsymbol{\alpha_t}$ leads to a linear superposition of co-polarised fields at each element and signals at each resolution cell:

The signal received at the 'beam' directed at $(\boldsymbol{\varepsilon}, \boldsymbol{\alpha})$ (i.e., with the appropriate phase weightings) at range $\mathbf{R}$ is

$$\mathbf{V_{beam}} = \textstyle\sum_m(\mathbf{V_{rec}}(\boldsymbol{\varepsilon}, \boldsymbol{\alpha}, \mathbf{R}, \mathbf{t}, \mathbf{m}).\exp(-\mathrm{j}\varphi_m(\boldsymbol{\varepsilon}, \boldsymbol{\alpha}))) \tag{3.11}$$

The ability to form signal models for any target shape by linear superposition of scattered signals indicates that Maxwell's equations can be solved for any combination of such targets, with sensitivity and resolution determined by the standard radar equations. The next step is to consider the uniqueness of such a solution.

3.2.2 The electromagnetic uniqueness theorem

The physical rules for propagation and their consequences for EM surveillance solutions are brought together by the EUNIT, which encapsulates the foundation for the function of radar. The EUNIT provides that signals received at an aperture exposed to electromagnetic waves emitted into and scattered and diffracted from within a volume containing a distribution of charges (found as polarised molecules, ionic compounds, free electrons in conductors and dielectric materials) are uniquely determined by (a) the exciting fields and (b) the distribution of charges. By defining and controlling the exciting field and measuring, continually, the scattered signals, the geometric distribution and behaviour of charges can be inferred, uniquely. For further detail on the EUNIT, see Bleaney and Bleaney, 1965 [7].

HSR aims to acquire all information encoded by targets in signals scattered in its direction, and to interpret it in terms of a unique description of the targets in the VoR. To achieve 'uniqueness' is subject to the signal to noise ratio, range sampling, resolution and ambiguity, Doppler sampling, resolution and ambiguity and the resolution and sidelobe performance of the antenna or antenna array. The interpretation is straightforward in the absence of multiple reflections; less so when they are present.

To place bounds on a unique solution, the first requirement is to know in detail the power, frequency, source position(s) and timing of the exciting (transmitted/irradiating) field. The EUNIT then provides that the signals scattered by each target are unique, but may require unachievable sensitivity (as with conductive shielding of one charge structure by another, or multiple scattering or diffraction effects). These may introduce desensitised regions or potential ambiguities in the perceived target field. The second requirement is to fully sample and measure the signals received at each point of a sufficiently large and sensitive receiving aperture, from which the positions of scattering objects or their reflections can be determined. There is no assumption about planarity of the surface, about the shape of a beam, nor about the sequence of observations, other than the spatial and temporal sampling constraints.

The EUNIT provides that, if the VoR is fully covered by the radar's persistent irradiation with an adequate receiving array, the effects of all targets within it may be found by suitably analysing the received signals. These are stored as the content of an associated computer memory, and can be discovered and analysed subject to the noise and power budget of the sensor and the dwell time chosen for the process.

Since data are acquired continually from all directions, the same memory content can be analysed by parallel processing resources over independently chosen time spans to achieve different surveillance functions. In this way, whereas a scanning radar searches space sequentially, and is limited by the speed of light and the narrowness of its beam, HSR acquires the full spatial data set to computer memory and searches there for evidence of targets.

In the case of radar targets whose motion and scattering of EMWs are different (based on resolution in at least one dimension, and the precision of the sensor), the EUNIT provides that it will be possible to resolve target signal components by processes of linear filtering, rather than by threshold detection followed by track filtering and discrimination.

A key aspect of analysis under the EUNIT is that while the form of detection used on BSRs relies primarily on power or amplitude of signals at a specific processing step, the information about target motion and behaviour is primarily encoded in the scattered signal phase. Doppler shift is identically the rate of change of scattered phase. Every radial movement results in a linearly related change in the received signal phase, and every tangential movement results in a change in the phase gradient across the receiving array. If observed continually, these trends can be measured and any information related to presence, position and motion can be extracted.

An example is the ability to resolve an airliner from a wind turbine. The response due to the motion of the aircraft's airframe is dominated by a single-line or marginally spread Doppler spectrum, whose frequency is proportional to the radial speed of the aircraft. The response due to the motion of a wind turbine, whose (usually 3, large) blades rotate repetitively at a rate between 0.1 to 1 Hz, consists of a continuing series of 'flashes', whose Doppler components are spread over much of the Doppler spectrum. The aircraft and turbine signatures are necessarily different under the EUNIT, and for a HSR that acquires all the relevant EMW signal information they can be resolved by appropriate linear filtering. Suitable forms of filter exist to resolve and separate these two forms and have been proven under a wide range of conditions. Different sets of filters may be used in parallel to classify an aircraft (helicopter, propeller, jet, rotor characteristics, etc.), to monitor much smaller targets such as birds at shorter ranges or indeed, to monitor the condition of wind turbines themselves.

Based on the ubiquitous and continual sampling of space and time, many surveillance applications may of course operate simultaneously in the case of HSR, rather than needing to search airspace differently in real, finite time, constrained by the speed of light.

3.2.3 Huygens' principle

Huygens' principle (HP) is a statement arising from Fresnel-Kirchhoff diffraction theory, that provides a simple basis for calculating the propagation of electromagnetic waves in the presence of scattering objects. As a wave progresses through free space and is defined over a wavefront at time t_0, its later configuration can be calculated by treating every point on the wavefront as a spherical source of signal energy, emitting the same signal energy as arrived at each point. Plane waves remain plane waves, spherical waves remain spherical and beam patterns are conserved. However, where the wave is disturbed, e.g., by the presence of a conductive or dielectric sphere, surface currents (either conduction or displacement) are induced, that reduce the amplitude of the forward wave and scatter the deficit as a spherically symmetric, outward-travelling field. Many sufficiently small spheres can be used to simulate non-spherical structures, and HP can be used to estimate the scattered fields as complex vectors.

These spherical 'scattering nodes' are designed to decompose a (not necessarily symmetric) structure into a computable model of its expected scattering behaviour, and to calculate the form of scattered signatures in time and space. The scattering cross-section of an object can be modelled by representing it as an assembly of spheres of the same material as the object and radius not greater than $\lambda/2\pi$, and distributed evenly over the surface of the object at intervals of $\lambda/4$.

Using the type of model described above, the structure and components of an air target can be modelled to evaluate the probable effectiveness of a staring radar (or a radar using a scanned beam) in resolving targets that conform with the chosen model from those (such as birds or clouds) that typically do not.

3.2.4 The reciprocity theorem

The reciprocity theorem is a fundamental in electromagnetic theory and provides that if a known signal is applied to an antenna feed within a VoR and signals propagate to a second antenna, the received signals will be the same whichever of the two antennas is the transmitter and which the receiver. In the staring radar case this is true for each single receiving array element, and to apply it to an extended array the effect of phase adjustments in beamforming is taken into account.

It also applies when a signal propagates via an intermediate reflecting surface. This effect is to be seen when dealing with scattering via two separate objects in forward or reverse order in the case of azimuth multipath.

3.2.5 The speed of light as a constraint

Maxwell's equations and the known physical constants they contain lead to a definite speed of propagation of EMWs. It is possible to communicate with spacecraft near the Moon's orbit in less than 3 seconds. It is possible to obtain a reflection from an aircraft 100 miles away in about one millisecond. However, when surveillance requires the interrogation of objects in many different directions sequentially, the speed of light becomes a critical limitation.

It also constrains how the EUNIT works in relation to electromagnetic waves and retarded potentials.

Radar sensors measure the distance to a target in proportion to the propagation time $\tau_{\mathbf{prop}}$ between transmission, scattering and reception, based on the constant value of $\mathbf{V_C}$.

For radar in which single transmitting and receiving beams are matched, with azimuth resolution $\mathbf{Res_{\alpha}}$, to observe targets in a particular direction within the VoR azimuth sector ($\mathbf{VoR_{\alpha}}$) both beams must dwell in that direction for at least the propagation time before moving on to the next resolution cell. To place $\mathbf{N_p}$ pulses on the target further increases the scan time and therefore the revisit time.

The scan period τ_{scan} therefore has a physical lower limit

$$\tau_{\text{scan}} > (\text{VoR}_{\alpha}/\text{Res}_{\alpha}) \times \text{N}_{\text{p}} \times \tau_{\text{prop}}, \tag{3.12}$$

and τ_{scan} is typically in the order of seconds. The minimum revisit time is the same, but a multifunctional scan tends to further divide the available time between spatial sectors and functions.

In principle a staring radar can make observations over the whole field of regard with a single transmitted pulse; however, to achieve the same sensitivity as the scanning radar over 360 degrees will take the same number of pulses in total and the same time. Meanwhile the target is continually revisited in the staring case, also generating Doppler and other information.

This is not a distinction in energy efficiency, but in the analytic nature of the observation – continually rather than intermittently sampling the encoded information for every resolution cell.

3.3 The staring radar power budget

The RRE allows engineers to estimate the effectiveness of any radar system in detecting targets of a given size (radar cross-section; RCS) at a given range, azimuth, elevation and speed, in the presence of Johnson (thermal) noise plus noise contributed by front end devices.

3.3.1 Signal power, noise, aperture, resolution, dynamic range and accuracy

The standard RRE (3.1) approximates the signal power received from a target of radar cross section **σ**, but assumes free space conditions and does not include reference to polarisation, multipath, diffraction, interference or other propagation effects, and only to the scattering pattern of the target at its particular orientation.

Signal to noise ratio: We can include the presence of random noise with the RRE to yield the staring radar sensitivity equation:

$$\mathbf{SNR = P_r/P_n = P_t.G_t.\sigma.G_r.G_{co}.G_{nc}.(\lambda^2/4\pi)/(4\pi R^2)^2/(B_n.kT.NF)} \qquad (3.13)$$

The power of random noise in the radar's receiving circuits is proportional to its bandwidth $\mathbf{B_n}$. For a 'monochromatic' pulse, the higher the range resolution the greater the bandwidth needed, and the greater the peak power required at the transmitter. We note that for a frequency-modulated transmission the range resolution is constrained by the modulation bandwidth, rather than by the (lower) noise bandwidth, which may be retained at the cost of multiple-delay demodulation channels, and an increase in the radar's occupied bandwidth.

Noise sources are similar to those for conventional radar; however, for staring radar more attention is focused on phase and gain noise as the factor that may limit dynamic range in high-clutter environments (see Section 3.4).

Aperture and directional resolution: The receiver array dimensions $\mathbf{Ap_h \times Ap_v}$ determine the aperture (close to the effective area $\mathbf{A_e}$), related to the effective receiver gain $\mathbf{G_r}$ and the angular resolution of the radar, both in azimuth and elevation ($\mathbf{Res_\alpha = \lambda/Ap_h;\ Res_\varepsilon = \lambda/Ap_v}$), and subject to details of the array geometry and channel elements.

Note that while the gain, $\mathbf{G_r}$, of an array of a given physical size is inversely proportional to the square of the wavelength, the total radar sensitivity is proportional to the receiving aperture, but not to the wavelength. The gain of a transmitter with a given aperture increases with frequency; however, the narrower beam reduces time on target for a scanning radar. Staring sensitivity after processing is independent of frequency, but directional resolution is not.

Range resolution is determined by and inversely proportional to the radar's operating bandwidth ($\mathbf{B_{op}}$), and for a monochromatic pulse, $\mathbf{B_{op} = B_n}$. Then:

$$(R_{rg} = V_c/(2^* B_{op})) \tag{3.14}$$

The **dynamic range** of a phased array receiver formed of **Nr** receiving channels tends to be greater than that of a single-channel receiver with a reflecting antenna, because the array solution uses many receiving circuits in parallel, each serving a small receiving aperture. The overall noise factor will be similar, but the total signal power handling capacity of the array, shared between many channels, is greater by the factor **Nr**. Overall dynamic range is enhanced by beamforming and coherent integration. However, performance may also be affected by interference (accidental or intentional), clutter and phase noise.

The **Doppler resolution** of staring radar is equal to the inverse of the CPI and the effective dwell time. So, for a dwell time of 2 seconds, the Doppler resolution is 0.5Hz. At an L band operating frequency of 1 300 MHz, this corresponds to a radial speed differential near 0.05 m/s.

The **accuracy** of measurement of range, azimuth, elevation is related to the resolution in each dimension by the signal to noise ratio:

$$\mathrm{Acc}(R, Az, El) \approx \mathrm{Res}(R, Az, El)/\mathrm{sqrt}(SNR) \tag{3.15}$$

3.3.2 Sampling space and time

A further set of mathematical constraints or design challenges arise because a radar design typically implies various sampling functions, arising from the pulse repetition interval, any scan interval, the beam width and the effective dwell time, and any beam or aperture divisions arising from multiple functionality. Received sinusoidal signals are either fully sampled and correctly measured in the course of reception, or may exceed the Nyquist limit in which case aliasing will occur.

Nyquist's criterion is that for a signal containing frequencies up to a bound of $\mathbf{F_{max}}$, it can only be accurately represented by a sequence of samples if the interval between samples is $\mathbf{I_{samp}}$ where

$$\mathbf{I_{samp}} < 1/(2\ \mathbf{F_{max}}). \tag{3.16}$$

The challenge for staring radar, in which frequency shifts are fundamentally important for target analysis, is to reconcile range and Doppler ambiguities that arise (see Section 3.3.3). However, persistent observation is available for every resolution element. This allows flexible control of the duration of the CPI; sensitivity and Doppler resolution can be increased, and continuity, can in principle, be maintained with overlapping CPIs.

First, we must follow up the application of Nyquist's theorem for signals that exceed the F_{max} bound, but nevertheless can be measured in terms of complex amplitude. In its basic form, Nyquist's theorem requires that for a sampling frequency $\mathbf{F_S}$ the signal frequency lies between $-\mathbf{F_S/2}$ and $+\mathbf{F_S/2}$. Its phase may then only change, between successive samples, by between π and $-\pi$ radians

This rule can be extended to the case of signals at higher frequency but whose phase rotation between samples (separated in time by the PRI) lies within a range $\Delta\varphi$, where:

$$(n-1).\pi < \Delta\varphi < (n+1).\pi \quad (3.17)$$

where **n** is a constant integer.

In principle this boundary places only a very high limit on acceleration, but in reality it is also limited so that Doppler ambiguities do not overlap. This limits mean acceleration within a CPI up to, for example, +/–60m/s^2 (6G) for long-range L band radar, or up to +/–800 m/s^2 (80G) at higher frequency but shorter range.

The observed frequency (F_{obs}) is then related to the true Doppler frequency (F_D) by

$$F_{obs} = F_D + \text{or} - n.F_S \quad (3.18)$$

Signals beyond +/– $F_S/2$ may be processed linearly provided that this more flexible rule is observed.

Using this Doppler information to determine the radial speed of a target, especially for applications such as ATC, with fast targets at long range, must recognise that for any Doppler bin F_D the actual target speed may be

$$V_r = (F_{obs} - / + (n.F_s)).V_c/2F_c \quad (3.19)$$

The value of **n** depends on the actual target speed V_r and on F_C and F_S (equal to the PRF).

3.3.3 Ambiguities in range and Doppler

The design of staring radar is subject to ambiguities that arise because of the finite speed of light (from Maxwell's equations). The pulse repetition interval places upper limits on the maximum range and the maximum Doppler shift that can be sampled unambiguously. Too short a PRI, and signals scattered from the further ranges in the FoR may arrive after or during the transmission of the next pulse, causing range ambiguity or desensitisation. Too long a PRI and the Doppler components will be sampled at less than the strict Nyquist boundary associated with the maximum radial speed of targets within the specification.

To avoid ambiguities altogether, HSR may operate within a combination of range and radial speed, determined by the operating frequency and the PRF, such that the product **LimH** = $8.V_{rmax}.R_{max}.F_C/V_c^2$, is less than 1.

Here V_{rmax} is the maximum radial speed, R_{max} is the maximum instrumented range, and F_C is the operating frequency.

For example, a HSR designed for maximum range of 10 NM and maximum radial speed of 800 knots at F_{op} = 1.3GHz meets this requirement, with LimH ~ 0.85, and there is no ambiguity.

However, many combinations of practical requirements exceed this value in terms of required operating frequency, range and speed. Ambiguity can then not be completely avoided; however, where the nature of targets is appropriate, methods are available to resolve a finite degree of violation of this rule.

It is notable that the operating frequency itself is directly linked to the Holographic Limit, which results in a preference for lower operating frequencies. In the example above, an X band radar with the same range would be limited to speed of less than 100 knots to maintain unambiguity.

In designing to avoid ambiguity, range ambiguities take a higher priority than Doppler ambiguities because of non-linearities or saturation that may occur during transmission or in the presence of intense clutter. We assume here that a sufficiently low PRF is used to avoid ambiguity in range. Therefore, for many surveillance requirements Doppler ambiguity will occur.

A target discovered in a range-Doppler surface[a] (excluding range ambiguities) may have several ambiguous Doppler frequencies; however, provided continual observation, the correct values of speed and state of acceleration can be determined over time through its actual range rate. Resolution of this ambiguity may be achieved not by explicit comparison of measured range rate after track filtering, but by analysis of signal amplitudes in adjacent range gates during one or more CPIs. For a HSR operating at 1 300 MHz with bandwidth of 1 MHz and range resolution of 150 m, a PRF of 1 kHz and CPI of 2 seconds, the unambiguous range of radial speed is +/−60 m/s or +/−116 kt. The value of **n** is needed to resolve the ambiguity.

In 2 seconds a target at 60 m/s moves by 120 m – i.e., it migrates by almost a full range gate. This change (known as 'Range Walk') is manageable in the course of linear signal processing, primarily because successive range gates in a HSR beam operate coherently, meaning that a phase measurement in one gate is the same as that for the same target in the neighbouring gates at shorter and longer range, and the signal ratios between successive gates are scalars that vary smoothly as the target progresses. The issue in this case is not one of actual measurement error but of demands on computing.

If the target scattering amplitude varies (by fading or multipath) during the course of its transit, as well as varying because of its transit from gate to gate, the fading effect will be common to the two returns, and range ratios will be maintained. By observing a sequence of at least 3 range cells in the beam, the fading effect can be compensated and the range migration measured directly. For targets moving at +/−500 m/s, each point on the range-Doppler surface has 9 possible values of radial speed, which can be resolved between pairs of processing intervals provided sufficient parallel processing threads. The correct radial speed can then be fed forward to the tracking function.

3.3.4 Sensitivity under range walk

With an extended CPI (say 2 seconds), high radial speed results in migration of the target between range gates. For speed up to 500 m/s and range resolution of 150 m, the target may move through several range gates within one CPI. Without explicit process provision the effective gain may be reduced by 8dB. To maintain sensitivity in the presence of such high speed, different versions of each range gate are formed, based on concatenating successive time segments in successive range gates, as described in Chapter 7.

Without continuity of observation, the choice between ambiguous radial speed values is under-determined due to the presence of unknown radial acceleration

[a]An effective tool in observing and assessing target and radar interactions is the range-Doppler surface – a 2-dimensional plot – in which relative intensities of signals in a HSR 'beam' are displayed with Doppler frequency on the x-axis and range along the y-axis.

during an unobserved interval. With continuity, time windows used for Doppler spectrum evaluation leave no gaps. They may be advanced by any number of pulse intervals, and a starting assumption is that 50% overlap will be efficient.

3.3.5 Coherence and decoherence

Perfect monochromaticity in a signal appears to allow the extraction of signals from random noise without any lower limit as the CPI is extended to longer and longer times, but in the real world the available gain may be limited by a process of 'decoherence' between reception, target scattering and transmission. An early question to ask about the useful extent of dwell time is whether the processing gain available from coherent integration of a target return, as observed by the radar, may be limited in cases where the target itself suffers a form of 'decoherence' over time.

Here, references to decoherence should not be confused with quantum decoherence, which is a key concept in describing the evolution of states in interactions of subatomic or quantum particles versus those of macroscopic bodies. It is not relevant in that sense to the coherence of radar signals and targets.

Aspects concerning reception gain, timing errors and phase noise will be addressed in Chapter 6. Here we consider whether decoherence may appear as a property of radar targets that constrains the potential of coherent integration, and one that degrades more severely with increasing carrier frequency.

For staring radar, where it is such a central concept and enabling characteristic, we need to be clearer about our definition of coherence.

Coherence is not synonymous with monochromaticity. It is the relative property of a waveform that allows it to be described by a regularly varying complex exponential function of time. Decoherence expresses the loss of such regularity.

Even a random waveform, if known and used as the basis for such a model (with as many terms as samples) would provide full coherent gain. When the analysis tool used is a Fourier transform, the model is a linear set of single sinusoidal terms, each of which is orthogonal to any other and represents a single, linear rate of progression of phase with time, within the Nyquist limit, at frequency intervals equal to the inverse of the CPI. More complex models allowing not just monochromatic fits, but fits of increasing complexity (such as frequency-modulated sine waves) are also effective. These represent discrete levels of acceleration, jerk, yank, etc.; i.e., they allow coherent integration with non-linear progressions of phase with time, and select for increasingly dynamic trajectories, requiring increasing computational capacity to explore.

The process of coherent integration remains valid provided that there is a regular progression of complex amplitude over the CPI, from one digital sample to the next, as described above in Section 3.3.2 with acceleration limits consistent with solid bodies and Newtonian rather than explosive dynamics. The effect of target dynamics on staring sensitivity will be discussed in Section 3.11.1.

Phase noise represents a lower bound on coherence and can have significant second-order effects in the presence of large (typically clutter) signals.

3.3.6 Photons, airspace and memory

Staring radar does not 'search' airspace in any physical, sequential sense except that of range: the irradiating wave is incident sooner at short range than at long, not because of a search pattern, but because of the speed of light. Signals from all locations within the VoR are acquired during the pulse interval, and time domain beam formation also occurs within that interval.

Radar engineering and design can benefit from a direct reference to their links with quantum physics.

A small HSR transmitter emits a pulse with a peak power of, for example, 2 kilowatts, 1 microsecond long, at a frequency $\mathbf{F_c} = 3$ GHz, at a pulse rate of 8 kHz. Physically it is a superposition of photons.

The carrier frequency determines the energy, $\mathbf{E_{ph}}$, of transmitted photons, via Planck's constant, **h**.

We have $\mathbf{E_{ph}} = \mathbf{F_c}.\mathbf{h}$, where $\mathbf{h} = 6.6 \times 10^{-34}$ Joule-seconds. The photon energy is then $\mathbf{E_{ph}} = 2{\times}10^{-24}$ Joules, the pulse energy, $\mathbf{E_{pu}}$, is 0.002 Joules, and each pulse consists of 10^{21} photons. The pulse duration yields a bandwidth near 1 MHz. In this case, 2 048 pulses are assembled for Fourier analysis.

A target object with radar cross-section of 1 sq. m, at a range of 10 km, encounters and scatters about 10^{12} photons from this pulse. The array aperture then receives about 3 000 photons from a single pulse and target, during a 1-microsecond sampling interval. During a CPI it receives about 6 million photons with coherent phases over the array, with total coherent energy of about 25 femtojoules per CPI. Johnson noise for the array is equivalent to about 0.5 femtojoules per CPI, yielding a signal to noise ratio of about 17 dB after beamforming and Doppler transformation.

This is consistent with the range equation and links the radar power budget with quantum energy states. With respect to the probabilities of any one photon interacting with each array element, the quantum rules mean, in effect, that each photon checks out every possibility of absorption by electrons in each element. In these numbers, this yields the result expected from Maxwell's (non-quantum) equations, which remain the basis for further modelling and calculation.

Each PRI, range samples from every channel are acquired and distributed to memory in a digital beamforming processor, so that the entire VoR is updated every 125 microseconds.

Subsequently, the whole VoR is available for access based not on the transit time of photons but on the memory access time. The photon transit time in the 10km-radius VoR is in the order of 100 microseconds, whereas the memory access time is measured in nanoseconds.

Ubiquitous or staring irradiation makes the staring array available for signals from any part of the VoR. Provided that memory and data communications capacity of the radar are appropriate, computing capacity may be added as necessary and functions performed in parallel without disrupting the physically limited operation of the radar. Much recent effort on multifunctional active electronically scanned array radar has considered subdivision of the radar array between various functions and waveforms. That

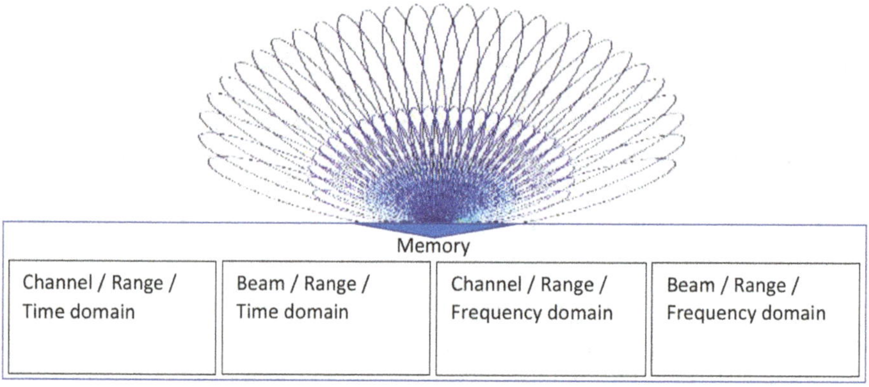

Figure 3.3 *Data can be organised by channel and range and by beam and range in the time domain. Fourier transformation gives the frequency domain. This is referred to later as a 'RAED' structure.*

concept is an outgrowth of the paradigm of sequential spatial search and risks greater delays compared with the staring configuration.

Figure 3.3 illustrates the wide receiving beam pattern of HSR and the connection from the VoR to the Channel/Beam/Range memory segments for target searches.

3.4 Multipath propagation and the EUNIT

The EUNIT addresses propagation using time-varying retarded potentials and propagating waves. With time-varying sources and scatterers the effects of reflection and shadowing, absorption, diffraction and multiple scattering need to be accounted for so that boundary conditions are resolved. The issue is whether the major targets (which form the primary boundary conditions) are represented by separable phase progressions, including propagation via multiple paths.

We shall address 'multipath' propagation in further detail in Chapters 6, 7 and 9. However, it is appropriate to note its physical origin here. It is significant for both scanning and staring radar.

In free space, with a small number of targets spread over a large number of resolution cells, the probability of overlaps in reception is slight, and the solution for each arises from measuring resolved range, Doppler, azimuth and elevation values. However, while the EUNIT and Huygens' principle still hold, the solution is more complex if secondary or multiple reflections occur – i.e., the signal travels from the transmitter and back to the receiver via a secondary reflector. Multiple propagation (or 'multipath') is a common occurrence with targets near the ground or sea surface.

Surface multipath occurs in two distinct ways. The first and more critical is that the signal transmitted from the radar to a target arrives by two paths that may interfere destructively, leaving the target without irradiation. This situation is not recoverable

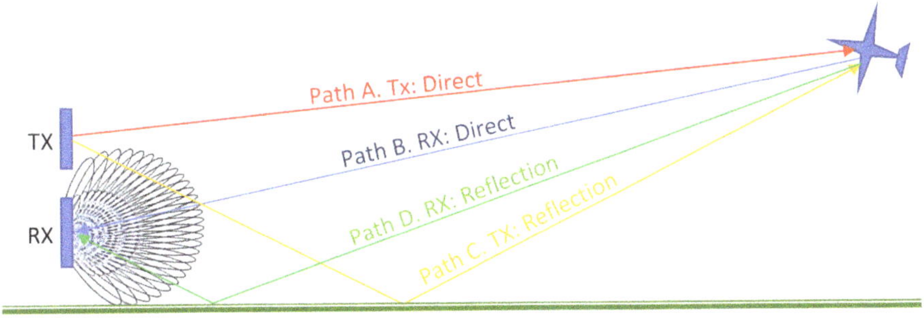

Figure 3.4 *Propagation paths available where a ground reflection occurs*

on the receive side. Its effect can be limited by treating either the ground surface with absorbing or reflective material or, more practically, by ensuring that transmitter side-lobes directed at negative elevation (below the ground surface) are always less in amplitude than those transmitted at the same angle above the surface. This can be achieved to a substantial extent by careful choice of phase and amplitude adjustments in the transmitter, but is more challenging the closer the target is to the horizon.

The occurrence of multipath at the HSR receiver can be detected directly, since at the range and Doppler frequency associated with a target it causes the power received at the array to vary with position in the aperture (which does not occur with single-path plane waves). A horizontal reflecting surface causes the power to vary with vertical position, and a close-in vertical reflector results in differences in the horizontal direction.

In Figure 3.4, A is the Transmit-direct path and B the Receive-direct path. C is the Transmit-multipath and D is the Receive-multipath. Interference occurs at the target between A and C. Destructive interference at the receiver occurs when path D interferes with path B.

For surface multipath the delay differences are typically small. The four paths AB, AD, CB and CD are coherent but are received with different phase progressions up the array: in general, a single-CPI measurement will not resolve them, but observation over successive CPIs may be sufficient. The signal information they transfer may support analysis of the distinct paths themselves and of the reflecting surface that causes them. Methods will be discussed in Chapter 7.

Multipath can result in loss of signal and measurement errors, but it does not invalidate Huygens' Principle or the EUNIT. In compliance with the EUNIT, it uniquely arises from and communicates, over time, information about the reflecting surface as well as the target.

Azimuth multipath is addressed in Chapter 9.

3.5 Mechanical and geometric effects

Mechanical and cost constraints aside, a radar sensor can function more effectively the larger the aperture of its antenna. For ground based radar, while the size of radar antennas can impose added cost, environmental and structural consequences, size itself is a less demanding constraint than for airborne radar (note the size of PAVE PAWS, Jindalee and especially Cobra Mist radars).

Ground-based, staring radar has the inevitable consequence that an adequate aperture must be provided simultaneously for surveillance in every VoR direction, whereas a scanning radar can provide coverage of the VoR with a single, rotating, flat-panel aperture, albeit intermittently and slowly. For staring radar with a 360-degree azimuth VoR a cost penalty therefore arises with the number of antenna elements required. However, these are not transmit-receive modules. They will be substantially cheaper than those in a AESA, and the overall space occupied and the structure required are unlikely to disadvantage the static, staring case.

For coverage in transmission the staring solution does not require a large aperture either for beamforming or for sensitivity. A much smaller array can be used, and assuming the VoR is broad in azimuth (say 90–360 degrees) and narrower in elevation (say 20 degrees), it can consist of a small number of elements vertically, each a simple radiator, depending on the state of polarisation. These few fixed, high-power elements are less expensive than a large number of phase-adjustable elements.

A staring radar also requires only fixed mechanical stability, rather than dynamic stability, with an associated cost saving and no Rotating Joint. There are of course constraints on the overall design of radars that may be remote, arising from requirements for power supply, security, communications, access for maintenance and stability under weather conditions.

In total, and subject to costs of processing that will be addressed in Chapters 6 and beyond, it should be expected that surface-based staring radar can be cost-effective when compared with a scanning configuration.

For airborne radar, the primary constraints on size arise from the airframe geometry, scale and flight characteristics, which in many cases override considerations that would otherwise govern the performance of the radar. In such cases, demand for high operating frequencies, high beam agility and high power at those frequencies may dominate the designer's freedom of choice. Airborne staring radar may have substantial applications, but in this book we consider primarily ground based applications, in which size and weight may be subordinate to functional performance requirements.

3.6 Beams and sidelobes

The design of radar antennas has been aimed at narrow beam widths (implying both high angular accuracy and sensitivity) with very low sidelobes (implying the ability to resolve target signals and positions with widely varying intensities). These in turn impact the necessary size of antennas. To achieve very narrow, high-gain transmit/receive beams requires accurately shaped and stable but movable reflectors, or arrays giving accurate knowledge of element positions and current complex gains. In practice sidelobe gains are non-zero, with sufficient gain that given a minimum

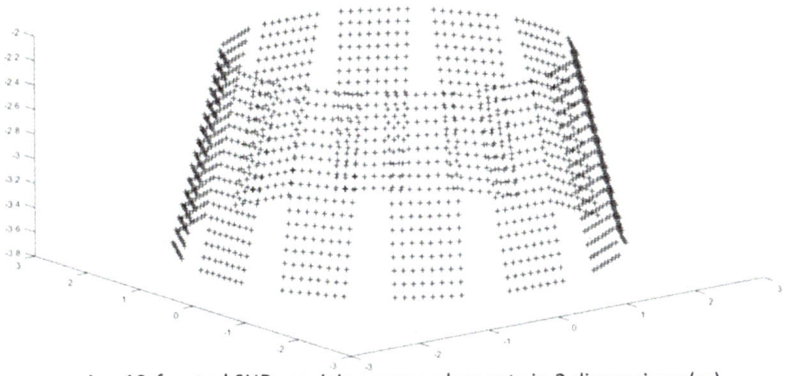

A. 12-faceted SHR receiving array; elements in 3 dimensions (m)

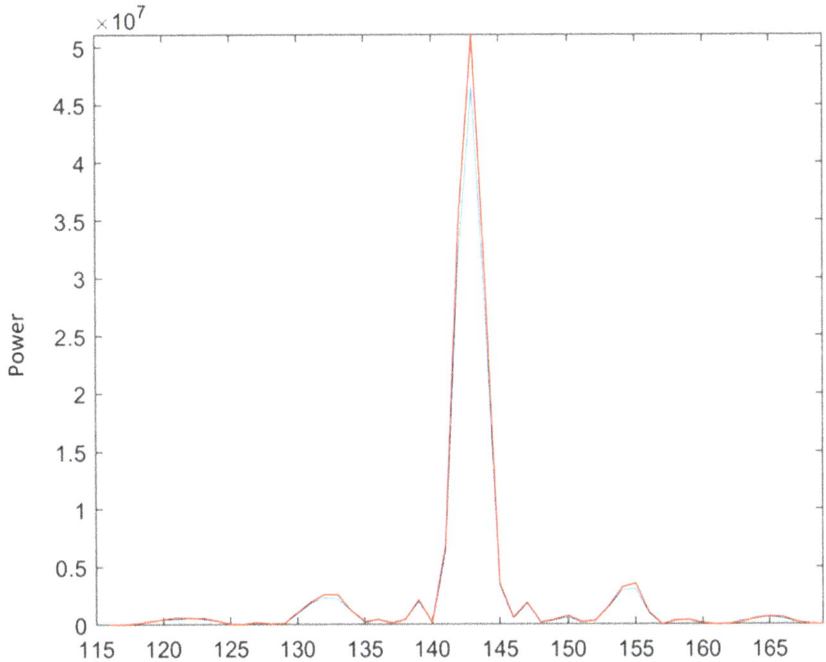

B. Unweighted azimuth beam pattern for 12-faceted SHR receiving array

Figure 3.5 (A), (B). 12-faceted array geometry and beam patterns, unwindowed

target is detectable in the main lobe, larger and closer targets will also be detectable in sidelobes.

Sidelobes are the result of the interaction of the actual antenna aperture and elements, weighted appropriately in the case of an array, with the wavelength, according to Fraunhofer diffraction theory. Efforts to minimise them are a major focus for antenna engineers and are expensive in performance terms in that the antenna array will need a large overall extent, high stability, calibrated elements and lossy windowing functions.

For a scanning radar of any type there is a time delay between interrogating a target through the main lobe and an unintended response from the same target when observed via a sidelobe. These time delays are uncorrelated with the PRI, the Doppler period and any fading characteristic of the target. Therefore, signals that are detected in any sidelobe are decorrelated from and incoherent with those received from the same target when in the main lobe, and the two have to be treated as separate targets, only distinguishable by non-coherent target and track processing. Sidelobe detections and sidelobe interference present fundamental and expensive limitations on scanning radar performance.

All antennas and arrays have sidelobes, but their significance for staring radar architectures is different from their significance in sequential scanning. The process of parallel coherent digital beamforming still results in residual sensitivity in unintended directions – i.e., in sidelobes. However, these sidelobes are formed numerically from the same simultaneously received signals as the main lobe; just with different, known complex weightings. They will be at the same range and Doppler shift and closely correlated in phase with the main lobe response. Therefore, in searching the range/beam/Doppler space, main and sidelobe returns can be co-processed coherently to determine a singular position and radial speed. Sidelobe misdetections can therefore be largely eliminated in the staring case. An array geometry and azimuth beam pattern for a multi-faceted HSR receiving array are illustrated in Figure 3.5.

In the case of interference via sidelobes, their different significance has a similar effect as for targets. In the case of staring radar, while sidelobes still occur, an interfering signal will be found in the closest main lobe and will be phase-correlated in its sidelobes. It is possible to identify the different lobe returns and find the correct direction of arrival, and various strategies are then available for their rejection, as we shall see in Chapters 6 and 10.

Where the source of potential interference is predictable, as in the case of a neighbouring node of a surveillance network, the solution is normally to operate on an orthogonal carrier frequency. However, we shall see in Chapter 10 that the staring concept presents alternatives that may contribute to the surveillance system's effectiveness.

3.7 Targets, the propagation medium and histories

Targets in the VoR form the boundary conditions that both enable and constrain the radar function. For radar to succeed, targets must embody contrasting properties with those of free space. They must contain features of a scale sufficient to diffract or scatter the wavelengths in use and must exist in a space that allows adequate wave propagation. For example, targets embodying scattering features smaller than a bee require operating frequencies at X band or higher to scatter efficiently, while robust propagation through many kilometres of rain needs signals at lower frequencies.

A scattering target's cross-section varies with its material and scale, its shape, its orientation with respect to the incident wavefront and to the direction of observation, and with the operating frequency and waveform. For spherical, metallic targets, the radar cross-section has a peak close to 4 times its optical cross section for a wavelength equal to the target circumference, converging to a ratio of 1 at short wavelengths and rapidly towards zero at longer wavelengths. Real targets may be modelled in terms of assemblies of small spheres, and linear, planar or curved surface components will scatter in ways consistent with Fresnel-Kirchhoff diffraction theory and Huygens' principle.

A complex target shape of many wavelengths' scale will yield cross-sections that vary rapidly with orientation, including both peaks and effective zeroes. A short target dwell time will yield an unpredictable value with a mean intensity close to the average, but a very low intensity (the effect of 'fading') has a finite probability.

The long and continuing dwell times used by staring radar tend to reduce the likelihood of missed detections due to target fading, and yield Doppler components related to the airframe shape and its continuing changes in relative orientation. The observed cross-section is a scalar value (in the case of linear polarisation), whose variation with time depends on the target's shape and motion, and also on multipath effects. It is challenging to decode in a quantitative way, but it offers substantial opportunities for analysis where long-term histories are available. A significant example is that observation of longer-term changes may assist in the analysis of aviation incidents (close approaches, collisions, impacts and dismounts). These opportunities will be addressed further in Chapter 8.

The effect of aspects of the waveform other than its carrier frequency depends on the detail of the shape and orientation of the target. A waveform can be chosen to favour returns from one target and configuration over another, and suggests capabilities in target recognition. The corollary is that this type of adaptive choice of waveform will probably be disadvantageous if the target type or orientation differs from the expectation.

A preferred strategy for a staring radar is to keep the irradiating signal as simple as possible, consistent with resolution and scattering requirements, so that any information encoded in the scattered waveform originates with the target, independently of the signal generation.

3.8 Target and clutter types, features and models

The physics of radar and its targets must accommodate a wide range of objects and materials.

Broadly these fit within a small number of classes. In Table 3.1, we outline the physical means whereby they can be distinguished by means of radar of an appropriate design.

Table 3.1 Target classifications consistent with the materials and models we have outlined

Target type	**Cross-section**	**Speed or behaviour**	**Ranges**	**Classifier**
Civil aircraft	1–1 000 sq. m	100–600 kt	1–60 NM	Prop/JEM/ traject.
Military aircraft	0.5–1 000 sq. m	50–1 000 kt	1–70 NM	Prop/JEM/ traject.
Helicopter	1–100 sq. m	0–200 kt (hover/ rotor)	1–60 NM	Rotor/hover
UAV (fixed wing)	0.001–10 sq. m	10–200 kt	0.1–20 km	Prop/RCS
UAV (rotary)	0.001–1 sq. m	0–50 kt (hover/ rotor)	0.1–10 km	Rotors/traject.
Clutter type	**Cross-section**	**Speed or behaviour**	**Ranges**	**Classifier**
Buildings	1–10 000 sq. m	0	1–10 km	Static/scale
Vegetation	1–10 sq. m	0–2 m/s (waving)	1–10 km	Location/phase
Sea clutter	0–1/100 sq. m/ sq. m	0.1–10 m/s (rolling)	1–60 NM	Elev/Vrad/ random
Wind turbines	100–100 000 sm	0.1Hz–100 m/s (rot.)	5–60 NM	Static/ Repetition
Birds	0.001–0.1 sm	2–20 m/s (irregular)	0.1–10 km	Traject./no rotor

The most significant, common feature that supports analysis under the EUNIT is that targets of interest are Newtonian solid bodies: they have a continuing existence in the VoR, and their motions are constrained by physical inertia and limited forces (motive or environmental due to air motion). Maxwell's equations and the EUNIT then place limits on how their scattering properties can change as a result of these motions, and therefore, subject to sampling, on changes in the encoded waveform information. Newton's laws do the same in the mechanical field.

Unmanned aircraft and birds, buildings and vegetation constitute a set of types generally observed at shorter range. Other types are targets and clutter observed by longer-range surveillance radar.

Certain aspects of staring radar favour lower operating frequencies such as S or L band. The smallest object feature dimension that scatters effectively at these frequencies is in the region of 3 to 6 cm (conductive) to 5 to 10 cm (dielectric), respectively. This is consistent with all these types of target, although small rotary drones are tending to the lower end.

Most types have either all-metallic surfaces or a combination of dielectric and metallic (drone motors, batteries, conductors). Their overall cross-sections range down to about 0.001 sq. m (−30 dBsm); however, many have features such as rotors whose cross-sections may be below 10^{-4} sq. m (−40 dBsm), which are found in association with drone airframes.

Staring radar can be equipped to detect and discriminate between the motions of rotor blades and those of birds' wings, considerably simplifying the problem of false reports caused by birds observed by counter-drone radar sensors.

3.9 The volume of regard and radar networks

The VoR is defined by the distance, in any direction, at which the specified minimum target can be detected with the specified probability. For a classic primary surveillance radar this is a circular volume whose height depends on the vertical beam pattern of the receiver. Such a radar, whose transmitter and receiver operate through the same antenna, is known as monostatic, and is typically ground based and situated in static terrain. Alternatively it may be aboard ship and subject to its own proper motion and that of the ocean surface.

A different VoR can be constructed using separate transmitters and receivers to form a 'bistatic' radar, and subject to constraints of time and frequency, multiple transmitters and receivers may operate as a 'multistatic' array. We shall encounter such arrays in Chapter 10.

Radar measurements may be 2-dimensional or 3-dimensional, and the resolution cells within which targets may be detected are scaled in terms of the bandwidth of the radar, and the antenna aperture compared with the operating wavelength.

In Chapter 10, we shall consider a coherent, staring radar network that may offer modes of operation in which spatial search by monostatic sensors, whose VoRs overlap and may support MIMO operation, but do not cross-irradiate, is replaced by a multiply-irradiated airspace in which multistatic operation is exploited efficiently.

Each node of such a network may operate with a local transmitter and receiver, and may use a different operating frequency. However, it is a particular benefit of staring radar that, in addition to detailed and precise target analysis, they may be allowed to operate at the same carrier frequencies and with synchronous timing and coherent waveforms. This approach to networked surveillance offers great spectral efficiency, non-cooperative multilateration of targets and also multi-directional target analysis, which will be addressed in Chapter 10.

3.10 Atmospheric losses and precipitation

For all surface-based radar sensors the radar propagation path passes through the atmosphere, which contains turbulent air, clouds and precipitation. To maintain surveillance performance for targets of interest, sufficient power budget is needed to support attenuation of forward waves by atmospheric gases and rain, and appropriate discrimination is needed against backscatter by wind vortices or fluctuations in rainfall. Attenuation and scattering coefficients for weather features are given in Chapter 8, Table 8.1.

There is a greater penalty to be paid at higher frequencies due to these sources of attenuation and clutter; however, higher frequencies may allow more precise weather characterisation, especially where polarimetric measurements are available.

3.11 Analytic solutions for targets and clutter

Information derived from a target by a staring radar is contained in the complex amplitude sequences received and typically described in Fourier transforms of pulse sequences, sampled at fixed delays. The radar generates effectively continuous, sinusoidal signals, whose complex spectra are determined by amplitude and phase modulations superposed on target Doppler shifts. The received signals are analytic functions of time. (Note they are not mathematically formal 'analytic signals', which have specific properties not relevant to radar Doppler.)

Staring radar aims to derive unique target trajectory solutions from an effectively continuous series of complex signal values received over an aperture. The values are the signals that arise from Maxwell's equations and the boundary conditions: they contain a combination of trajectory and target information which, under the EUNIT, is unique.

If the trajectory can be described by a convergent polynomial (which in the absence of explosions it will be), then the complex amplitude sequence received will be an analytic mathematical function; typically it is differentiable, is not discontinuous and can be represented by a superposition of sinusoids.

Fourier analysis is a first and very efficient step in detecting coherence in signals that arise from near-ballistic motion. Fourier transformation focuses on sinusoids, based on their orthogonality, with resolution inversely proportional to the input dwell time, and without loss of signal energy. In other words it is reversible. It is efficient at focusing near-monochromatic signals, while random noise during the sampling period is resolved into its separate, complex frequency components with Rayleigh-distributed amplitude and random phase (or Gaussian-distributed in-phase and quadrature components).

Coherence is a richer property than monochromaticity, and here we return to the term 'analytic'. In the case of scattering from a near-ballistic target, accelerations result in non-linear motion but not discontinuous positions. The phase progressions generated by such motion remain coherent to the extent that phase deviations from linearity can be traced by a convergent polynomial: acceleration brings finite second-order terms, jerk brings third-order terms and so on. To find these coefficients may be a demanding computing task, but the deviations do not represent loss of signal energy or information; they take more general searches and more time to identify.

Analytic solutions based on staring radar have two advantages: first, the actual trajectory dynamics are observed and can be determined by coherent processing, and secondly the CPI can be shifted or extended as necessary to optimise the response.

For a staring, coherent radar we are free to choose the length of the CPI, and fine Doppler distinctions can be enabled by extended dwell times. Persistent dwell provides the potential for retrieval of a continuous time domain and Doppler history, and an extensive resource for classification and target recognition. We shall be concerned with the physical constraints on this ability to assess the nature and behaviour of targets and clutter.

The benefit is that it becomes possible to determine activity within the VoR in an analytical way – i.e., a way that is uniquely determined by the physical targets themselves and can be calculated by the solution of explicit equations. We have prepared for such solutions in earlier sections of this chapter, and here we can introduce issues that may benefit from such analytical solutions as applied to SRC and ATC radar systems described in Chapter 5, and with respect to target dynamics and coherence.

3.11.1 Doppler effect with target dynamics

An apparent decoherence can arise from radial acceleration of the target. In its simplest form a coherent signal (as scattered by a target approaching radially and at a constant speed) has a monochromatic angular frequency (Doppler shift) equal to its rate of phase advance, and this is efficiently captured during Fourier transformation. However, radial acceleration results in variation of the frequency and a spreading of signal energy across at least a part of the Doppler spectrum. Although the amplitude of the Fourier transform loses integration gain if the frequency varies during the CPI, coherence itself is not lost. Signals lose coherence if the phase shift from one cycle to the next becomes random, as with thermal noise, or fails to meet a Nyquist-related criterion. For a physical object under acceleration, these conditions will only apply at very high acceleration rates or carrier frequencies, as described in Section 3.5. Under normal dynamics, phase changes will be strongly correlated pulse-to-pulse, and an analytic solution for coherent integration will exist. Examples are illustrated in Figure 3.6.

The use of Fourier transformation for coherent integration on a target in such a trajectory results in reduced amplitude in the frequency domain not because energy has been lost but because it is spread conservatively between successive FT bins. The sequence of phase and amplitude in these bins then retain the dynamic information.

The high-resolution plots show that the in-phase and quadrature components (red and yellow, respectively) are not randomly phased or located in frequency but follow an ordered sequence.

The F_D sequence is where information describing the dynamic motion is encoded; the radial motion can be reconstructed in detail from that sequence.

The range of resolvable analytic trajectories is large, and their solution will incur a potentially large computational burden. It might be unmanageable if a non-conditional and wide-ranging process were applied to all resolution cells for all dynamic cases. However, as referred to above, the spreading of energy through the Fourier transform does not dissipate signal energy but does spread it in frequency. Provided that cell discovery can be extended to total coherent energy (e.g., using a Vector Histogram; see below and Chapter 6: Cell Discovery and HSR Signal Metrics), sensitivity is maintained. In that case selection of a sufficient subset of cells for a wider range of processing and dynamic search may be achievable.

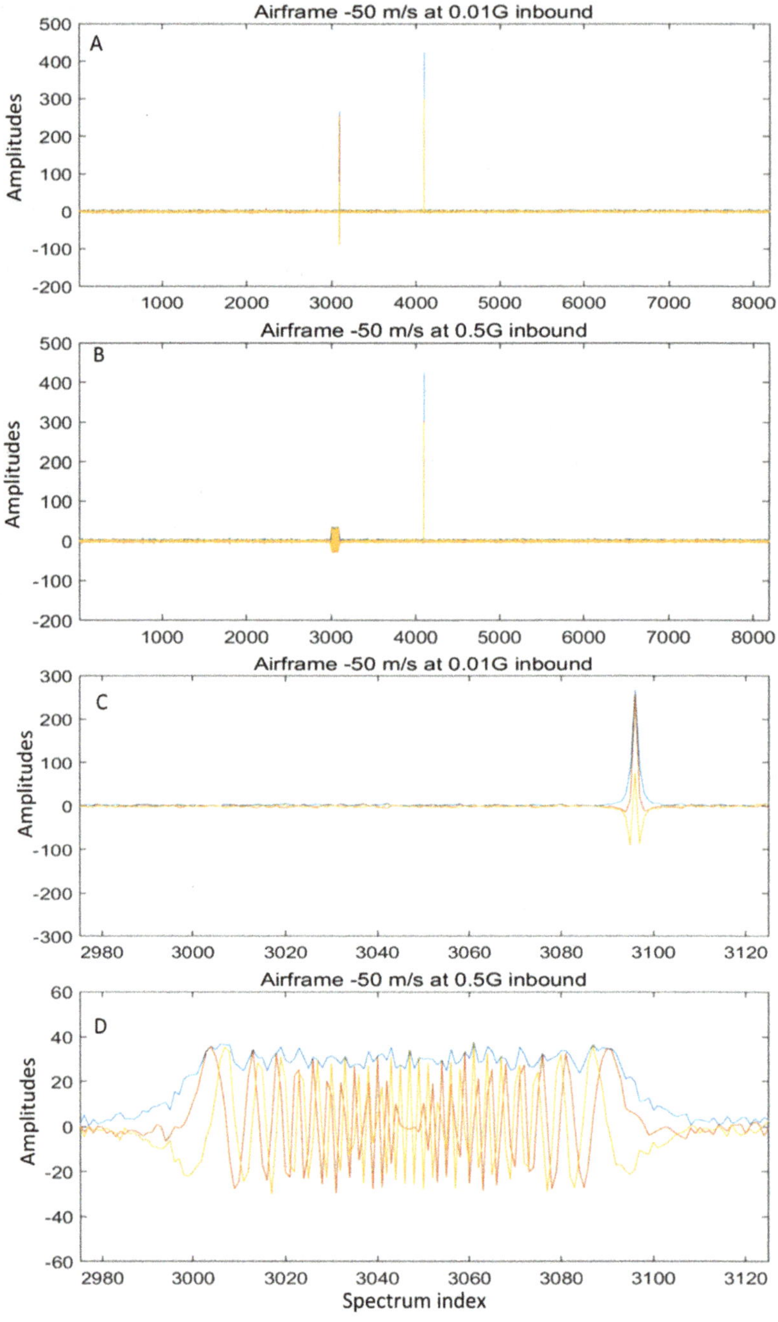

Figure 3.6 *A – D. Doppler spectra for (A) near-constant radial speed, (B) a constantly accelerating profile; (C) and (D) show the effect of low and higher acceleration at expanded resolution*

Target dynamics, subject to Nyquist-consistent sampling, do not result in decoherence. They represent departures from the simple case; a constant radial target speed, for which a Fourier transform is ideally suited. To maintain coherent gain is a matter of process and computing capacity rather than a physical constraint, and it is only necessary that a sufficient range of dynamic cases is considered. Once an active cell is discovered, with estimates of range, speed and acceleration, much more computing power can be applied to exploit the complex signal sequence.

Appendix 1 contains a range of dynamic target behaviours and the resulting FFT outputs.

3.11.2 Target modulating features

Many targets of interest include features that (rather than phase modulating the scattered waveform as in the Doppler effect, where a constant radial speed results in an F_c -proportional Doppler shift) are encoded directly as amplitude or complex modulation. These effects can be modelled in the same way as the Doppler effect, using a Huygens decomposition of spinning components attached to and moving with the airframe.

Figure 3.7 shows the 8 192-point spectrum modelling a quadcopter approaching at 10 m/s, with 2-bladed, 15 cm rotors operating at 10 000, 10 400, 10 900 and 11 300 RPM. Each rotor generates 4 components, the outer pair related harmonically to the inner, and all symmetric about the airframe component at Fourier bin 4 020. The sequence is limited by the Doppler shifts of the blade tips. These spectrum features illustrate the scattering and modulating characteristics of a drone, modelled using the method described above.

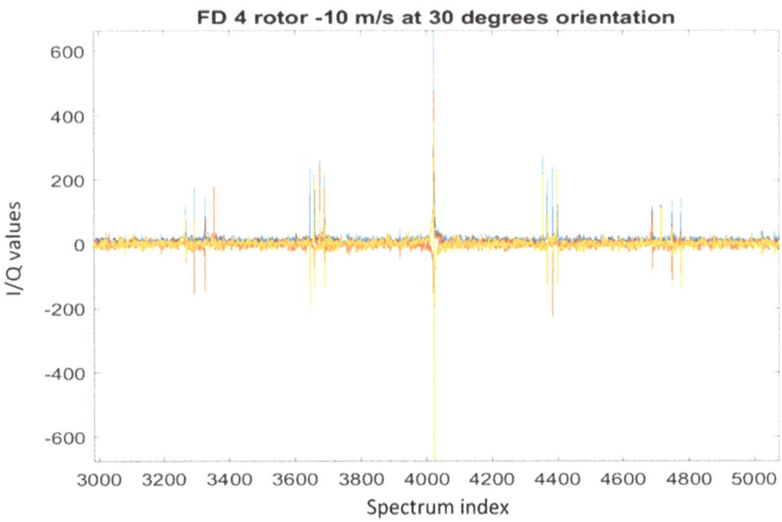

Figure 3.7 Spectrum scattered by both Doppler and modulation effects

3.11.3 Resolution cell analysis for large VoRs

Cell analysis to this point has been based on measurements up to the point of detectability; i.e., can the minimum target be detected under the conditions of RCS, speed, dynamics, multipath, etc.? Once a cell has been discovered to contain signal energy representing targets other than pure static clutter, or to be affected by large static objects, or by interference, or by a circuit failure, its raw data history can be selected for further intense analysis.

Examples of cell contents that may need distinct treatment are those that contain targets at high acceleration rates, large clutter targets where noise is degraded with phase noise, directional sources of interference, moving clutter from surface traffic, moving clutter from wind turbines, etc. In some cases these may be pre-defined based on known sites and conditions; however, operating conditions may vary and a basis for making cell-specific process selections will be necessary.

The whole VoR may contain over a million resolution cells. The form of processor best equipped to treat this large space is currently a Graphics Processing Unit, or GPU. A single-chip GPU may contain over 1 000 32-bit parallel, floating-point arithmetic processors and a substantial memory, and will be able to apply over 8 TeraFlops to process incoming data. To accommodate all the cells it may be necessary to apply several or many GPUs to the whole VoR, and this can be achieved both by separating the overall range window into separate swaths to be processed by paralleling of GPUs and by assigning the processing of multi-facet sectors to specific subsets of azimuth directions. However this high speed is only properly accessible if applied in a regular, unconditional stream. There will be a point at which a smaller number of cells (but still potentially a substantial number, in hundreds or thousands) need to be assigned for condition-based processing in the nature of target analysis by CPUs. Cells can be directed into an appropriate stream based on universal analysis following cell formation, discussed in Chapter 6.

Channel signal data arrives initially in the range/time domain and can be Fourier-transformed directly into the frequency domain. Time domain and frequency domain outputs for each cell are formed via Paths a and b below, and frequency-domain versions for each channel by Fourier transformation via Path c:

a. complex gain-weighted sums of all channel outputs in the time domain,
   $\underline{\mathbf{V}}_{\mathbf{TD}}\,(\mathbf{R}, \alpha, \varepsilon, \mathbf{t}) = \sum_{\mathbf{h,v}}(\underline{\mathbf{V}}(\alpha,\varepsilon, \mathbf{h}, \mathbf{v}, \mathbf{t}).(1/\mathbf{G}(\alpha,\varepsilon, \mathbf{h}, \mathbf{v})').\exp(\mathbf{j}\phi(\alpha, \varepsilon, \mathbf{h}, \mathbf{v}))$
   where $(\mathbf{G}'.\exp(\mathbf{j}\phi))_{\alpha,\varepsilon,\mathrm{h,v}}$ is the fixed complex gain at α, ε for each element **h, v** of the array, and **V** symbolises a voltage level as opposed to velocity.
b. Fourier transformation of the time domain series for each beam and range cell:
   $\underline{\mathbf{V}}_{\mathbf{FD}}(\mathbf{R}, \alpha, \varepsilon, \omega_D) = \mathrm{FT}(\underline{\mathbf{V}}_{\mathbf{TD}}(\mathbf{R}, \alpha, \varepsilon, \mathbf{t}))$,
   where a fast Fourier transform algorithm will typically be used.
c. Complex gain-weighted sums of all channel outputs in the frequency domain.

Paths (a + b) and (c) should give the same results unless different window functions are required, but may apply to different signal tests and provide options for different lengths of CPI. The channel/beam data structure via paths (a), (b) and (c) is illustrated in Figure 3.3.

The complex signal time sequence and frequency spectrum represent the information content of each cell, identified in parallel by range, azimuth and elevation $(\mathbf{R}, \alpha, \varepsilon)$. If an assumption is made that targets are represented by peaks in the Doppler spectrum, then detection may be achieved in a similar way to a BSR, but with the option of using the extended Doppler spectrum instead of the range variable to determine a CFAR threshold.

The continual interrogation of each cell by a staring radar yields a continuing sequence of complex signal data about the content of each cell, which, under the EUNIT and subject to the power budget of the radar, contains information on every aspect of target scattering, including position, size, motion and dynamics. As a result, in Chapter 6 we seek a method to discover those cells that require further processing on the basis of departures from statistics typical of no-target cells.

Discovery and a choice of processing on the basis of thresholding raises an issue of the form of processing capacity. Parallel processing engines such as GPUs provide very high rates of operation, but they do not easily accommodate conditional direction and selection of processes. It is important to carry out as much of the processing as capacity allows in a non-conditional manner – i.e., in parallel on all resolution cells, reserving specific cell-oriented processes for fast DSP, FPGA or CPUs.

However, this issue is so central to the operation of staring radar that we seek a mode of operation that avoids, as far as possible, conditional process control but allows for distinct but parallel processing paths.

An efficient, non-conditional sorting method is a Vector Histogram, which will be described in Chapter 6. In summary, it replaces a threshold process, in which a number is compared with a pre-determined value to decide a future process and memory allocation, with one in which location address and content are transposed. The future process applies to the content of each content-derived address, determined non-conditionally by the prior process. Its cost is that it will require greater memory provision than the result of a process such as a threshold.

Cell characteristics that can be selected for different processing streams are discussed in Chapter 6. They include cells with abnormally low or high noise distributions, with saturation in the time domain, with evidence of high static clutter or phase noise, with repetitive clutter signatures and cells with normal target/clutter contents.

Cells containing targets in dynamic trajectories that stretch the ability of Fourier analysis to provide coherent gain can be detected and are amenable to analysis, including speed resulting in Range Walk, accelerative Doppler Walk or potentially Azimuth Walk, as will be seen in Chapter 6.

3.12 Spectrum selection and occupancy

At a number of points in this chapter, the choice of operating frequency of a radar sensor has been raised.

The choice of frequency for a staring radar will be subject to the same regulatory regimes as for scanning radars, but somewhat different criteria for performance and cost. For staring radar, where the analysis of targets is a key objective, the selection is driven in part by the features of targets that are subject to analysis, and in part by ambiguities in the solutions. For small airborne targets, features on the scale of 5 cm or less may be of interest at ranges up to 10 km. For larger targets, speed ambiguities below 100 kt may be of concern, at ranges up to 60 nautical miles.

No designer of radar systems has the luxury of assuming exclusive occupation of the radio spectrum. The avoidance of mutual interference between radar and other spectrum users, whether within the same overall system architecture or not, has required the use of orthogonal carrier frequencies, and considerable effort in terms of design to meet regulatory requirements, combined with the awareness of increasing spectral congestion. Wide spectral bandwidths have the advantage of yielding high spatial resolution and potentially improved clutter discrimination; however they militate against the availability of spectrum for other users, whether for radar or for other functions of radio systems. Wider spectral bandwidths also imply higher levels of random noise in receiving circuits and potentially higher power requirements in transmitters.

Specific allocations are provided for surveillance radar by Ofcom and the CAA in the UK, by the FCC and NTIA in the United States, and by national organisations in other countries hosting radar sensors. The regulations are coordinated overall through the International Telecommunications Union. The availability of spectrum has been both an enabler of high-performance surveillance and a constraint on the design and deployment of radar sensors. Mutual interference can take the form of noise and saturation through direct beam interactions, but also indirectly where a target irradiated by one transmitter is within the receive beam of a neighbouring system. Coordination of operating frequencies for scanning radars has been critical in both of these senses.

The primary radar operating bands in the UK are as shown in Table 3.2:

The constraints on spectrum occupancy by radar are related to its angular and range resolution, target characteristics and size constraints on the radar. Staring radar has characteristics that permit certain improvements in interoperability of

Table 3.2 UK primary radar operating bands

Operating band	Minimum (GHz)	Maximum (GHz)	Typical application
W	76	90	Automotive collision
Ka	18	27	Automotive surveillance
X	8	10	Short-range aviation
C	5	6	Military
S	2.7	3.1	Medium range aviation
L	1.215	1.400	Long-range aviation
P	0.225	0.39	Remote sensing

radar stations which, in combination with a relatively low primary requirement for spectrum, may offer improvements in overall spectrum efficiency. We shall return to this topic in Chapter 10.

3.13 Conclusions on staring radar physics

This chapter aims to outline how radar signals, scattered by objects of interest within a VoR, communicate descriptive information about those objects, their positions and motions, and how they can be analysed and understood remotely.

We have set out the physical laws governing the function of radar, and introduced the premise that for objects large enough to scatter EM waves of a given wavelength, and for radars with sufficient power and resolution, solutions can be found for all such objects.

A key distinction between holographic surveillance radar that stares into its whole VoR and those that do not is that the interrogation of the whole volume is mutually coherent, whereas in other cases it is not. In later chapters, and to maintain awareness of this distinction, reference will be to the Coherent Volume of Regard (CVoR) in place of VoR; for scanning radar the CVoR is limited to the transmit/receive beam width during its transient dwell time.

The conservation laws under which radars operate remain the same, but it is clear that, by interrogating the whole volume continually and coherently, the EUNIT provides that whole information about the target and its trajectory is accessible if signal data are placed in sufficient computer memory with sufficient speed of access and process.

Beam scanning with a short dwell-time, however, does not meet the criteria of the EUNIT. Under the EUNIT, as we have seen in Chapter 3, all observed targets and trajectories are encoded in the scattered EM signals; however, long gaps mean that rapid, transient target manoeuvres can escape observation. Many trajectory solutions cannot be excluded analytically and must rely on statistical track suppression.

The rate at which a BSR can acquire information about its whole CVoR is limited by the speed of light, its beamwidth and the dwell time required in each pointing direction. Having chosen a scan rate and beam width, the radar's function is constrained, its ability to interpolate target positions becomes its primary challenge, and its ability to meet the requirements of the EUNIT is lost.

By irradiating a whole CVoR and processing signals arriving in all directions continually, a staring radar has the option of parallel, function-selective processing based on coherent, complex information received by direct but remote contact with each target throughout an extended period.

Not only does HSR permit the recovery of sensitivity apparently lost by spreading transmitted signal energy, continual target interrogation extends that sensitivity to offer a very much wider, and effectively unlimited field of surveillance information, target capture and behaviour discrimination.

References

[1] Feynman R.P. 'Ch. 1, Electromagnetism'. *Lectures on Physics*. **2**. Addison-Wesley Publishing Company; 1977-02.

[2] Feynman R.P. 'Ch. 40, Statistical Mechanics'. *Lectures on Physics*. **1**. Addison-Wesley Publishing Company; 1977-02.

[3] Feynman R.P. 'Ch. 25, Linear Systems'. *Lectures on Physics*. **1**. Addison-Wesley Publishing Company; 1977-02.

[4] ITT authors. 'Ch. 41, Information Theory / References'. *Reference Data for Radio Engineers*. Howard W. Sams & Co; 1977.

[5] Feynman R.P. 'Ch. 44, The Laws of Thermodynamics'. *Lectures on Physics*. **1**. Addison-Wesley Publishing Company; 1977-02.

[6] Smith G.S. *An Introduction to Classical Electromagnetic Radiation*. Cambridge University Press; 1997-08-13.

[7] Bleaney B.I., Bleaney B. *Electricity and Magnetism*. Oxford University Press; 1965.

Chapter 4
Applications of holographic staring radar

Chapter 3 described the physical origins of radar target information and the need for linear data processing to retain and decode it. In this chapter, its applications are examined, including both mainstream and more niche uses. We start with a summary of the main differences between staring and scanning radar that lead to different outputs in terms of performance and whole-life costs.

Since the initial development of high-speed, wide dynamic range, analogue to digital convertors, there has been an increasing emphasis on numerical processing of radar echoes. Advances in digitisation have been matched by software and computing hardware to create more complete and higher-performing systems. In more recent times, improvements in radio-frequency hardware technology have introduced great complexity, resulting in some cases in prohibitively expensive, specialised devices. Conversely, the cost of digital signal processing, in real and relative terms, has considerably reduced, and the flexibility of software functions has grown. This issue of the cost of radar hardware versus the cost of digital signal processing is one of the major drivers behind the improvement in radar system performance more generally, but also the development of new and alternative radar concepts that have the common characteristic of continuous illumination (or 'persistent', not to exclude pulse modulation), and come under the umbrella term, 'staring radar'.

By continuously illuminating and receiving from a Coherent Volume of Regard (CVoR) we obtain more information that offers the prospect of enhanced precision of measurements. Not only are there more observations; within a wide range of dynamics, target echoes are coherently encoded from all motions in the Volume of Regard (VoR), as described in Chapter 3. Each sample adds not only to the precision of an estimate but also to the resolution and dimensionality of the process.

A staring radar illuminates a large region on transmit and uses enough narrow beams to observe the whole region on receive. The ability of such staring radars to look and dwell everywhere all of the time can be likened to the difference between searching a darkened room with a narrow-beam flash lamp versus switching on a floodlight with 'holographic' laser coherence, and has similar advantages. In radar terms, this means that multiple receive beams simultaneously cover the illuminated region and targets in each beam, in parallel, synchronously and coherently, rather than the serial-sectors approach that is fundamentally part of a scanning radar.

This parallel data acquisition and synchronous processing in a staring radar can take advantage of the lower costs associated with evolving digital signal processing, shifting the emphasis away from highly bespoke hardware technology and on to general but highly capable parallel signal processors. The continuous illumination and reception of echoes, leading directly to many more observations of targets over much greater dwell times changes the challenge of tracking from sparse statistics to fully sampled solutions. These characteristics are crucial to the performance advantages that staring radar can bring about and are central to both old and new applications alike.

Persistent illumination of an area of interest means that echo signals can be integrated over long and selectable periods and enable fine Doppler resolution. This facilitates the detection of targets that have very low radial velocities and also provides a method of separating multiple targets in a manner analogous to fine range resolution (which, of course, can also be a feature of a staring radar). Many radar applications are based on Doppler discrimination for target detection. Continuity of illumination and reception not only allows fine Doppler resolution but the dwell time for any radar resolution cell can be tailored to target types and surveillance circumstances. Detection, taking place in parallel in place of the series mode inherent to a scanning radar, aids continuity and robustness of tracking. Track updates can be extremely rapid; in principle as high as the pulse repetition frequency. By placing more of the radar system gain on the passive (receive) side of the system with a broad illumination beam there can be a reduced likelihood of interception by electronic surveillance measures (ESM) and a fully digitised receive array antenna can allow adaptive digital beam forming to minimise interference or deliberate jamming. Fine Doppler resolution also provides a means of discriminating different types of targets from one another by virtue of their different motion characteristics. Given these qualities it is unsurprising that staring radar is finding a raft of applications, sometimes in areas of traditional radar operation and sometimes as a reaction to environmental changes and new target types. Here, we consider just a sample of current and emerging applications that fit well with the attributes of staring radar, and the challenges and requirements they impose.

4.1 Airspace challenges

Airspace and especially low-level airspace is undergoing fundamental and rapid changes that will present ever increasing challenges to radar surveillance systems. Over the next 10 years, manned air traffic has been predicted to double. In addition, over the same period, the number of small commercial and hobbyist unmanned aircraft, or drones, is set to increase enormously such that their number will more than match that of manned aircraft. Overall, this is a quadrupling of targets of interest in a short space of time. Further, drones will occupy predominantly low-level airspace with altitudes of up to just a few hundred metres and many below 100 m. This represents a substantial alteration to the use of the skies, which may have all manner of consequences. For example, these are altitudes that are utilised by birds. Birds are extremely sensitive to habitat alterations, and it is unclear what the consequences for bird populations might be. There are other examples of how low-level airspace is already being occupied. Wind turbines can reach a total height of over 300 m. The number and magnitude of wind farms is likely to increase as alternatives to fossil fuels are developed. Windfarms are located both on land and out to sea, and can generate significant and problematic clutter for traditional radar systems, seriously affecting their false alarm performance and that of air traffic control (ATC) systems. This occurs to such a degree that wind farms very often experience problems in obtaining planning approval; they are unsynchronised with surveillance radars and generate variable clutter, false detections, tracks and track breaks. This is where we begin our examination of staring radar applications.

4.1.1 Wind farm mitigation

Ever since the first wind farms appeared, there have been concerns as to the quantity and quality of clutter they generate and the effect this may have on airport and air-route surveillance radars and hence the effect on air traffic safety. Fundamentally, a single wind turbine represents a large static scatterer in the form of the wind farm tower and three large moving, intermittent scatterers in the form of the turbine blades.

Many efforts have been made to resolve this problem, either by tracker modifications, by radar or wind farm siting or by adaptive nulling in the direction of a turbine, in each case to avoid signal information encoded by the turbines. The solution based on staring radar is that the turbines yield signals that encode the positions and every movement of each turbine, so that if observed so as to meet the EUNIT their radar returns can be separated from those of aircraft, and suppressed. This operates by acquiring and using more information about the turbines and their movements, rather than less.

Figure 4.1 shows a typical wind turbine and its constituent components. To a radar, the tower will cause a large echo response near zero Doppler. Each blade delivers a large and varying echo response, with a Doppler span dependent on the operation of the turbine in the prevailing weather conditions and for a given illuminating radar specification.

Figure 4.2 illustrates the HSR response to a single moving 3-bladed turbine. Leading edge Doppler returns are to the left; trailing edge returns to the right of the spectrum centre, at index 128.

The key aspects of performance required of radar in the presence of wind turbine generators (WTGs) within the ATC environment are:

Item of performance	Without wind farm	Over wind farm	Regulator
Probability of detection	>90%	90%	CAA
False alarm rate	<3 per 4 seconds	<3 per 4 seconds	ESASSP
Latency	<2 seconds	<2 seconds	CAA
Report interval	<2 seconds	<2 seconds	CAA

The horizontal 'stripes' in the figure are the Doppler sequences from the wind turbine blades. The majority of the radar echo from the turbine is centred around

Figure 4.1 A wind turbine tower, nacelle and rotor, within a wind farm

zero Doppler, superposed on intense tower responses with small radial displacements and suppressed, near-zero Doppler. Wind farms generate spectral responses from the turbine blades that also exhibit significantly higher Doppler values. It is this combined response that competes with echoes from aircraft and complicates or even defeats detection and track processing. The results, in terms of the air traffic controller's display, are varied. Targets still detected may be incorrectly located and detection of targets may be missed altogether. This can cause target tracks to break or false tracks to be created in the area of the windfarm.

So-called 'in-fill' radars have been developed to address this problem, including staring solutions. For example, Aveillant Ltd. (a Thales company) has a product, the Theia 16A (A for aircraft), that has been designed to tolerate clutter caused by windfarms. It has an instrumented range of up to 5 nautical miles and provides staring surveillance over a quarter of a hemisphere. It operates in the L-band portion of the electromagnetic spectrum and has an operating bandwidth of 2 MHz. Target updates can be as high as four times per second and the data are reported in standard Asterix Cat 34/48 for ease of integration into an ATC display.

Figure 4.3 shows an example Theia 16A mounted on an ISO container. The radar is housed in the top unit under a protective radome in which the transmit and

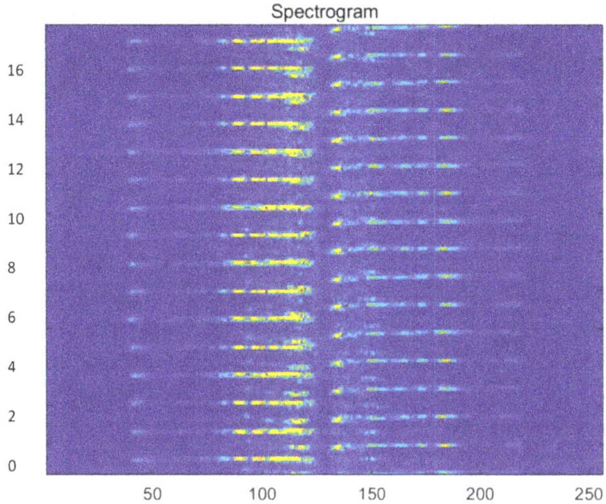

Figure 4.2 Spectrogram modelling a series of 45-m-blade wind turbine Doppler radar returns

receive arrays are housed side-by-side. It has been in operation for several years near East Midlands Airport and is demonstrably effective.

These radars can be fabricated from simple subarrays that are cheap to make and to house in a suitable environment. Early digitisation leads to high dynamic range and capacity for extensive linear signal processing including Doppler filtering and coherent gain. Designs ranging from 64 to 8 192 channels have been produced, tested or planned. A key engineering challenge is computing capacity, necessary to communicate and process data representing the large number of VoR cells resolved in range, azimuth and elevation, and fine Doppler bins.

Figure 4.3 Theia 16A in operation in support of East Midlands Airport

4.1.2 Unmanned air vehicles (UAVs / Drones)

As we remarked earlier there is a massive increase in the number of drones being used both for a variety of commercial applications as well as by a significant and enthusiastic hobbyist community. Drones have a combination of characteristics that make them especially challenging to be reliably detected by radar systems. They fly at low altitudes, often below 100 m. They fly slowly relative to their manned counterparts. They are physically small and consequently have extremely small radar cross-sections (RCS). The RCS of a small hobby drone is in the region of −20 dBsm to −30 dBsm. This value of RCS makes stealth targets look positively enormous. To make mattes even more complicated, these are also the attributes of birds. In other words, the echoes of birds and drones will have a similar magnitude. This means the radar has to have the additional capability of discriminating between birds and drones if an accurate picture of the skies is to be achieved. The flight trajectories of birds and drones tend to be very different with birds, usually, exhibiting a more erratic flight profile, and using different means to achieve lift and motion. These physical differences will be manifest as differences in a time series of radar echoes and provide a means of discrimination. The continuous illumination of a staring radar means that such characteristics can be observed without interruption providing continuous measurements to enable birds and drones to be correctly classified.

Thales-Aveillant has also developed a drone surveillance radar called the Gamekeeper 16U (U for unmanned aircraft) for counter UAV applications. The Gamekeeper radar has a coverage range of 7.5 km and is able to detect targets with a radar cross section as small as −20 dBsm to a range of 5 km.

The key aspects of performance required of radar in the counter-UAV environment are:

Item of performance	Typical requirement
Range	5 km
Sensitivity	0.01 m^2 (−20 dBsm)
Classification	90% UAV vs. birds
Dynamic range	>80 dB
Report interval	<2 seconds

Figure 4.4 illustrates the Aveillant Gamekeeper 16U radar and its operating housing. It has some similarities with the SRC concept introduced in Chapter 5, but operates in a different frequency band. Gamekeeper incorporates transmitter, receiving array and processing, prior to display, within its housing.

Figure 4.5 shows an example spectrogram formed from data recorded using a Gamekeeper 16U radar system.

These spectrograms show time running along the horizontal axis and Doppler frequency on the vertical axis. There are clear differences between the time evolution of the Doppler frequency content of the bird and drone echoes. The bird exhibits a near sinusoidal response indicating it is flying quite tight near-circles, in this case possibly consistent with the behaviour of a raptor. Conversely, the drone shows a

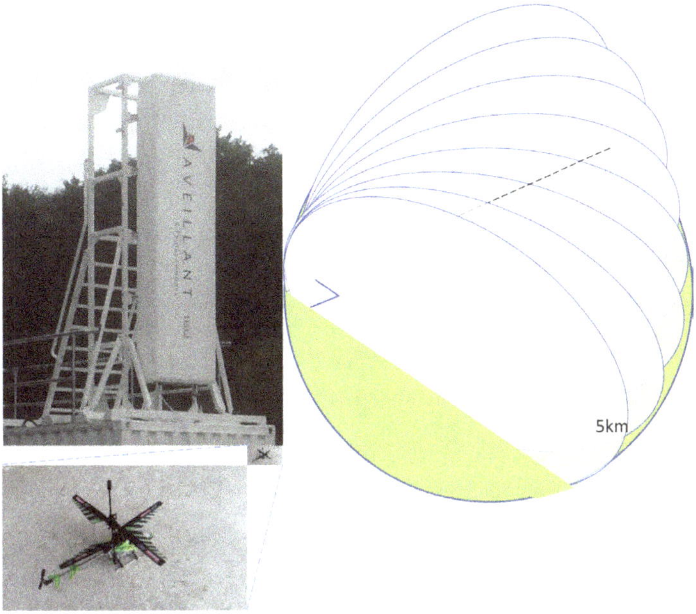

Figure 4.4 Aveillant (a Thales company)'s Gamekeeper 16U drone surveillance radar: a small drone and an illustrative coverage diagram

main body signature that illustrates a turn from an initial positive velocity to one that is negative, and showing much more variable speed. Further, and even more characteristic of drone echoes, is the presence of Doppler sidebands caused by propeller rotation. A wide variety of signal processing techniques from standard pattern recognition through to advanced AI methods could be employed to discriminate birds from drones; the challenge here is to develop a technique that can deal with the wide range of environmental conditions. These include, for a given site, the time of the day which can be a big influence on the bird population, through to the time of year in which seasonal variations play their part. Equally, at different radar locations there will be different bird populations, so that species-specific classification

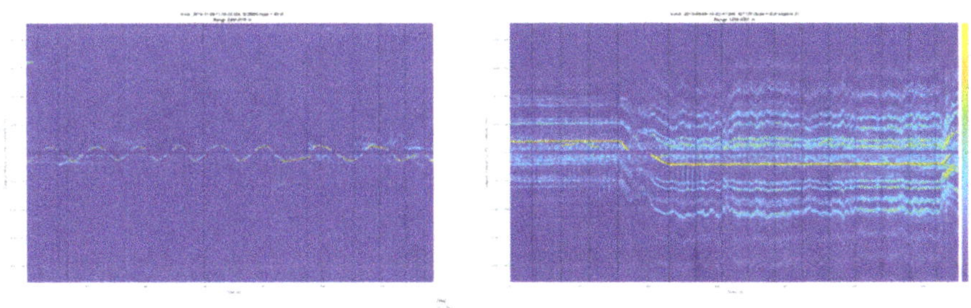

Figure 4.5 Example spectrograms for (a) a bird and (b) a drone

would be preferred over stochastic trajectory measures. There are also a wide variety of drone types; fixed wing, rotary wing, quadcopters, hexa and octo-copters and so forth. All these variables inevitably result in performance uncertainties and challenges in expressing performance in statistical terms.

4.1.3 Air surveillance integration

The occurrence of clutter caused by wind farms and the arrival of small, low, slow aircraft such as drones are significant issues for air traffic safety, and staring radar is being developed for application in ATC. In principle, both airport surveillance radars (ASR) and air route surveillance radars (ARSR) can be satisfied by a single design of radar, although in recent times these have been segregated between the S-band or the L-band portion of the electromagnetic spectrum, respectively. ASRs provide surveillance in the vicinity of airports, and ARSRs provide surveillance between airports. ARSRs tend to operate over longer ranges and a longer wavelength has advantages, especially in relation to weather, whilst the available bandwidth at L band is smaller.

Aveillant (a Thales company) has demonstrated the flexibility of its staring radar design, using common radar hardware in providing coverage for drones to 5km and aircraft to 60 nautical miles, including wind turbine mitigation, drone classification, and an array geometry suitable for the construction of common-frequency networked operation. Aveillant offers the Theia 64A, a staring radar that uses similar

Figure 4.6 Aveillant Theia 64A on the beach at Muckleburgh, Norfolk, demonstrating mitigation of 80-turbine Sheringham Shoal wind farm

technology building blocks as the Theia 16A. It operates at L-band and is illustrated in Figure 4.6.

Its larger receive array results in narrower beams and, therefore, enhanced azimuth accuracy compared with the smaller array, with range increased to 20 nautical miles.

Two-dimensional receiving arrays naturally lend themselves to both elevation and azimuth determination and, hence, these systems can provide a 3-D picture of air activity, important for defence applications.

In Chapter 10, the issue of wide-area surveillance is addressed by radar networks in the context of increasing focus on spectrum occupancy and efficiency.

A longer-range variant has also been demonstrated, operating at over 60-nautical miles, in preparation for a staring radar with a 360-degree azimuth CVoR, and is illustrated in Figure 4.7.

Figure 4.7 Thales-Aveillant 4-faceted staring array research tool for 360-degree coverage

In expanding the function of staring radar for 360-degree surveillance, a key issue is that if 3 or 4 flat panel arrays are used in combination, they will yield variable performance around the circle, and will require at least 40% additional equipment and cost than that suggested by resolution near each panel boresight. A circular or multi-faceted array can improve that aspect. Such departure from a planar array makes the usual, efficient Fourier transform method of beamforming inapplicable, and for a large array forming many beams simultaneously the alternative represents a heavy processing burden that may appear out of proportion to the rest of the radar. Nevertheless in studies to date, whilst computing becomes a significant part (up to 20%) of the radar cost, providing the necessary capacity is not expected to threaten its cost-effectiveness overall (see Chapters 5 and 7).

Sidelobes in this design may exceed those normally achievable by flat panels with appropriate aperture windowing; nevertheless their formation in coherence with all the main beams deployed by the array greatly reduces the performance challenges that they represent. This is discussed in detail in Chapter 7.

The key aspects of performance required of HSR in the ATC environment are:

Item of performance	Typical requirement
Range	60 NM
Sensitivity	1 m^2 (0 dBsm)
Classification	Aircraft vs. various noise and clutter
Dynamic range	> 80dB
Report interval	<2 seconds

4.2 Imaging complex targets

Radar imaging is well-established and highly effective in the case of synthetic aperture radar (SAR). In that case the radar sensor moves according to a known or accurately measured trajectory and the sequence of complex amplitudes it receives as a function of time, range and its own position encodes the shape of the objects in its CVoR.

Except for the fact that in SAR it is the radar that moves and the VoR scatterers that are static, by reversing those two roles, and allowing the trajectories of targets to vary, the function of staring radar is extremely similar. In fact the well-known inverse synthetic aperture Radar (ISAR) method can be seen as a subset of staring radar, usually applied to single targets.

To form an image of a target its different scattering centres must be treated as independent objects. Their trajectories relative to the radar are used to establish its central motion, after which the relative positions of those scatterers can be established.

It may seem a paradox that what appears to be a vulnerability of staring radar (i.e., that targets executing dynamic trajectories appear to decohere and threaten the process of coherent gain normally achieved by Fourier transformation) in fact is the precise encoding of dynamics and shape that the EUNIT provides in the form of complex amplitude sequences. A target executing an exactly circular trajectory around a radar, for which every Doppler component is zero, in fact minimises the encoded information only providing a radial profile, whereas an agile target provides copious information about its motion and shape.

This is a broad subject, which will not be pursued further here, but it expresses the large potential of staring radar. Precise requirements are expected to follow as the imaging opportunity develops and is recognised in government and industry.

4.3 HF Radar

4.3.1 Over the horizon radar

Over the horizon radar (OTHR) radars generally transmit and receive in the HF part (3–30 MHz) of the electromagnetic spectrum and have been around since the early developments of radar. OTHR was developed to overcome the limitations of higher frequency operation in which range limits were determined by the position of the horizon. OTHR radars fall into two categories; the first are ground wave systems which use very low frequencies which propagate via coupling to the earth's surface and diffracting around obstacles. The second type of system uses the ionosphere to bounce radiation so that it is directed to an area of interest beyond the local horizon.

There are many examples of these types of radar with, perhaps, JORN in Australia and ROTHR in the United States being the best known. Figure 4.8 shows an image of the JORN OTHR antenna. Note the large spacing between elements dictated by the long wavelengths used in the HF band.

The use of low transmission frequencies and the desire for very long detection ranges, inevitably leads to large equipment sizes. Indeed, OTHR radars can occupy areas as large as a few square kilometres. Detection ranges can be as high as several thousand kilometres and targets as small as fighter aircraft can be observed. OTHR radars may 'floodlight illuminate' an area of interest and also form multiple beams on receive. They also exploit advanced processing concepts such as multiple input multiple output (MIMO) to enhance angular resolution and accuracy. Very sophisticated processing is used to observe moving targets against a background of stationary clutter. However, the ionosphere is not stable and compensation has to be applied to correct for this. Equally, detecting targets over the sea surface has to be conducted against clutter that is moving and hence has a finite and variable Doppler extent. The staring nature of OTHR means that very fine Doppler resolutions are possible, providing both integration gain for sensitivity and target detection through differential movement. Most OTHR systems are large, have required long development programs and are owned by governments and implemented/operated by industry. The JORN radar is currently the subject of a contract between the Australian government and BAE Systems.

Figure 4.8 The JORN radar antenna

The key aspects of performance required of radar in the OTH environment are:

Item of performance	Typical requirement
Range	250 NM
Sensitivity	0.5 – 5 m^2 (–3 dBsm)
Classification	–
Dynamic range	>80 dB
Report interval	<12 seconds

4.3.2 HF Radar for ocean monitoring

The HF part of the electromagnetic spectrum (3–30 MHz; wavelengths 100–10 m) has also long been used as a tool for monitoring the properties of the ocean surface and typically uses a staring geometry. In this way the only moving component is the sea surface and variable long dwell times can be used to interrogate aspects such as wave speed. HF radar systems are cost-effective tools to monitor coastal regions at ranges of up to around 200 km. They are mainly used in applications such as search and rescue, marine traffic information, oil spill monitoring, mapping of sea state parameters (significant wave height, wave period and direction, wind direction), tsunami early warning, ship detection and tracking and in the calibration and validation of numerical simulations.

Difference between the speed measured by radar and the known speed and direction of a vessel through the water allows determination of ocean currents; a key component in the determination of weather and an ingredient in wider climate studies. The HF part of the electromagnetic spectrum is used as the associated wavelengths are a good match to those of the sea surface and advantage can be taken of Bragg scattering, in which regularities in surface wave patterns yield a form of resonance with incident radiation. Exploitation of the Bragg effect yields information about wavelength, speed and direction of the sea surface. A minimum of two radar systems is needed to resolve the two-dimensional current field from radial velocity data. The two radars must be spaced sufficiently far apart but look at the same ocean area such that there is a significant angle between them. This separation distance will be a function of the range to the patch of sea under interrogation for each radar system. Globally, around four hundred systems are in operation, with further installations planned. In fact, there are emerging networks of cooperating HF radars used to improve the quality of information monitoring and its subsequent exploitation.

The key aspects of performance required of radar in the ocean monitoring environment are:

Item of performance	Typical requirement
Range	200 km
Sensitivity	High
Classification	Sea state, wind direction, wavelength
Dynamic range	>80 dB
Report interval	<12 seconds

The US High Frequency Radar Network (HFRNet) derives radar data from 31 organisations, including measurements from Canada and Mexico. HFRNet currently holds a collection of data from over one hundred and fifty radar installations. In recent years, in Europe, HF radar networks have been growing steadily with over sixty systems currently deployed and many in the planning stage. In Asia and Oceania, more than one hundred and ten radar stations are in operation. Figure 4.9 shows the antenna structure for the WERA HF radar system used to improve the reliability of ocean forecasting for the Port of Rotterdam, in support of both efficiency and safety.

Figure 4.9 The WERA HF radar system at the Port of Rotterdam

4.4 Radar for autonomous vehicles

It is a strategic goal of the automotive industry to be able to offer self-driving vehicles that, ultimately, should have no need for a steering wheel. Currently a variety of sensors are being examined with the aim for the vehicle to be able to probe its environment with sufficient clarity and accuracy to enable the generation of instructions that can safely control a car from the start to the end of a journey. These include cameras, lidar and radar. It is likely, in the shorter term, that a combination of sensors will be needed in order to provide true autonomy. However, cars need to function in all weathers. Radar is the only sensor that can offer true all-weather performance regardless of whether it is night or day. This implies there may well be a reliance, in certain foul weather circumstances, on radar as the prime or possibly even, in extreme conditions, the only operating sensor. As a result, extensive research has been under way for at least three decades, including development of low-cost, yet highly sophisticated radar sensors. In work carried out in this field several years ago the target price for the components of a multi-function, short range radar was $50.

These radars operate at two regulated transmission frequencies centred at 24 and 77 GHz. The 24 GHz systems are used for shorter range tasks, sensing objects in the immediate vicinity of a vehicle and for specific duties such as blind-spot monitoring, lane-keeping assistance and for parking. The 77 GHz radar systems are used for longer range tasks such as sensing objects further away for safety when pulling out of a junction, and specific duties such as automatic distance control and automated braking.

The radars themselves are the subject of intense development but have the common design themes of wide bandwidth for high resolution, multiple input, multiple output array antenna functionality for narrow receive beamwidths, waveform design and a staring mode of operating. In other words, they illuminate a broad area on

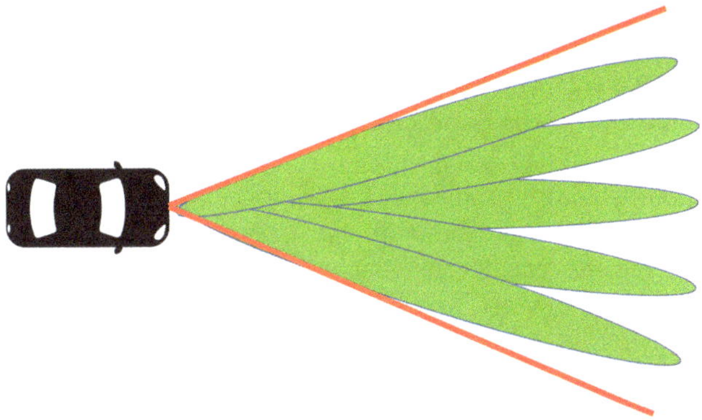

Figure 4.10 *The illumination beam and multiple receive beams of a MIMO vehicular radar system*

transmit and use the MIMO technique to generate, digitally, a series of beams on receive to fill the illumination area. Bandwidths are as high as 3 GHz and may eventually fill the entire allocation of 5 GHz that the regulators have made available at 77 GHz. A bandwidth of 3 GHz provides for a radial range resolution of just 50 mm, suitably fine for resolving details of objects that can be used not just for detection but also for recognition. Recognition is a key component of understanding the environment surrounding a vehicle, and some degree is necessary for autonomy. These radars use high Doppler resolution to provide a detailed picture of movement within an observed scene. This is done using a frequency-modulated continuous wave (FMCW) transmission signal with extended dwell periods to yield fine Doppler resolution. Note though, the combination of very fine range and Doppler resolution leads to challenges on solving range-walk – where a target moves through a range resolution cell during the processing period and Doppler walk for a target whose radial speed moves between Doppler resolution cells during a processing period. As described in Chapter 3, this does not invalidate the coherence of the Doppler phase sequence but requires appropriate processing to retain sensitivity for each dynamic case.

Figure 4.10 illustrates schematically the floodlight transmission beam with multiple receive beams generated via MIMO for a typical advanced vehicular radar concept. Potential targets comprising the view ahead of the vehicle can be better separated in the multiple receive beams enabling improved detection and recognition.

The key aspects of performance required of radar in the automotive environment are:

Item of performance	**Typical requirement**
Range	200 m
Sensitivity	0.1 m^2 (−10 dBsm)
Classification	Vehicle, pedestrian, fixed clutter
Dynamic range	>80 dB
Report interval	<0.1 second

A number of manufacturers offer vehicular radar systems, including NXP, Inras, Fujitsu, Bosch, Valeo, to name just a few.

Figure 4.11 shows an example from NXP in which both the transmitter elements and the receiver elements can be observed on the cream panels. Typically, the transmitter will send a signal that is sequenced across the transmit elements such that on receive larger 'virtual' arrays can be formed using the MIMO technique. This facilitates the finer resolution that comes with the larger virtual antennas. Digitisation of all the receive elements enables the generation of multiple beams thus filling the entire illuminated field of view.

We refer again to MIMO techniques in Chapter 10, in relationship to the demands of wide area surveillance combined with congestion of the radio spectrum.

The NXP TEF810x fully integrated 77 GHz transceiver has three transmitter and four receiver arrays to support the MIMO processing and has an instantaneous bandwidth of 2 GHz, which can be extended to 4 GHz using a technique known as chirp stitching. Each receive chain includes a 12-bit ADC, sampling at an effective rate of 40 MS/s. The ADC is followed by a programmable decimation filter with variable decimation factors from 1 to 16. The TEF810X is a single-chip automotive FMCW RADAR transceiver for short-, medium- and long-range RADAR applications, covering the full car RADAR frequency band from 76 to 81 GHz. The mm-wave front-end part consists of a waveform generator offering flexible chirp control, three transmit chains featuring binary phase control and output level stabilisation, and four receive chains with high input compression and low noise figure. The highly integrated nature of these radar systems lends them to mass production at a scale where costs can be extremely low making them affordable for all vehicles of the future.

Whilst there are further advances that need to be made to ensure viable all-weather autonomy for truly self-driving cars it is intuitive that look ahead in a staring mode of operation will be the best way to gather and interpret the core information needed to generate the instructions for safe control of a vehicle.

Figure 4.11 NXP TEF810× fully integrated 77 GHz transceiver

4.5 Passive radar

Another large class of staring radar is passive radar. We discuss passive radar only briefly as it is a well-known technique that is reported extensively elsewhere [1]. Passive radar exploits transmissions of opportunity and, hence, only requires a receiver. Many of the most commonly available and most used transmitters of opportunity emit continuously and over broad regions, often spanning 360^0 in azimuth. Perhaps the most popular transmitters of opportunity are those used for radio and television broadcasting. This is because they offer the highest powers and hence provide for relatively long-range detection, principally of aircraft targets and using just a single receiver. Figure 4.12 shows, schematically, the passive radar concept. By using broadcast transmitters of this type passive radar is fundamentally a staring radar and techniques described in this book may apply. However, there are some important differences between passive radar and monostatic staring radar that need to be considered.

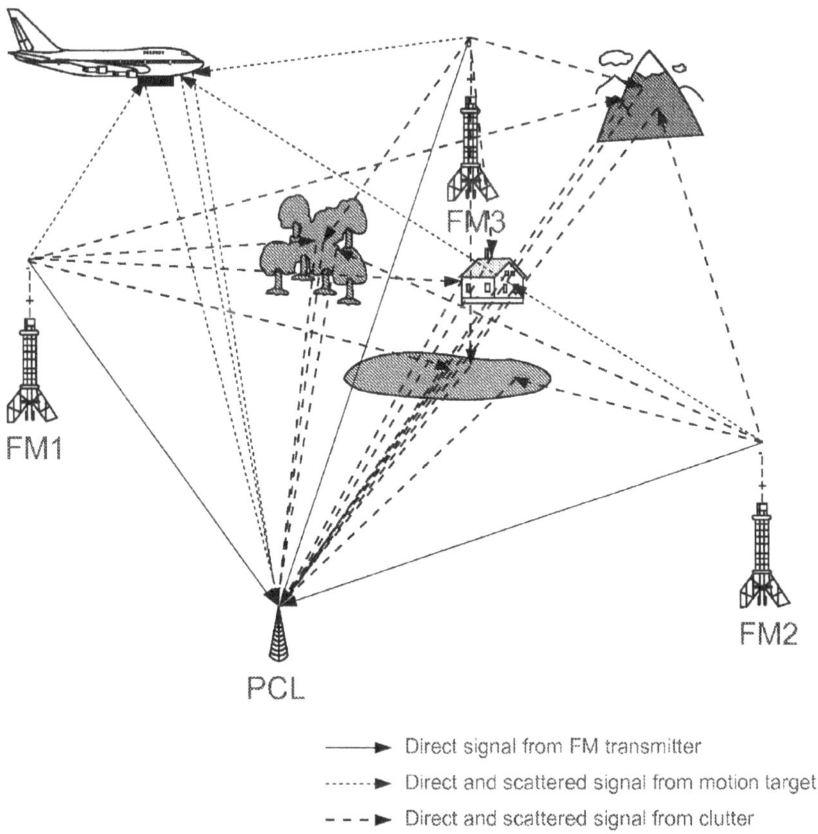

Figure 4.12 Illustration of passive radar

The key aspects of performance, required of radar in the passive environment, as applied in the ATC environment are:

Item of performance	Typical requirement	Regulator
Coverage	Surface to 50 000 ft.	Not yet established
Sensitivity	1 m^2 (0 dBsm)	
Classification	Threshold-limited	
Dynamic range	–	
Report interval	<2 seconds	

In essence, passive radar has some fundamental differences from staring radar as described in the other chapters. The basic processing technique will have things in common, with the exception that in Staring Holographic Radar the waveform of illumination is known in advance, whereas passive radar lacks the inherent coherence in transmission that provides the basis for holographic scattering, and will require complex waveform recovery processes, for each transmission used, to extract encoded trajectory information as described here. Note, passive radar is typically bistatic or multistatic (depending on the number of transmitters or receivers used). Passive radar has no control over what is emitted by the transmitter nor over where it is emitted. Nevertheless, with sufficient knowledge of the signals and their coverage very creditable performance can be achieved. For passive radar, as for the networked radar scheme outlined in Chapter 10, two signal components arrive at the receiver. One of these components surveys the VoR whilst the other propagates directly from the transmitter. This is used as a reference but must not overload the receiver. Matched filtering of the reference signal with the surveillance signal, with appropriate delays to the region of interest, yields a range-Doppler surface from which targets can be detected using the approaches introduced earlier. However, it is important to note that the direct signal will also be detected in the surveillance channel and has to be suppressed in order not to act as a limit on detection range. Passive radar also will detect clutter at or close to zero Doppler frequencies in the same way as more conventional monostatic staring radar. In many developed systems, the receiver consists of a fully digital array so that multiple simultaneous beams can be formed and large volumes of space surveyed, once again relating to our staring radar discussion.

Hensoldt, The Fraunhofer Institute, Thales and Leonardo are just a few of a large number of organisations that offer passive radar systems commercially. An example using VHF and UHF transmitters of opportunity and the receiving antennas for both bands can be seen in Figure 4.13. The smaller ring of 'plates' comprises the UHF antenna and the larger ring of dipoles, the VHF antenna. The circular nature of both antennas offers a 360° field of view in which air targets can be detected and tracked. The received signal at each antenna element can be digitised, enabling full digital beam forming to be employed, including adaptive beam forming. The small size of these antennas compared to a typical phased array at these frequencies means that the receiving beams generated are relatively broad. Smart processing techniques are used to remove ambiguous targets that result.

Figure 4.13 The Hensoldt TwinVis passive radar system

The Hensoldt TwinVis is a passive radar that has a number of operational advantages. For example, as the receiver is remote from the transmitter, it cannot be located from its own transmission and subjected to jamming. Further, it avoids further congestion of scarce radio spectrum since it relies on third party transmitters. Equally, the bistatic geometry is one that has potential to 'see' stealthy targets which have been designed only to defeat monostatic radar. The TwInvis system can be used for both military airspace surveillance as part of a ground-based air defence capability and is also offered for ATC applications.

Passive radar, as with conventional monostatic radar, can come in many forms. The concept has also been used for indoor monitoring within buildings. For example, the ways in which people use and move around buildings can be tracked subject to source within or near the building (a further realisation of the EUNIT). Passive radar's ability to monitor movement through walls has been developed for use by the emergency services. It can also be used for fall detection for the elderly and infirm.

It has even been used to monitor animals both for agricultural applications and in big game parks. Passive radar offers an alternative to more traditional approaches to radar design and is likely to find more widespread application as the technology further evolves. Its vulnerability may lie in the user's lack of influence over what is transmitted, and where it is incident in the VoR.

4.6 Other applications

In principle, there is no limit to the applications that can be fulfilled using staring radar, provided they can be developed to meet the terms of the EUNIT. Maritime surveillance of other vessels and objects at sea; high spatial resolution radar imaging of the ground; inverse synthetic aperture imaging; missile scoring, tracking of artillery, rockets and motors, etc. Weather radars including evaluation of phenomena such as tornadoes and wind shear are all examples where the high Doppler sensitivity of a staring radar can be exploited. Many radar applications make use of Doppler shifts to differentiate legitimate targets from clutter, and it is the acquisition of continuing series of complex amplitudes that give staring radar its particular strengths. Doppler effects are used to provide detailed classification information about targets, derived from their bulk and internal motions. It is small wonder that staring radar is rapidly gaining traction as an attractive but in some ways disruptive addition to the stable of traditional radar systems, enabled by low cost, high capacity computing.

An early application of airborne staring radar (1960s onwards) has been its use in measuring the depth of polar ice sheets. The upper and lower surface topography of ice sheets and glaciers have been surveyed in Antarctica, Greenland, the Himalayas, the Andes and many other sites of climate interest, but the reflecting characteristics of the bed also provide information on its shape and its thermal state and, therefore on factors affecting its stability.

Chapter 5 will focus on configuration examples to illustrate various HSR applications.

References

[1] Griffiths H.D., Baker C.J. *An Introduction to Passive Radar*. Artech House; February 2017.

Chapter 5
Configurations for HSR

The discussion of staring radar will benefit from identifying a small number of examples of HSR configurations. Without limiting the discussion to these examples, they will provide a simple way of referring to specifics, and a basis for comparing approaches.

This chapter provides reference points with which the relationship between physics and the engineering of HSR can be explored in Chapters 6–10. Engineering defines what methods are available and affordable, and how the device will work; the physics shows whether the device can work, what can be the benefits and how closely engineers' designs may approach the ideal, and together they determine what will emerge for the purposes of the user.

Each example will imagine a practical construction allowing sensors of different sizes and capabilities to be built, and expectations of functionality and costs to be considered.

The HSR examples are, first, an early proof of the concept, illustrated in Figures 5.1–5.3, which is followed by two generic configurations of ground-based staring radar; one transportable, the other a larger, fixed installation. Each is intended to establish a specific CVoR; to decode coherent, complex signals scattered by objects within the CVoR and received across the observing array aperture; to deduce the presence, motion and scattering characteristics of those objects, and to report those that conform with the requirements for surveillance.

5.1 HSR configuration examples

Many forms of radar exist that stare into their Volume of Regard (VoR), including the historical forms described in Chapter 1. The examples introduced here are primarily concerned with radar as an air surveillance instrument, and how coverage can best be provided for appropriately scaled CVoRs, with assessment and reporting of numerous targets and trajectories.

A Proof of Concept radar (PoC) focused on the requirement to suppress radar clutter due to rotating wind turbines while maintaining sensitivity to aircraft overhead. Two were built by Cambridge Consultants, and were tested successfully by Aveillant Ltd at locations in the UK and in Texas.

The two generic configurations take the form of single HSR sensors, scaled to cover single CVoRs of different extents in range and azimuth, and to monitor different scales of target. Figure 5.4 illustrates a short-range configuration (SRC) planar staring receiver array, intended for small targets at shorter ranges, to provide surveillance in the neighbourhood of critical infrastructure, sensitive sites and venues, etc. Figure 5.10 illustrates a more extensive and non-planar receiving array configuration (ATC), to achieve 360° coverage, at ranges suitable for air traffic control, up to 60 nautical miles. More extensive and networked coverage is considered in Chapter 10, in which stations of this second kind may operate in the form of a coherent network.

5.1.1 Common features of staring radar

These configurations share a number of aspects and features that arise from the concept of staring radar, rather than from specific performance requirements.

5.1.1.1 Interrogating the CVoR

In the staring solution there is no scan sequence. The entire CVoR is irradiated at the pulse repetition frequency (PRF); signals are conditioned and digitised near the point of reception for each array element. All element signals require conversion to high-resolution digital form, after the operating frequency band has been filtered to match the radar's transmitted and spectrally occupied bandwidth and to avoid interference, but before functions such as beamforming and detection.

Interrogations of spatial resolution cells take place not sequentially, in airspace, but in parallel, in computer memory, after data are acquired from the entire CVoR. To support multiple functions, all signals can be retained in raw I/Q form for extended target and clutter analysis.

All spatial sectors and ranges can be resolved simultaneously, at the PRF, into cells defined by range, elevation and azimuth. To achieve solutions under the EUNIT, signal data must be received as complex amplitudes and stored. HSR brings forward uniform, linear, coherent pre-detection signal processing for all resolution cells, to discover those that contain significant waveforms of any type. That subset can then be submitted for more detailed analysis, which may then be data-conditional, of the time- or frequency-domain features observed. Dwell times and CPIs can then be divided, extended or concatenated to obtain necessary performance in target capture, analysis and discrimination.

5.1.1.2 HSR components

The examples in Sections 5.1.2–5.1.4 depict ground-based sensors, equipped with:

1. A transmitter whose outgoing radiation is broadly bounded to cover the CVoR;
2. A receiving array oriented to digitise and condition received signals scattered from within the CVoR;

3. Highly parallel data processing and data communications forming 3D resolution cells:
4. A programmable, stable timing and signal control device.

These elements have distinctive functions in a number of ways:

1. **The transmitter** uses a linear power amplifier and a transmitting antenna in the form of a narrow vertical array illuminating all the desired range of azimuth and elevation positions within the CVoR.
2. **Receiving array** elements occupy known positions, each with line of sight into the CVoR, and each with an accurate, coherent time reference to the transmitter. Arrays can be modular assemblies. They require accurate calibration to allow for thermal and ageing variations in the complex gain of each channel, and high-resolution digital signal conversion. The function of each array channel needs to be monitored. Receiving elements must be protected from damage and preferably (if mechanically possible) from saturation by the transmitter.
3. **Received signals** are communicated to the radar processor. Complex amplitudes are sampled for every range resolution cell and for every receiving channel, and are combined to resolve the range, azimuth, elevation and Doppler spectrum of signals correlated over the array.
4. **Resolution cells** are interrogated after application of a set of linear filtering processes (pulse compression, beamforming, Doppler filtering, etc.) and their signal contents evaluated. Those whose outputs qualify against a set of thresholds during the most recent CPI are 'discovered' as candidates for conditional processing, target tracking and reporting.
5. The **timing and signal controller** needs to generate highly stable and fairly simple waveforms, rather than focusing on agility or complexity.

Based on these common features, a staring radar is equipped to interrogate objects within the CVoR.

The signal data acquired are not constrained by a specific beam pattern, scan timing or dwell period; they are stored in extensive, fast-access memory, so that any surveillance application can be activated, within the processing resource, including target detection and tracking, clutter suppression, target analysis and discrimination, target dynamics, multipath analysis, etc., without the need for re-acquisition, array or time multiplexing, or waveform adaptation.

Signal data are initially received channel-by-channel, in time domain sequence; beam-wise data are formed in the time domain, followed by Doppler transformation of each resolution cell. If beamformed, time domain data are needed, the reverse order of beam-forming and Doppler transformation may be used, yielding four views of the incoming encoded data: channel-wise in the time domain, beam-wise in the time domain, channel-wise in the frequency domain and beam-wise in the frequency domain. These different views are useful for resolving different signal sources and types, as will be seen in subsequent chapters, and form what will be

Table 5.1 SRC and ATC outline design parameters

Aspect	SRC	ATC	ATC vs SRC	Comment
Peak transmission power	1 kW	40 kW	40: +16 dB	ATC net gain 4 dB
Transmit gain	15	5	0.33: −5 dB	Simple antenna
Pulse compression	1	64	12: +14 dB	Chirp
Target range	5 km	60 NM	−54 dB	
PRF	6.67 kHz	1 kHz	−8 dB	
Target cross-section	0.01 m^2	1 m^2	100: +20 dB	
Receiver effective area	0.5 m^2	9 m^2	15: +12 dB	
CPI	0.3 seconds	2 seconds	6.7: +8 dB	2 048 point

referred to as a RAED memory structure. This comprehensive, coherent and linearly inter-related overview of all CPI signals is unique to staring radar.

The first example of HSR illustrates the PoC form built by Cambridge Consultants Ltd., and tested by Aveillant Ltd. between 2012 and 2014 to confirm basic functionality and performance in initial applications, and to justify and pursue further development and manufacture of this form of radar.

The SRC and ATC concepts outlined below are focused on ground-based, static implementations of a surveillance radar. SRC is transportable and potentially mobile, while ATC should be a larger fixed structure. Marine and airborne variants are conceivable, but those additional complexities lie outside our scope.

The primary constraints on the physical realisation of the SRC and ATC examples are safety with respect to transmission and servicing, and positioning with respect to access, neighbouring structures and terrain, within the requirements for sensitivity, accuracy and reporting rate.

Operational applications of the staring radar principle can be based on commercial-off-the-shelf components, assembled as modules to form a wide range of sensor configurations. In this chapter two further distinct theoretical examples are introduced: 'SRC' to operate at S band (3 GHz) and configured to monitor a 5-km, 90° sector for the presence of small, slow and low drones in the presence of both static clutter and populations of birds; and the second, 'ATC', configured for 360° coverage to 60-nautical miles, and operating at L band.

To clarify the difference between SRC and ATC configurations, key parameters are given in Table 5.1.

Either configuration assumes a rigid structure, mechanically stable under any necessary environmental conditions, with externally facing assemblies for transmission and reception both of the radar signals and of data communications, and internal assemblies for power amplification, power conversion, system control and for data communications and data processing.

5.1.2 Proof of concept HSR

The PoC HSR was designed both to exercise and prove the fundamental radar function, but also to resolve targets of different types – in this case, aircraft and large, moving wind turbines. It is pictured in Figure 5.1.

The PoC HSR example operated at L band. Its transmitter used a single-microwave patch dipole with a broad field of view in both azimuth and elevation, and the receiver used a planar antenna array, pictured, unenclosed, in Figure 5.2.

Figure 5.3 considers two aircraft crossing a wind farm (A). The expected output for BSR surveillance is illustrated in (B). Evidence shows that HSR offers user-required output, as (C).

Its sensitivity was specified to detect an aerial target of RCS 1 m^2 at a range of 5 NM, in the presence of wind turbines with cross section up to 10 000 m^2.

The PoC CVoR covered a sector of 90° in azimuth and 90° in elevation. The planar receiver was an array of elements, each matching the whole CVoR. Each array channel included simple amplification, filtering and conversion.

In-phase and quadrature signal components for each element were passed continually from the array to a central processor, where digital time-domain pulse sequences were processed to yield high-resolution Doppler spectra for each range cell. Receiving beams were formed to measure both azimuth and elevation target positions.

Figure 5.1 PoC planar HSR array in a trial of air surveillance in the presence of wind turbines

Figure 5.2 Proof of concept HSR in operation within a wind farm near Iraan, West Texas. The main enclosure houses the receiving array and the smaller radome houses the transmitter.

An example of the function of the PoC HSR when integrated with an operating air traffic control radar is shown in Figure 5.3.

The PoC radar was used in East Anglia, Texas, Scotland, Kent (UK), Coventry and Northumbria to demonstrate that the persistent observation provided by HSR could be used to mitigate wind turbine clutter.

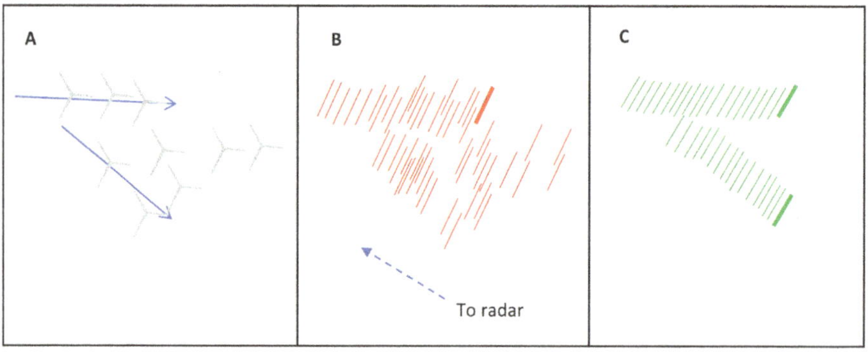

Figure 5.3 Illustrating successful HSR wind turbine suppression. (A) 2 aircraft overfly a wind farm. (B) scanning radar reports both aircraft and turbines. (C) HSR tracks aircraft while suppressing turbines

5.1.3 Short range configuration (SRC) outline

To illustrate functions at small scale, SRC models a transportable radar sensor for detection, classification and reporting small targets to ranges of several kilometres. The primary function from which to develop a user requirement for SRC is that of monitoring the presence of drones (unmanned air vehicles – UAVs) to 5 km, with radar cross-section of 0.01 m^2 (−20 dBsm), over an azimuth quadrant.

The SRC example occupies a single enclosure, including transmitter, modular receivers and processor. To match the operating wavelength to the scale of components of small drones (~25–50 mm), and with cost and portability constraints, we consider operation at S band (3 GHz). This example can use commercially available components and commercial-off-the-shelf communications and computing equipment.

The SRC would transmit from an array 0.5 m tall × 0.1 m, and its receiving array (illustrated in Figure 5.4) should be 1 m tall × 0.5 m in width. The single sensor enclosure array dimensions may be 1.8 m tall × 0.5 m.

The transmitter is intended to cover a narrower (20°) vertical beamwidth.

5.1.4 Air traffic configuration (ATC) outline

Air surveillance at ranges up to 60 nautical miles requires considerable upgrading from the SRC configuration.

The ATC configuration loses target sensitivity because of the greater maximum range requirement, effecting a 4th-power reduction in the returned EM signal intensity (power). The PRI is increased, but so also is the dwell time, yielding the same coherent gain. The transmitted power is increased and longer pulse compression is provided, while the transmitted power is dispersed over 360° instead of 90°.

ATC is specified here for targets with 100× the radar cross-section specified for SRC with, for example, 40 kW peak power, 64× compression, a vertically narrower transmit beam pattern, a larger receiving aperture and longer dwell time are sufficient in theory to compensate for 22×-longer range. ATC is expected to cover 360° in azimuth and ranges up to 60 NM (111 km) for targets of RCS greater than 1 m^2.

5.2 SRC outline resources, structure and functions

An SRC radar is expected to be transportable and to detect small, slow drones at up to 5 km, to report at more than 2 Hz, and also to be able to discriminate reliably between drones and birds, which may have a comparable or greater radar cross-section.

The outline begins from considerations of hardware: transmission, antennas and beams, the target, operating frequencies, signal conditioning, etc. Staring radar acquires signal information encoding target characteristics and trajectories throughout the CVoR, passing data for processing to discover, capture, locate and analyse the encoded targets.

Table 5.2 Example requirements for SRC configuration

Item	Minimum	Maximum	Comments/Caveats
Operating frequency	2 700 MHz	3 100 MHz	S band: Scatter from drone components
Maximum range	5 km	10 km	Critical infrastructure
Minimum range		200 m	
Range resolution		75 m	
Range accuracy		10 m	
Minimum target		0.01 m^2	Hobby drones
Target discrimination			1 – Doppler spectra and 2 – trajectories
Maximum elevation	20°		At maximum range
Minimum elevation		0.5°	Surface multipath
Transmit boresight		+10°	
Azimuth resolution		14°	Boresight
Azimuth accuracy		1.4°	20 dB SNR
Elevation resolution		7°	Boresight
Elevation accuracy		<0.7°	20 dB SNR

Its function is bounded by the hardware used, but is defined by the radar's selectable dwell time and the various processes through which the evolution of complex amplitudes, through time, can be decoded in terms of target characteristics and motions.

The hardware specification of a SRC sensor should include items consistent with Table 5.2.

5.2.1 SRC physical configuration

SRC would operate at S band with dimensions as described in Section 5.1.3. The front face forms a protective radome, but other faces should be largely conductive to minimise sensitivity to external electronic and radio signals. SRC should be mounted above ground at sufficient height that the transmitted signal will not exceed safe limits in spaces with either public or operational access.

For sector coverage as envisaged for the SRC form, a flat array gives a simple receiving configuration, and is illustrated in Figure 5.4. The SRC model uses a receiver array of 16 rows of 8 elements. A manufacturable structure might comprise 8-modular 'slates', each containing 4 × 4 array elements.

In this form of radar, the function of beamforming in a specific direction requires parallel access to all elements with useful gain in that direction. In the case of SRC, with a planar array, this means that outputs from all 8 slates are required at the single beam-forming processor.

As a planar array, SRC receiving beams can be formed conventionally by means of 2-dimensional fast Fourier transformations (FFTs). It will require a processing rate of about 30 billion complex floating point operations per second and is well within the capacity of modern graphics processors operating 32-bit arithmetic. FPGAs and various RISC machines would also be capable of accommodating this

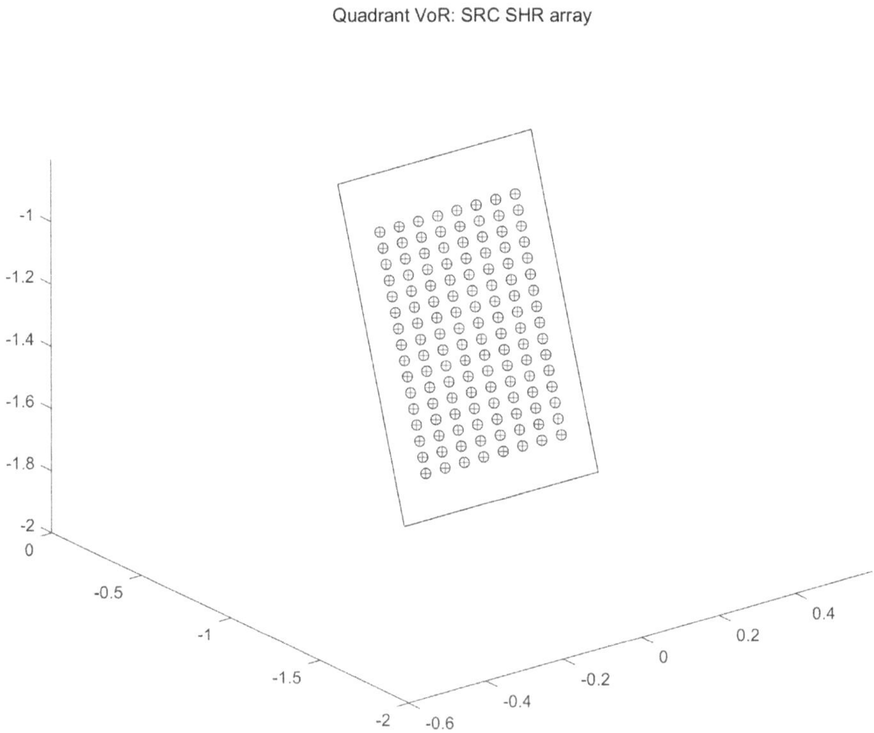

Figure 5.4 *A planar S-band receiving array for SRC configuration. Dimensions in metres.*

rate; however, they tend to be oriented to fixed-point arithmetic and at this time a GPU would be the processor of choice.

A method capable of evenly spaced beams, but also with variable azimuth resolution, would use beam-by-beam, element-weighted vector sums, with a process capacity requirement of 200 billion complex multiplies per second, also, as a non-conditional process, within the capacity of a GPU.

In Figure 5.4, the receiver is an array whose boresight direction lies above but near the horizon. The height axis is based on −1m at the top of the receiving array. SRC is intended to cover an azimuth sector of 90° for protection of infrastructure or public sites against small UAVs.

For SRC, the processing burden could be accommodated by multicore CPUs operating at say 3 GHz, plus a graphics processor such as recent Nvidia and AMD products providing up to 8 TFlops (TF) of 32-bit floating-point arithmetic.

5.2.2 SRC transmission

The SRC transmitter model provides a coherent, pulse-modulated but band-limited S band irradiation throughout the CVoR. Its transmitter output power is set at 1 kW, using a vertically focused antenna array with 90° azimuth field of view and a minimum pulse repetition interval of 30µs with duty cycle less than 5%.

To achieve the intended minimum and maximum ranges, SRC models a pulse length up to 1µs, and the repeated waveform consists of components covering both range windows and ambiguity cases. SRC's range could be extended by using a second transmit pulse capable of 16× compression. Figure 5.5 illustrates the I/Q transmission waveform envelopes.

The SRC transmitter provides a narrowed vertical beam pattern, illustrated in Figure 5.6.

Receive beams are formed in both azimuth and elevation within this vertical envelope.

5.2.3 SRC receiver channels and range cells

The SRC example receiver uses an array of 8 × 16 channels. While the front-end functions of radio-frequency filtering and conversion for a small array may be centralised, for larger arrays they are properly carried out locally at the array. A cost-effective form of assembly for SRC, and one that allows for the development of different configurations at reasonable cost, may be to combine 16-element slates, approximately 0.2 m^2. Each slate performs all functions from radio-frequency reception to output of high-resolution, serial digital, in-phase and quadrature signals for each of its 16 channels. SRC would then be built from 8 slates, in 2 columns of 4.

Range resolution for SRC is modelled at 75 m, with pulse width of 500 ns and effective bandwidth of 2 MHz, occupied bandwidth (−23 dB) of nearly 4 MHz. With instrumented range of 5 km, the receiver generates 100 range gates at up to 30 kHz. A PRF of 8 kHz can provide appropriate coherent gain, range and Doppler ambiguity, reporting rate and sensitivity to dynamics (see Table 5.1).

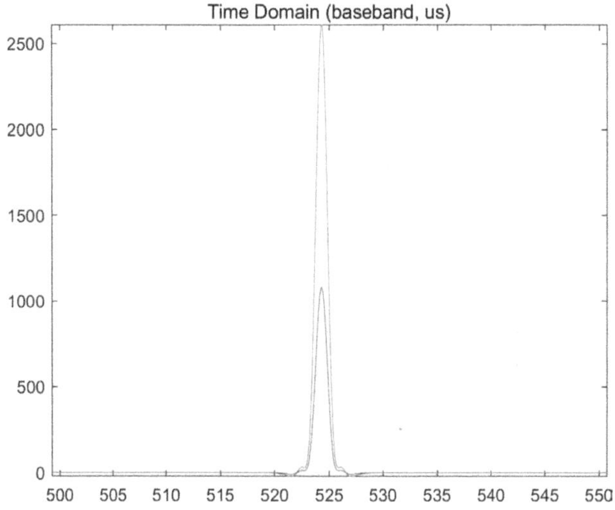

Figure 5.5 SRC time domain transmission (I & Q baseband vs time)

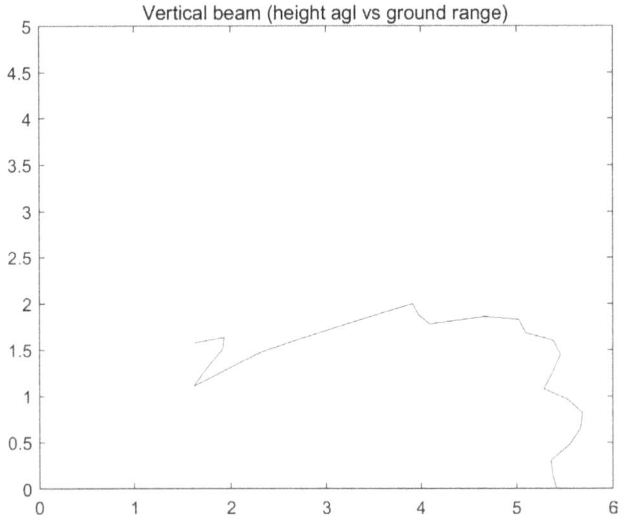

Figure 5.6 Vertical beam range plot for SRC (units: km)

The pulse repetition interval (PRI) of 125 μs yields unambiguous range of over 18 km.

In the example, each channel generates 16-bit I/Q samples at up to 3 MHz continuously, and for each 16-element slate the raw data rate is estimated at 1.5 Gbps. SRC would therefore generate 12 Gbps. Fibre-optic communications can provide the necessary capacity.

Overall the receiving array generates 1.5 GBytes per second in complex 32-bit signal values, to be communicated to the radar processor. Each CPI generates 0.5 GBytes of data in 0.3 seconds.

5.2.4 SRC Azimuth and elevation beam-forming and RAED data access

The SRC array consists of 128 receiver channels, generating a 2-dimensional beam array (nominally 16 elevation × 8 azimuth). For this flat array the directional resolution varies from 7° in elevation × 14° in azimuth, on boresight, to 10° at 45° elevation and 20° at +/− 45° in azimuth.

For both SRC and ATC configurations memory structure and access will be important. A structure in which both channel-by-channel and beam-by-beam data are accessible in quantity will be necessary. Similarly, access will be needed to both time domain and frequency domain data. An effective design approach will be to consider a 4-way data structure for access to Range/Azimuth/Elevation and Doppler data. This is labelled a RAED structure (illustrated in Chapter 3, figure 3.3 and Chapter 6, figure 6.1); it will be needed for data access in processes outlined here and in Sections 5.2.5, 5.3.4 and 5.3.5, and will be referenced further in later chapters.

As we have seen, beams on reception may be calculated by means of Fourier transformation or by direct complex weighted vector sums (CWVS). Fourier-calculated beams vary in separation across the field of view and require about 30 billion complex multiplies per second for SRC. The direct CWVS method, mentioned below for ATC, requires about 200 billion complex multiplications in the SRC case, for even beam spacing.

The output for all cells and for either method would occupy 840 MBytes in memory for 1 CPI.

5.2.5 SRC Doppler transformation

After beamforming, coherent integration in time can be achieved by Fourier transformation, with a standard window function across the CPI, generating Doppler spectra for each resolution cell.

Doppler spectra are generated more than twice per second. To obtain these Doppler spectra for each of 51 200 cells at that rate requires about 10 billion complex multiplies per second; well within the non-conditional capacity of a graphics processor.

The results of the primary Doppler filter are used, as will be described in Chapter 6, to determine which cells contain signals of various kinds – possible targets, clutter, interference, phase noise, etc., leading to cell discovery, target capture, classification, tracking, reporting and condition updates.

The SRC Doppler filter process requires 2.4 billion complex multiplies per CPI. It provides input to a generic and non-conditional sorting process referred to as a vector histogram (VH). This is described in detail in Chapter 6, and its output is used to direct downstream, potentially conditional processing.

5.2.6 SRC airspace partitioning

Beam-forming and Doppler transformation are linear processes and can be applied in either order, yielding different views of the signal space. For signals that only achieve significance after Doppler transformation it may be appropriate to derive a range-Doppler view in the first instance; however, where signals such as broadcast interference are present, as we shall see in Chapters 6 and 10, it may be necessary to derive beams in the time domain, while retaining access to channel-wise data, prior to Doppler transformation.

For SRC, in cases where high-amplitude clutter results in excess phase-derived noise that varies with direction, Doppler beams will be needed at an early stage.

5.2.7 SRC operation and processing

To describe the mode of operation of SRC we replace the sequence for scanning radar (point, transmit, receive, detect, track, report) with a process comprising twelve activities, several of which take place in parallel.

1. Stream 1 of operation, Transmit, is to maintain a series of pulse transmissions into the CVoR with minimum timing, frequency, phase and power deviations.

2. Stream 2, Calibrate, is interleaved between data acquisitions to obtain calibration data for the receiving array at predefined intervals.
3. Stream 3, Acquire, is synchronised with Stream 1 to receive, amplify, filter, frequency convert, digitise and transform to in-phase and quadrature components all the signals received at every array element and every delay value between transmissions and maximum time delay, as they arrive from Stream 1.
4. Stream 4, Transfer, synchronised with but delayed from (Stream 3.) by 1 PRI is to transfer all signal samples to memory in a highly parallel computing engine.
5. Stream 5, Storage, assigns signal data within high-speed memory for access by parallel downstream processes, managing memory and longer-term storage (Channel-wise, Time domain output).
6. Stream 6, Beamform, delayed from Stream 5 by 1 PRI is to beamform and explore every resolution cell (in range, azimuth and elevation) in the time domain to discover consistent noise, the presence of potential static or moving clutter or the presence of other disturbance or failure in operation (Beam-wise, Time domain out).
7. Stream 7, Channel Doppler, delayed from Stream 5 by 1 CPI forms Doppler spectra for each channel (Channel-wise, Frequency Domain output).
8. Stream 8, Beam Doppler, initiated when all resolution cells have been acquired in the time domain via Stream 6, through a CPI, is to Fourier transform each cell, forming Doppler spectra for each (Beam-wise, Frequency Domain output).
9. Stream 9, Cell Discovery, when all Doppler signal data are available (1 CPI from first PRI, plus Doppler FFT), Vector Histograms are formed from the output of Streams 5-8, to discover the presence of returns falling within the specified limits of Signals of Interest in a Discovered Cell.
10. Stream 10, Cell clarification and Target Capture, delayed to completion of Stream 9 (1 CPI plus DFFT plus VH): Prepare cell suppression; capture targets; based on VH multi-threshold results.
11. Stream 11. Report; report Targets of Interest (ToIs).
12. Stream 12. Maintain. Select channel and beam histories for Selected Cells; apply selected processing to assess, recover, classify, develop analytic tracks; store, develop metadata; manage storage.

The process is illustrated in Figure 5.7.

5.2.8 SRC summary

This section follows the scale and function of the smaller generic example of HSR. It is transportable, is of relatively short range, is oriented towards very small, slow aerial targets, and is of a scale suitable for use in maintaining security of critical infrastructure or sites of public assembly.

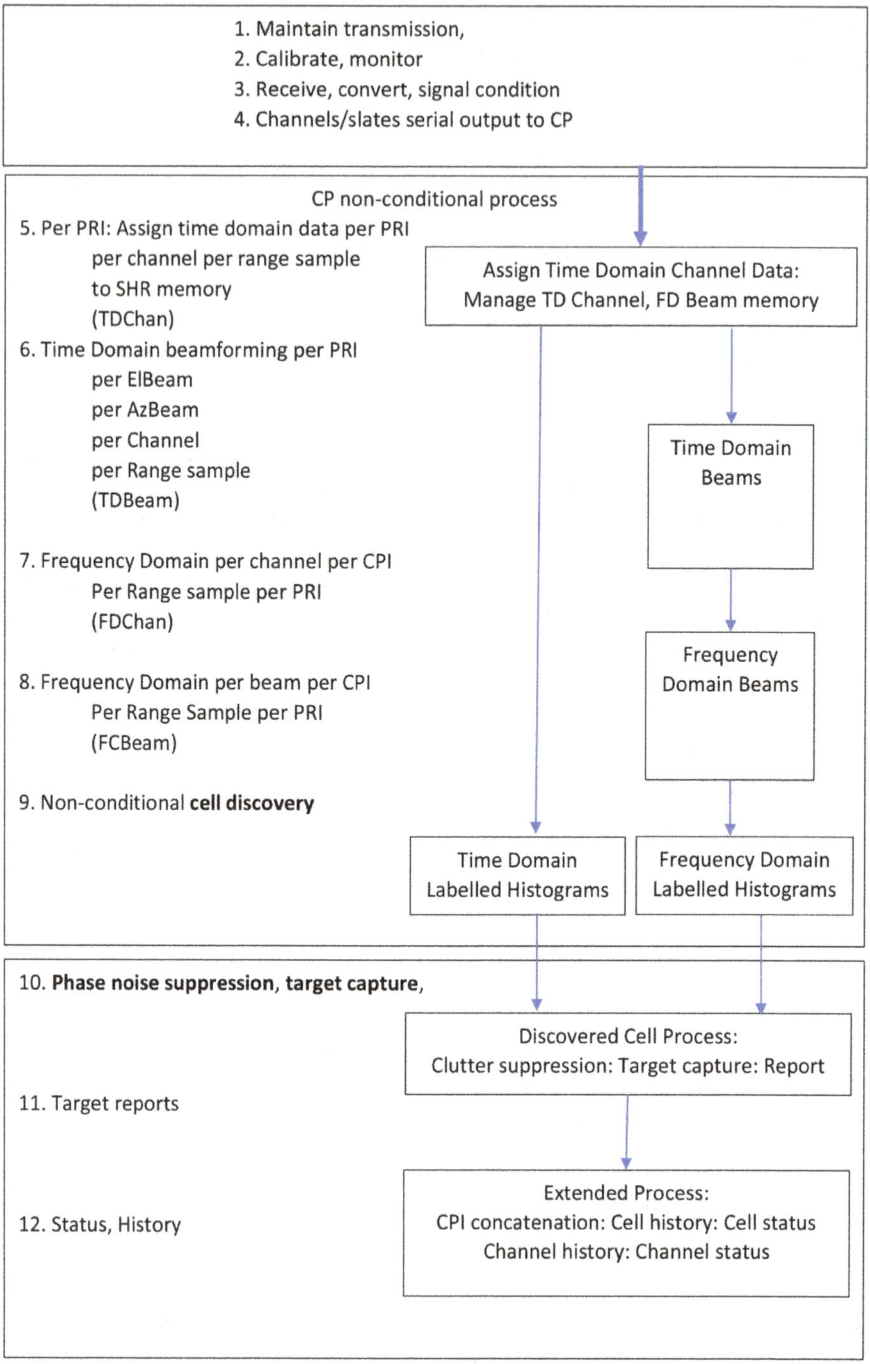

Figure 5.7 Operating sequence for a SRC configuration

5.3 ATC resources, structure and functions

The ATC configuration for staring radar shares many aspects with the SRC version, but on a larger scale and with additional features for use under circumstances of longer range, 360° coverage and a focus on safety rather than security.

The performance specification of a suitable ATC sensor should cover items consistent with Table 5.3.

Table 5.3 Example requirements for ATC configuration

Item	Minimum	Maximum	Comments
Operating frequency	1 215 MHz	1 400 MHz	L band for long range
Maximum range	60 NM		Civil Terminal Area
Minimum range		1 NM	
Range resolution		200 m	
Range accuracy		50 m	
Minimum target		1 m^2	Civil aviation
Maximum elevation	60°		
Minimum elevation		0.5°	
Transmit boresight elevation		+10°	FL 600 @ 60 NM
Azimuth resolution		2.5°	
Elevation accuracy		1 000 m	

5.3.1 ATC physical configuration

ATC is oriented towards large-area surveillance, and therefore longer range and effectively hemispherical coverage. When the CVoR extends to 360° in azimuth, with a non-rotating antenna, the physical configuration is complex, but different options are available. Two examples are:

1. a square or triangular truncated pyramid (as exemplified by the PAVE PAWS early warning radar, illustrated in Figure 5.8), and
2. a multi-faceted, conical array, illustrated in demonstration prototype form in Figure 5.9.

The configuration chosen for ATC is based on consideration, in outline, of these two alternative array geometries and appropriate beam-processing methods.

The key distinction between these array forms is that the pyramid separates coverage of the CVoR between planar faces whereas the multi-facet array beams are formed across all faces in view from a target at each azimuth. The applicability of different staring array solutions depends on the balance between the resolution and tracking performance of each and between the costs of different component inventories and computing burdens associated with multiple beamforming by each form.

The pyramid lends itself to efficient FFT for beamforming with flat panel arrays, but it forms beams that vary substantially in resolution and pointing intervals. Meeting minimum resolution requirements will lead to more expensive radar arrays.

Figure 5.8 PAVE PAWS Phased Array Radar, RAF Fylingdales (Creative Commons Attribution-Share Alike 3.0)

Figure 5.9 4-facet array demonstrator

A multi-facet array can provide uniformly spaced beams, with matching resolution, but will incur a greater computing burden.

This choice between ATC array geometries is complex, and is an important driver for processing. Static-phased-array radars tend to be rectangular (see various naval radars) or 3-faced pyramidal (PAVE PAWS), but a simple conical array example has 12 relatively narrow faces. In either case the ATC receiver configuration is assumed to consist of 4-element L band slates.

5.3.1.1 Resolution

A powerful reason for selecting a geometry using a minimum number of planar arrays is that to complete the beam-forming calculations the efficiency of FFTs is attractive. However, this approach has costs, since to make use of FFT efficiency the faces must be planar. The resolution on boresight is given by the ratio of the wavelength to the array aperture, and for resolution of 2.5° the aperture must be 19 wavelengths, or, for ATC, 4 m wide. In any one direction a single face must be used, and for a four-sided array each must accommodate targets within a 90° sector. Towards the edges of each sector, azimuth resolution is degraded and beam intervals increase by a factor increasing to 1.4 compared with the resolution on boresight. Therefore to meet the omnidirectional resolution requirement the aperture must be increased by the same factor, to 5.6 m, requiring 48 elements horizontally in each face. With 16 elements vertically the number of elements per face is 768, and with 4 faces the total is 3 072, or 768 4-element slates. For a 3-sided pyramid the factor is 2 rather than 1.4.

For the 12-facet array, for which an aperture dimension greater than 4 m is achieved by any 5 contiguous facets, resolution better than 2.5° is achieved in any direction with a complement of 1 536 elements, or 384 slates, exactly half that for the square array.

5.3.1.2 Costs of array options

For a radar in which the total cost is dominated by the array itself, this is a substantial saving; instead of requiring 3 072 elements the 12-facet array will require half as many, approximately halving the component cost.

For the square planform ATC the use of FFTs for beamforming would require in the region of 500 GFlops (GF) – well within the capacity of a single graphics processor, at a unit cost of several thousand pounds, including processing of the extended array to achieve the required resolution near the 45° azimuth directions. However the square form requires tracking and positioning algorithms designed to take into account the variable beam position intervals and resolutions, and so loses simplicity and mathematical symmetry.

By separating the radar into a larger number of narrower faces, but beamforming across faces that do not form a plane, the efficiency of Fourier transforms is lost. However, this approach reaps the benefit of being able to form beams in any direction with less than 3% variations in resolution. A subset of faces will be used for any beam, but as the forming set shifts from say faces 4–8 to faces 5–9, 80% of the channel signals are re-used, limiting any discontinuity in beamforming.

The cost of computing beams in this way is that of the processing burden. The process requires many numerical operations, but they are simple. The beams are formed using beam-by-beam CWVS calculations, across say 240 azimuth beams, each formed across 5 facets, with 32 elevation beams and 1 000 range gates, with a pulse repetition rate of 1 kHz. This beam-forming process will occupy about 10 TF. With process overheads, this might require up to 4 additional graphics machines, at a cost of a few tens of thousand pounds; a fraction of the array cost. At this scale, even larger arrays will also be feasible for longer range radars.

With respect to processing costs and benefits, the benefits of evenly spaced beams with uniform resolution outweigh the cost of the additional processing

Table 5.4 Different pointing and resolution results for 4-face pyramid and multi-facet array, (A) assuming the same number of array elements or (B) meeting the same resolution performance

	Pyramid	Multifacet	Cost factor
A. Azimuth resolution (0°, 90°, 189°, 270°)	4.77°	2.37°	
A. Azimuth resolution (45°, 135°, 225°, 315°)	6.95°	2.31°	
A. Azimuth pointing interval (0°, 90°, 180°, 270°)	4.77° (/2)	1.5° (2)	
A. Azimuth pointing interval (45°, 135°, 225°, 315°)	6.95° (/2)	1.5° (2)	
A. Channels	1 536	1 536	200 per channel
A. Complex multiplies/sec	200 GF (FFT)	14 TF (CWVS)	10 000 per TF max
B. Channels (for same worst-case resolution)	4 096	1 536	200 per channel
B. Complex multiplies/sec	1 TF (FFT	14 TF (CWVS)	10 000 per TF max
B. Cost factor	830 000	440 000	Channels + Proc.

capacity and the apparently efficient but less spatially-symmetric FFTs. The array then yields a fully flexible, ubiquitous radar.

For these reasons, and taking into account the pointing and resolution figures illustrated in Table 5.4, the choice for ATC is a 12-facet, truncated cone whose receiving array layout is illustrated in Figure 5.10. The transmitter will be a 4-faced vertical array above and at the centre of the receiving array. In operation it should occupy a 6-m tall enclosure intended for mounting on a tower between 10 and 20 m above ground. For this function operation at L band (1.2–1.4 GHz) can provide performance to 60 nautical miles with targets above 1 m^2 radar cross-section.

Reservations exist about moving away from the well-attested array format of planar faces and FFT-based beamforming, but Figure 5.18 will show that a smoothly varying, static, coherent beam set can be formed by a multi-facet array. The processing burden appears to be well within the capacity of current parallel processing platforms, and affordability should continue to increase subject to Moore's Law.

From Table 5.4 it is clear that while the process burden for the multi-facet array is greater by a factor of at least 10, the cost of that capacity is small compared with the additional array hardware required for the pyramid to achieve the same minimum azimuth resolution. The choice is therefore in favour of the multi-facet array.

5.3.1.3 ATC array geometry

The ATC receiving array consists of a number of sloping, planar facets, each carrying 128 receiving array elements, operating at L band. The array is close to 1.8 m tall and 5 m across, taking this form of a truncated, faceted cone. This is consistent with

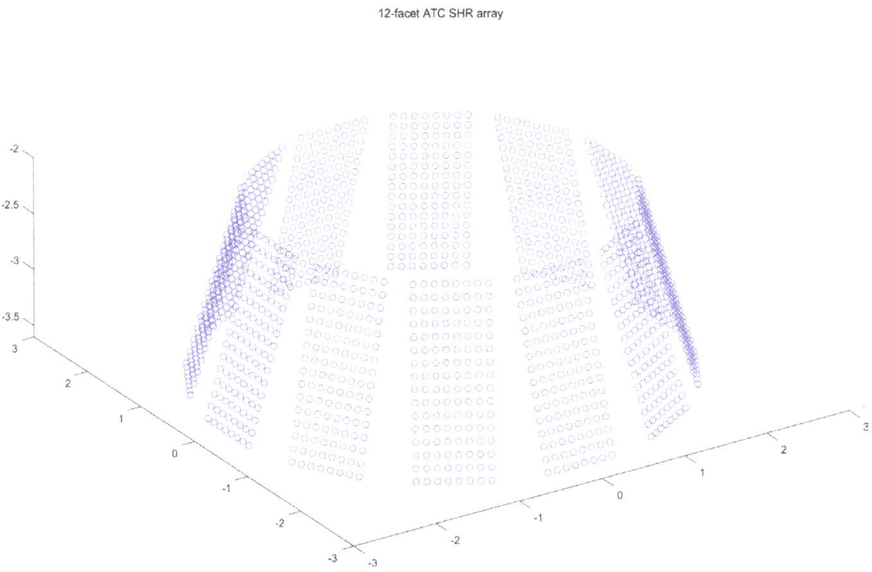

Figure 5.10 Receiving array outline for ATC staring radar configuration

a substantial assembly mounted on a tower according to conventional surveillance radar practice; the structure needs to be rigid, but not having large moving parts will be a simpler and less vulnerable installation than for beam-scanning radars.

Figure 5.10 illustrates the geometry of the 12-faced receiving array for a 360 azimuth degree radar capable of operation as a primary surveillance radar for aviation.

The receiving array, in this case operating at L band, with beam patterns as illustrated in Figure 5.18, offers azimuth resolution better than 2.5°, meeting the standard 3-nautical mile separation criterion to a distance of over 60 nautical miles. For some high-resolution applications, larger arrays may be required.

The transmitting antenna can be located above the array or can be located separately provided adequate timing and waveform control.

The multi-faceted array may be populated with 4-channel receiver slates; each can be controlled by a Field Programmable Gate Array. These can communicate via standard digital communications cabling or fibres and switches with a set of parallel computers, illustrated in Figure 5.17, that coordinate and process the signals acquired by each sector of the array.

5.3.2 ATC transmission

The ATC example models a 'chirp' frequency-modulated pulse with limited peak power, illustrated in Figure 5.11. The intermediate-frequency waveform is shown at higher time resolution in Figure 5.12, and the frequency/time plot is given in Figure 5.13. A dispersive digital filter is used to re-compress the pulse to a 1 μs envelope on reception, as shown in Figure 5.14.

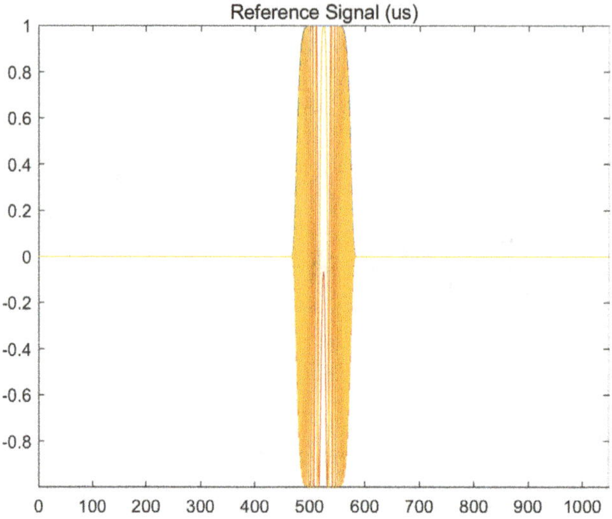

Figure 5.11 Extended FM chirp signal with flat peak amplitude

The pulse, pre-compression, is 110-μs long, with an effective bandwidth nearly 1 MHz, yielding length at half height near 1 μs, and occupying 2 MHz.

Dispersive pulse compression allows the effective power of the transmitter to be increased (by 100× in this case), and also for the transmitted spectrum to be controlled accurately. For a chirp pulse, the regulatory recommendation bounds the spectrum shape at the −40 dBc level, demanding accurate waveform control. The compressed envelope is shown in Figure 5.15.

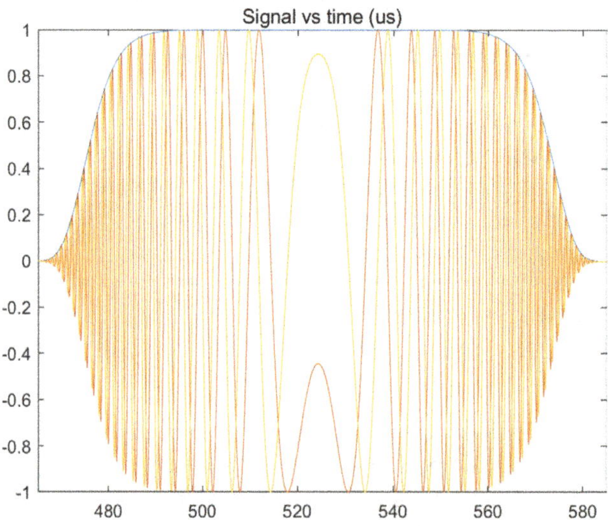

Figure 5.12 FM chirp (at IF) expanded

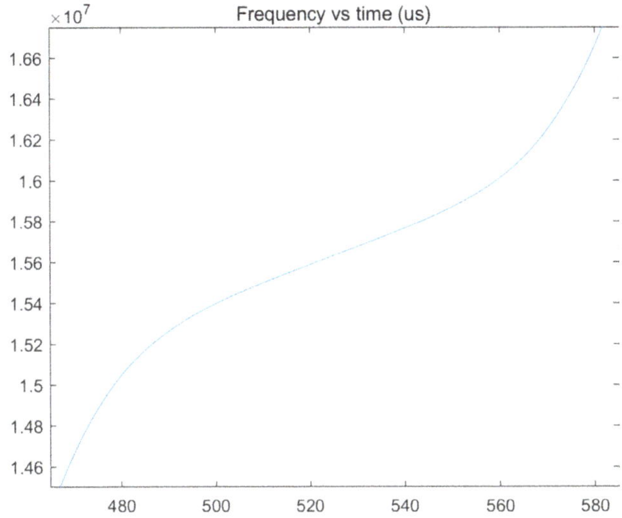

Figure 5.13 ATC example chirp frequency/time profile

The ATC transmitter model yields coherent, pulse modulated and band limited L band irradiation throughout the ATC CVoR. Its total peak transmitter output power is 40 kW, using four vertically focused, 8- or 12-element, phase-profiled antenna arrays, together providing a 360° azimuth field of view, a minimum pulse

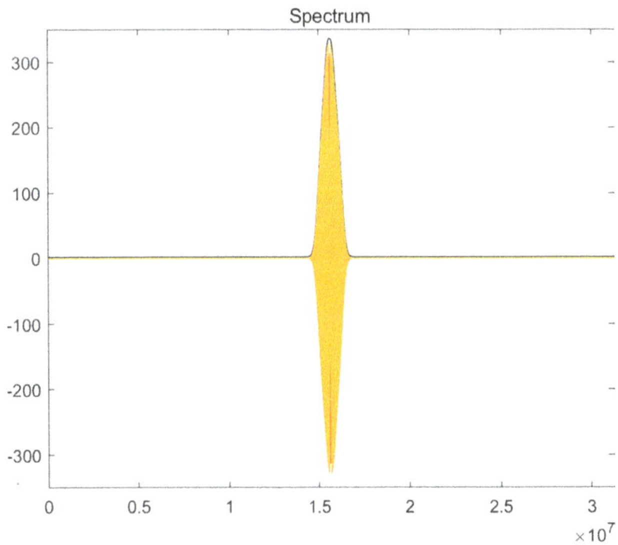

Figure 5.14 The effective bandwidth of the chirp is close to a Gaussian envelope nearly 1 MHz wide

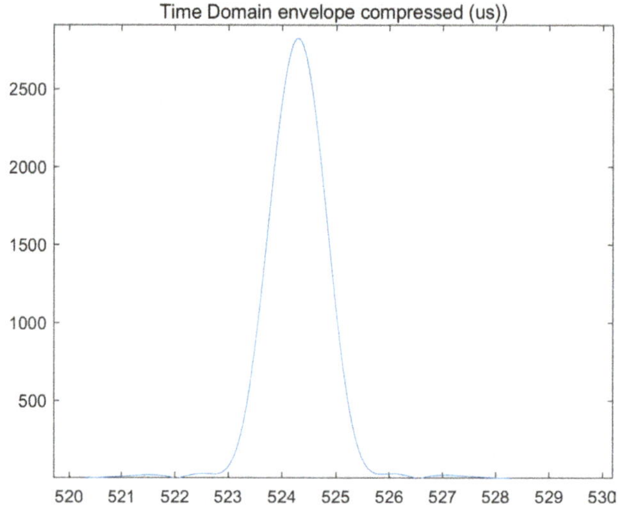

Figure 5.15 *The pulse after dispersive compression (μs)*

repetition interval of 500 μs, and a duty cycle limit less than 20%. The vertical beam pattern is similar to that for SRC, and is illustrated in Figure 5.16.

The extended transmitted pulse in many configurations will saturate the receiver or cause increased phase noise up to an effective range of 15 km. To achieve the required maximum and minimum ranges, ATC would use a sequence of two pulses to achieve from the minimum to the maximum range; a long chirp followed at

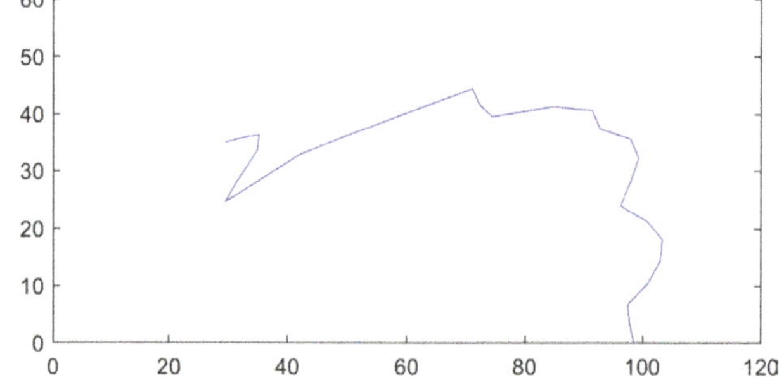

Figure 5.16 *Elevation pattern for ATC (km scales)*

maximum range by a short pulse similar to the SRC waveform, but at L band, to operate from 200 m to 15 km.

5.3.3 ATC receiver channels, range cells and data communications

The ATC example receiver is an array of 16 × 96 channels, arranged in twelve facets, each with 16 × 8 array elements, as illustrated in Figure 5.10. For smaller arrays such as SRC the front end functions of radio-frequency filtering and conversion may be centralised; for larger arrays, they are properly carried out locally at the array. A cost-effective form of assembly is to combine modular 'slates'. Each slate performs all functions from radio-frequency reception to output of serial digital, in-phase and quadrature signals for each of its channels, and is built into the multi-facet array. The channel locations in the array must be known and fixed within a small fraction of a wavelength.

Range resolution of this ATC example is modelled at 150 m, with effective bandwidth of 1 MHz, and regulated occupied bandwidth (−40 dB) nearly 2 MHz. With instrumented range of 135 km, the receiver generates a maximum of 900 range gates at a PRF of 1 kHz, providing appropriate coherent gain, range and Doppler ambiguity, reporting rate and sensitivity to dynamics.

Overall the ATC receiving array would generate 88 Gbps in complex 32-bit floating-point signal values, to be communicated to the Central Radar Processor (CRP) via Ethernet switches and optical fibre links.

Each CPI would generate 24 GBytes of channel data in the time domain for each 2-second CPI; each channel also generates 24 GBytes of Doppler spectra per CPI.

5.3.4 ATC beamforming

The process of beamforming for ATC requires that data from 5 array faces be combined for each beam in azimuth and elevation. ATC includes 12 parallel beamforming processors, each catering for beams covering 30° and receiving data from 5 faces. The interconnections are illustrated in Figure 5.17, in which the spokes are optical fibre links. The rectangle at each array condition signals for each array facet; those in red feed one sector radar processor (SRP) to form receiving beams between +/−15° azimuth.

5.3.4.1 Process rates

For ATC there are 6.5 million resolution cells; 900 in range, 180 azimuth (at 2° intervals) and 30 elevation directions (3° intervals), each requiring signals from 5 × 16 × 8 array elements, at the PRF, 1 kHz. The worst-case rate for the beam-forming process is 14 trillion complex multiplies per second, or 0.6 trillion complex multiplies, per azimuth sector, per second, or 1.2 trillion per CPI for each of 12 processors.

5.3.4.2 Data volumes

Each ATC CPI generates 24 GBytes of beam-wise data, both in the time domain and in the frequency domain, per coherent processing interval.

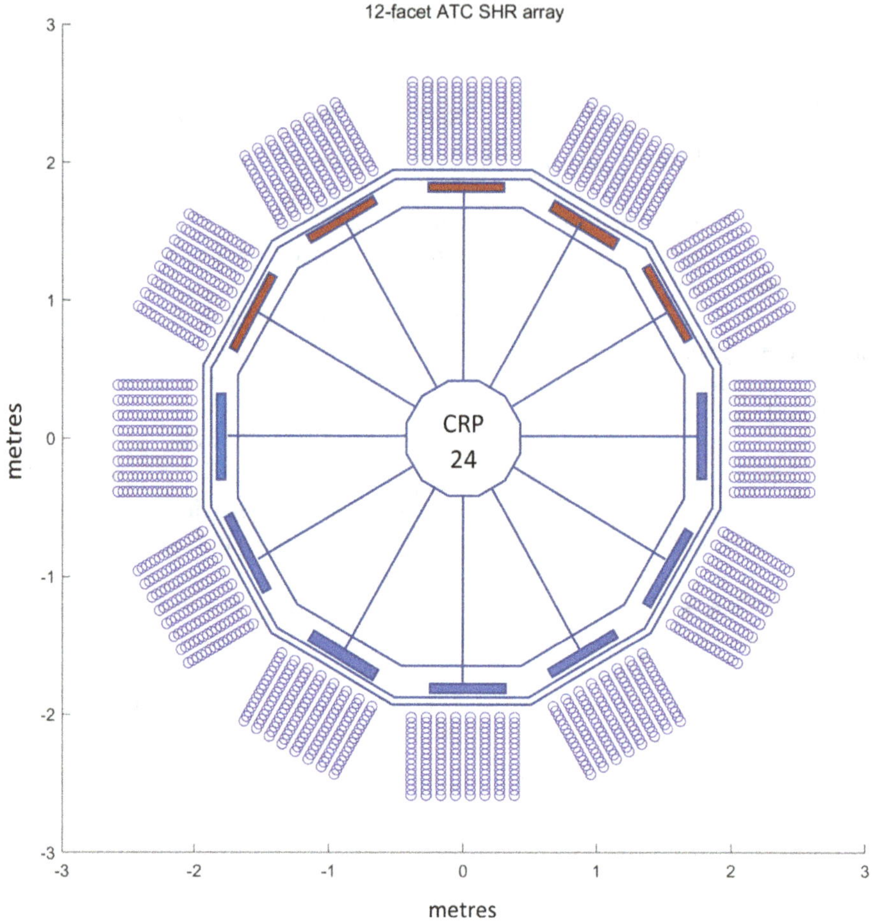

Figure 5.17 *In this example, each of 12 sector processors is linked with 5 facets of the receiving array. Twelve cell processors perform cell discovery and target capture.*

It is this process that sets the order of process capacity required for the ATC HSR example. A practical form of processor uses a number of CPU cores plus a highly parallel graphics processor, as prescribed for the SRC CRP, but in this case each 30° azimuth sector needs to be equipped with a SRP with capacity for beamforming, focused on throughput, and for downstream processing, focused on memory capacity.

Each of 12 ATC sector processors is linked with the array by high-speed serial data communications, allowing each to cater for an azimuth sector using data from a portion of the array. For the ATC example, each such processor forms azimuth beams over a 30° angle, taking data from 5 facets to form complex weighted vector sums.

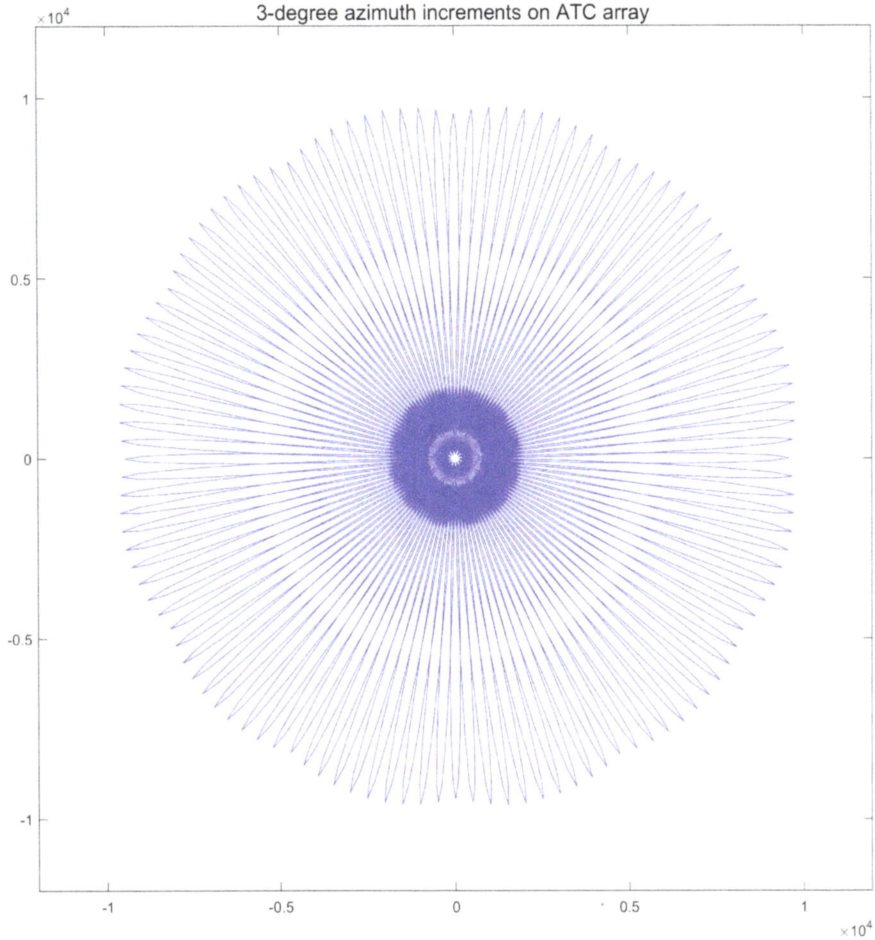

Figure 5.18 Three-degree-interval beams at 12° elevation for ATC. Gain ripples by about +/−0.2 dB.

Figure 5.18 illustrates a modelled azimuth beam pattern on receive for the ATC array, with one hundred and twenty 3°-spaced receive beams at 12° elevation. Beams can be set at about 1.5–2° intervals, but a coarser spacing is used here to aid visualisation.

ATC beams provide elevation resolution of about 5°, and accuracy typically 0.5°.

This configuration is preferable to the pyramid array with four independent faces, first because the azimuth aperture available for each beam is the same within about 2%.

The ATC model yields an azimuth beam width close to 2.5° around the azimuth circle; the multi-facet array and weighted vector sum, though it incurs a high rate of computations, allows beams to be formed in arbitrary directions, including regular series. Again, to achieve the same resolution at the 45° point a square pyramid array would need to have twice the installed hardware.

5.3.5 ATC Doppler and downstream processing

The ATC array model generates a total of 96 GBytes of signal data for each CPI, between four different views of the data. To this must be added the downstream data volumes.

5.3.5.1 Signal data volumes

For ATC this total volume is divided between the 12 sector processors, occupying up to 8 GBytes of signal data each. Downstream processing may require several concatenated CPIs, and 20 seconds' frequency domain data (10 CPIs) would occupy 40 GBytes in each processor. This is substantial but is not an unusual capacity requirement.

5.3.5.2 Downstream processes

The results of the four RAED data views are used, as will be described in Chapter 6, to determine which cells contain signals of various kinds – possible targets, clutter, interference, phase noise, etc., leading to cell discovery, target capture, classification, tracking, reporting and condition updates.

The modelled downstream processing continues, before target capture, to use the CVoR signal information to the maximum extent in identifying and suppressing phase noise (see especially Chapter 6), in identifying and suppressing moving clutter (Chapter 8), in accommodating range and Doppler walk (Chapter 7), in accommodating multipath (Chapters 7 and 9) and to accommodate coherent radar networking (Chapter 10).

These processes will require a considerable level of conditional processing, in addition to the non-conditional VH form of multiple thresholding. In addition the VH process for a cell, for 2 data views, is expected to require 2 MByte. For 5 M cells this would require 10 TB of storage, and between 12 SRPs this amounts to 800 GB each, which could be challenging. However, the more complex signal conditions requiring multiple-threshold (VH) processing, in cases of intense or repetitive clutter, or even of intense interference, tend strongly to apply to resolution cells near the surface, and between specific range boundaries. If the requirement for VH processing is limited to the lower 4 of 32 elevation beams, and to 50% of the range coverage, the memory demand reduces to 50 GB per sector processor.

Primarily for this reason it is probable that for each azimuth sector the capacity of a single SRP will be insufficient, and for ATC we expect that each sector will require 2 such processors.

Commercially available processors are available with CPU capacity for conditional calculations of about 20 GF (billion floating-point operations), plus non-conditional, parallel processing of at least 8 TF (trillion 32-bit floating-point arithmetic operations per second). These machines cost between € 5 000 and € 10 000 each, leading to a processing hardware cost in the region of € 200 000. This might result in the cost of processing hardware becoming comparable with half that of the ATC receiving array itself but a smaller fraction of the cost of a square pyramid.

5.3.6 ATC airspace partitioning

The ATC configuration for HSR shares many features with the SRC version, but on a larger scale, with CWVS beam forming, a different azimuth interrogation plan and with additional features for use under appropriate circumstances.

Low level resolution cells tend to yield more complex signals than those at higher altitude. At low elevation angles each cell will be monitored for targets of interest, but also for its status in terms of the presence of clutter, both static and moving; repetitive or random; the effects of phase noise, its complex gain, and its correlation with neighbouring cells.

5.3.7 ATC operation and processing

The parallel operating sequence is illustrated for ATC in Figure 5.19 and follows a similar sequence to that given in Section 5.2.7. Here the CVoR is expected to include volumes or areas with severe signal conditions such as intense clutter and interference, and signal data are prepared in four different views; again, a RAED structure:

1. Channel-by-channel in the time domain (to monitor array channel performance and status),
2. Channel-by-channel in the frequency domain (allowing phase noise suppression),
3. Beam-by-beam in the time domain (supporting time-varying clutter suppression) and
4. Beam-by-beam in the frequency domain (to discover target-consistent signal content).

For ATC, a channel-wise, frequency-domain view of the data is generated at Stream 7, allowing phase noise suppression and other process approaches.

5.3.8 Coherent staring radar networks

A future application of staring radar will be to provide high-performance, wide area air surveillance, with additional surveillance capabilities and designed to occupy a minimum of the radio spectrum.

This is envisaged as an extended implementation of staring radar in the ATC or similar configuration. It has been the subject of extended study prior to proposed commercial application.

The concept and accompanying considerations of functionality and risk are described in Chapter 10. A key aspect is the need to coordinate the operation of a number of ATC radars under comprehensive, accurate control of timing and target reporting.

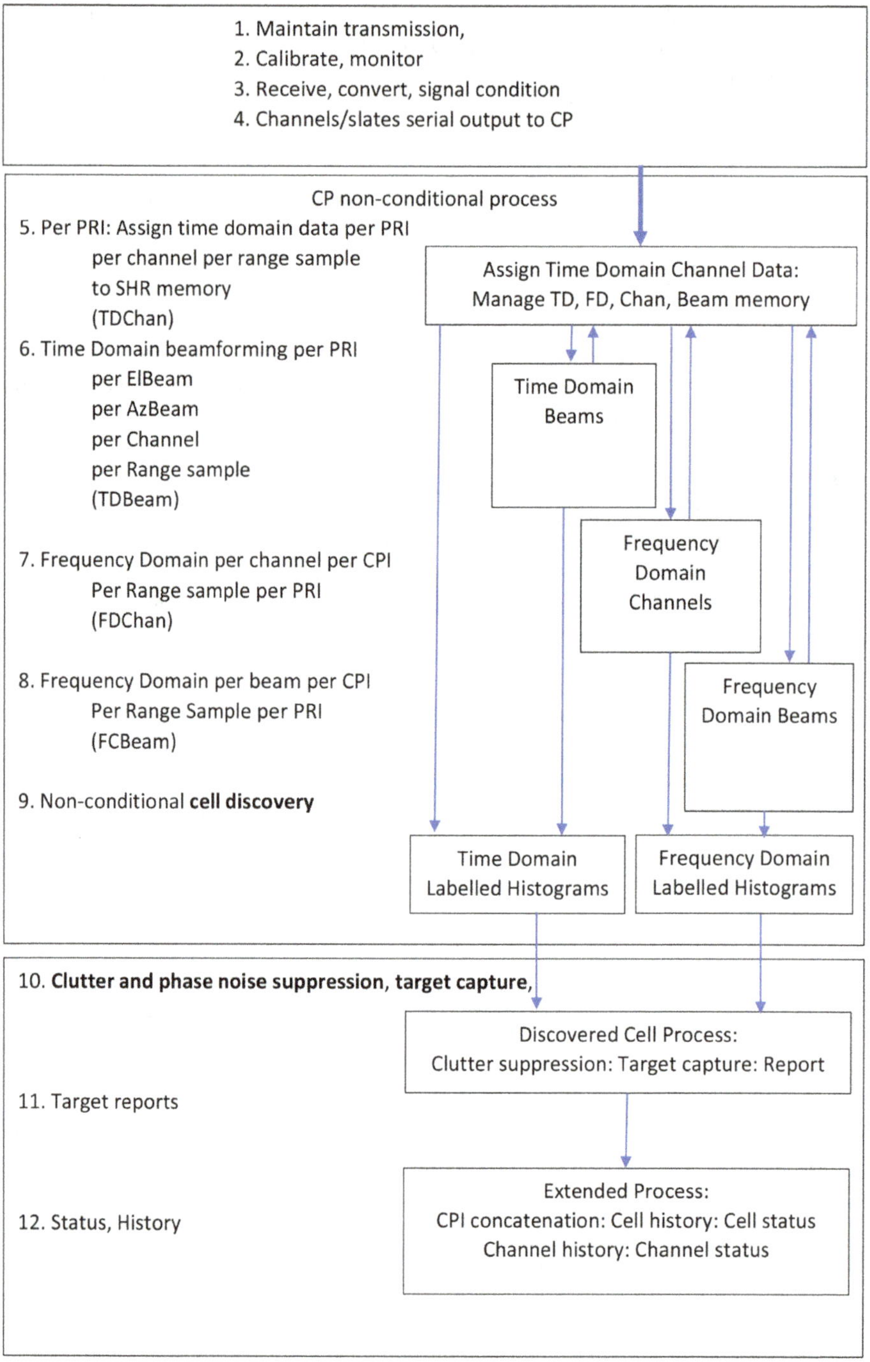

Figure 5.19 ATC conceptual operating sequence

5.4 Modular structure for surveillance HSRs

In a staring radar as we have seen the functions of transmission and reception are separate. Illumination by the transmitter is omnidirectional, and signals received by array elements from any direction are observed in simultaneous receiver beams. This requires substantial signal- and data-processing resources; much more extensive than for a scanned-beam processor, whose necessary throughput is fundamentally limited by a singly pointed beam and the speed of light.

The staring CVoR may be circular, centred on the radar or it may cover a smaller azimuth sector centred on the receiver where more specific coverage is required. The SRC example above surveys a sectoral CVoR; the transmitter uses a single vertical array column to illuminate it, and the receiving array uses 128 elements, each with a matching CVoR. For ATC a higher-power and broader-beam transmitter is used, but the receiving array is more extensive, and in both cases digital signal processing is required, with substantial capacity at the channel level and after beamforming.

In either case, staring holographic radar lends itself to a modular form of construction, in which standard components, whether in the transmitter, the receiver, the data network or in the computing function can be assembled at different scales for different coverage and functionality. Note that no rotating joint is required.

The SRC receiving array may be constructed from 4 × 4-element sub-array modules at S band that can provide for the assembly of a planar SRC receiver configuration, as seen in Figure 5.4. ATC may use a 2 × 2-element L band version for its more complex array, illustrated diagrammatically in Figure 5.10, for hemispherical coverage. We have referred to these modules as 'slates'.

5.5 Surveillance information

This chapter has identified two examples of staring radar that in principle allow the EUNIT to be exploited and the nature of targets explored by means of numerical models. Challenging aspects can also be discussed in sufficient detail and evaluated. Staring means that very much more information is acquired by the radar in the form of all the complex amplitude sequences it receives. Very much more processing is needed to resolve the different sources, trajectories and target behaviours.

The SRC and ATC example configurations allow these functions to be explored quantitatively; however, they do not represent designs. The increasing capacity and reducing cost of computing power provide that these large volumes of data and information can be affordable for radars whose functions go beyond target detection and tracking into classification and prioritisation.

Chapter 6 will focus on the measurement of signal characteristics and the discovery of resolution cells with contents of interest.

Chapter 6

Cell discovery and HSR signal metrics

Chapter 3 indicated the physical origin of distinctions between the EM signal information that is available under the electromagnetic uniqueness theorem to radar that stares continually into its coherent volume of regard (CVoR), and what is available to radar that dwells for short periods, separated by seconds, on any target in the course of scanning a non-coherent VoR. The EUNIT is well established, and we shall rely on it, with evidence and modelled results, as defining certain limits of what staring radar surveillance can or cannot achieve.

Staring radar as described operates by receiving signals at many channels of an array, by forming many beams, synchronously and coherently, between the channels and at many range gates, and then applying analysis such as fast Fourier transforms (FFTs) to explore the Doppler domain. Electronically scanned radars have the ability to dwell on targets for variable times, but this comes at the expense of other parts of the VoR. Staring radar is able to extend target dwell times without that cost and has the flexibility to use coherent processing intervals (CPIs) appropriate to any downstream process. It can then determine whether these signals qualify as representing targets of interest by evaluating the complex amplitude sequences at each spatial resolution cell formed in range, azimuth and elevation.

Staring radar has the opportunity to discover the presence and attributes of targets of interest over the extended CPI and may use a range of approaches to determine their presence.

The data received during each CPI are time-domain series of complex signal amplitudes for each element of the receiving array and for each irradiating pulse propagated throughout the CVoR. As these data are received, they are formed into time-domain beams, into element-wise frequency spectra and into frequency-domain beams. In the simplest case a single length of CPI is assumed, but this is not a necessity, and 50% overlapping CPIs are preferred. The process is illustrated in Figure 3.3 and generates a RAED (range/azimuth/elevation/Doppler) data structure, illustrated in Figure 6.1. The memory occupied by this structure is substantial, especially for the larger ATC configuration, and will be referred to as 'Cell Memory'. For SRC, it is estimated at 32 GB, and for ATC, it is 1.7 TB, shared between 12 beam processors.

In the data structure, resolved cells' complex amplitudes, defined in terms of range ($\mathbf{R}'$), horizontally and vertically located array elements ($\mathbf{H}$ and $\mathbf{V}$), azimuth (α) and elevation (ε) beams, in the time ($\mathbf{T_D}$) and Doppler ($\mathbf{F_D}$) domains, are arrayed in

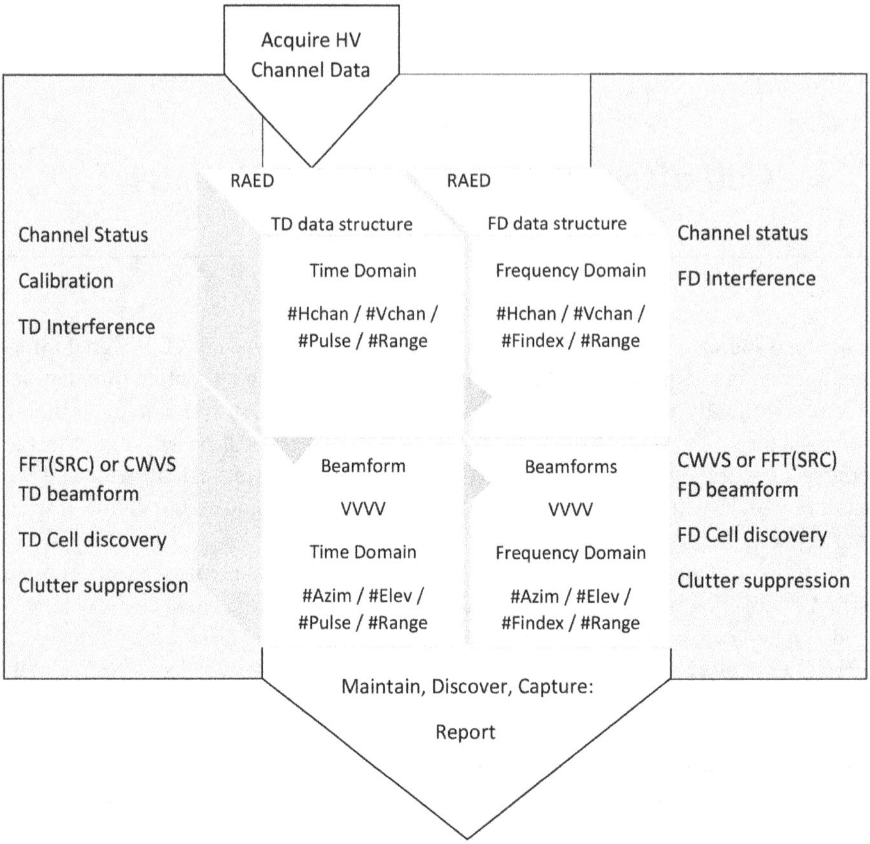

Figure 6.1 RAED data structure and functions

T_D sample time and as **F_D** spectra. This is termed the **RAED** structure, as referred to in Chapters 3, 5, 6, 7, 8 and also in Section 6.4.

The continuity of data acquisition for an HSR means that the boundary between pre- and post-detection, and thresholds themselves, need not be singular. This suggests versions of track-before-detect and classify-before-detect processes, and we consider a process of discovery, during which signals acquired in a resolution cell may indicate sources of varying significance, some of which can be classified in terms of target types, while others are significant in terms of the conditions of operation such as static clutter, mobile ground clutter, avian or turbulent 'litter', unintentional or intentional interference, or as symptoms of system malfunctions. Each of these will yield signals that contain substantial information about their source and potential responses but may not yet satisfy criteria for targets of interest.

We shall use the term Cell Discovery, referring to measurements of cell signal contents as the basis for system and environmental monitoring, as well as target capture. The radar process can be configured for different types of target or trajectory of interest, with appropriate dwell times and processes of analysis to achieve the necessary probabilities of detection, correct classification and constraints on false reporting.

We shall introduce a method of assessing the information content of a resolution cell, a 'vector histogram' (VH), which may offer more robust responses to signals than single detection thresholds, while avoiding the conditional process implications of single thresholds. This is a method that, in the place of Doppler filters followed by a yes/no decision as to whether the total signal content qualifies for tracking, can answer more questions in determining their significance; do they indicate good system operation? Do they indicate external interference? Can that interference be countered? Do they indicate a target of interest? How are the constraints of processing capacity to be met?

6.1 Channel, array and system status

Each channel of a staring receiver array contains components whose gain may vary with frequency, amplitude, temperature and ageing. As part of the operation of a staring array, the complex gain and noise performance of each array channel must be monitored, providing the basis for downstream processing, yielding measures of the condition of each channel under steady conditions and of the radio conditions under which the radar is operating.

6.1.1 Calibration requirements

To make the necessary tests, the array requires an effective method of calibration, by which known signals are injected at a known point in each channel, and their effects measured at a later point in the process. Since each channel in the array is separately filtered, amplified and gain controlled, frequency-converted, digitally converted and filtered and transformed into in-phase and quadrature components, calibration measurements must be made and assessed at a late stage in signal conditioning but prior to beamforming and cell processing.

A constraint on the calibration process is that the staring radar function exploits continuity, but the normal operation of receiver channels needs to be interrupted during calibration, ensuring that signals input during calibration are common to all channels. A method is needed whereby the reception function can be interrupted without defeating the objective of continuity. In practice, the whole process is only quasi-continuous, in that reception is typically disabled during actual pulse transmission, and short gaps may be introduced either between pulses or between CPIs when input conditions can be modified without severe discontinuity. Calibration is not required at millisecond intervals or even 1-second intervals, and it also may be accommodated as part of condition monitoring functions, with longer intervals. An option may be to reserve a late delay window at intervals for injection of what may be an adaptable calibration waveform, which may include thermal noise, minimum detectable signal tests, A/D and filter matching tests, saturation tests, bandwidth tests and, most significantly, tests of each channel's complex gain, prior to beamforming.

The transmission waveform also requires monitoring in power, frequency and phase stability, independently from the measurement of receiver characteristics.

Once the complex gain of each receiver channel is known, in the absence of externally received signals its noise figure can be measured and, with the transmitter in operation, its dynamic range may be determined. These values can be used in reporting the functional status of the radar.

6.1.2 Noise and interference

Under normal operating conditions and at delays well separated from the period of transmission, in the absence of targets each range cell's output is expected to be dominated by thermal and circuit input noise. The stability of this value can be checked at several points in the processing chain and may indicate a continuing state of health of each receiver channel.

Following transmission, signals greater than the noise level of each channel may arise either from large items of fixed clutter, from moving clutter such as wind turbines, from large targets of interest at short range, from interfering radio stations or radar systems, from jammers or potentially from instabilities in the channel circuitry itself. These cases may limit the available performance of the HSR and will be considered below.

To determine the presence of substantially or intentionally interfering sources, signals obtained via the receiver channels can then be evaluated in detail, both in the time-domain and in the frequency-domain, channel-by-channel or in different beam directions, where targets of interest are expected.

Before discussing cell discovery itself, we should address the process of radar detection and how it is normally achieved.

6.2 Target detection against noise and clutter

Conventionally in radar operation, a target is detected when the amplitude of its radar return exceeds a pre-determined threshold value. The strength of its return is determined by the standard radar range equation (Chapter 3, Eq. 3.1) where the actual scattering cross section of a target at any instant is discoverable from the transmitted power, the signal amplitude, the range, and the known gains of transmitter and receiver at the azimuth and elevation of that target (unless affected by multipath as discussed in Chapters 5, 7 and 9, or by lack of information about elevation). The amplitude of the return is then compared with the threshold to separate detectable targets from non-detectables, followed by the initiation of track processing.

This threshold is determined by a statistical process, which may be one of several alternate forms. Most are designed to maximise sensitivity while maintaining the rate at which random processes generate unwanted detections at a level within the capacity of the subsequent data processing chain. These are known as constant false alarm rate (CFAR) algorithms and are intended to provide against false detections either from thermal noise itself or, when combined with phase noise, from static clutter returns. Examples are discussed in Section 6.2.2.

Signal conditioning and the processing equipment may constitute an expensive part of a radar system, and in that sense the focus on optimising the immediate target signal content (intensity, range and direction) for instant detection and measurement from a single pulse, or a few, has been effective in supporting trajectory tracking in the presence of noise and sparse clutter.

6.2.1 Inherent noise in radar receivers

Thermal noise is constrained by the radar receiver front-end design. It is related to the bandwidth of the front-end filters, the operating temperature and the noise performance (noise factor; Nf) of the active input circuits up to the point of digitisation. The design should be such that the noise figure of the first amplifiers (the low-noise amplifiers) dominates noise arising at later stages in the design, especially those at which gain is applied and prior to digitisation. Beamforming then leads to the formation of resolution cells in which signals represent targets at distinct positions in range, azimuth, elevation (for a 3D radar) and Doppler shift.

Thermal noise itself, although forming a potentially obscuring background, can provide a stable and useful statistical reference. A staring radar has the opportunity to develop and exploit signal information as a continuing process, in which a steady and measured level of random thermal noise, far from ultimately degrading signal information, itself provides the necessary reference. It also provides an essential means of ‘dithering’ digital conversion processes of finite resolution.

The ability to assess the amplitude and complex spectrum of noise, and of significant departures from the known, expected distribution either in the time or the frequency domain, provides essential information both about the radar itself and the environment in which it operates.

Detection normally occurs by the comparison of signals that may or may not contain target information with a threshold. The threshold is calculated from a mean value of amplitude, of power, or a derived quantity such as the logarithm of the power, from a number of cells that are assumed not to contain a target. The threshold and the method by which it is set directly affect the most important element of surveillance performance – P(d), the probability of detection of a target of interest. Typically just a single threshold is used, determined by a CFAR process, and we will return to this later. The following sections are the illustrations of signals of different forms; noise, clutter, interference and targets themselves, in both time- and frequency-domains. For maximum flexibility, results are primarily based on Huygens numerical modelling. Figure 6.2 illustrates 270 milliseconds of received in-phase noise components for each of 64 calibrated SRC array channels. Low-level, low-frequency clutter has been attenuated by high-pass filtering in the time-domain.

After the calibration the noise levels may be expected to vary by only a few per cent in amplitude between channels and to give a good indication of the health of each channel. These signals are described as ‘random’, but they clearly share common aspects.

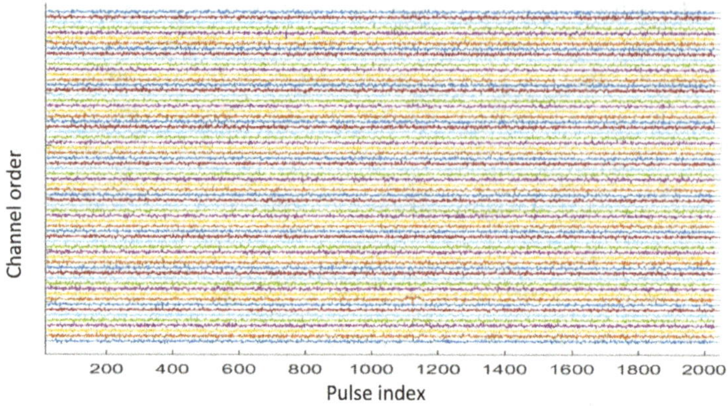

Figure 6.2 *Thermal noise modelling a 64-channel staring radar at a specific range, for 2 048 PRIs*

Persistent observation of 2 048 successive pulses at a chosen pulse delay yields a complex noise spectrum of 2 048 frequency indices, of resolution near 4Hz in this SRC example (near 0.2 m/s equivalent at the operating frequency near 2 700MHz).

The modular spectrum in Figure 6.3 is derived from channel noise signals in Figure 6.2 and resolved in range, azimuth, elevation, after partial suppression of near-static clutter signals. The residual low-frequency clutter is seen at Doppler bin 1 025. The left side of the spectrum covers Doppler shifts for approaching scatterers and the right side for receding scatterers.

Detection of targets in the presence of noise can be achieved by measuring the distribution of relevant, non-target, random signals, and setting the detection threshold relative to that distribution. A threshold-to-RMS noise modulus margin of between 12 and 20 dB would be typical. The challenge is to determine the

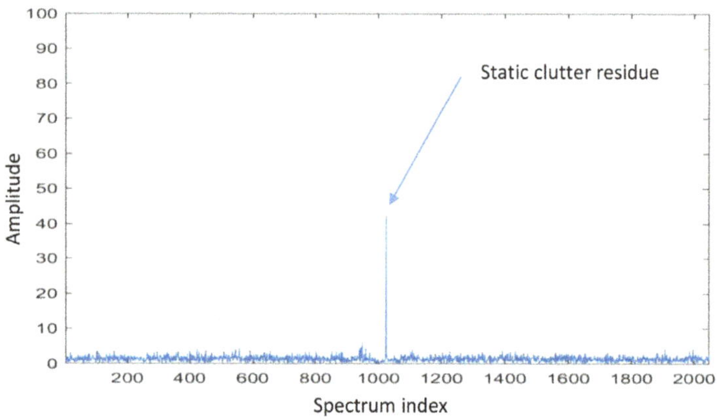

Figure 6.3 *Fourier spectrum for one resolution cell, after partial clutter suppression*

meaning of 'relevant, non-target, random' signals, and to manage the effects of clutter. Clutter in the form of static targets that may be large, but are not expected to move, may be suppressed by a 'clutter map' giving adequate desensitisation for zero-Doppler objects. CFAR methods are a standard feature and are described in Section 6.2.3.

We shall explore how cell output signals can be assessed in the case of an HSR, and whether this should differ from the standard methods used in the course of radar scanning. A substantive distinction is that a noise spectrum such as that shown in Figure 6.3 is generated for every CPI and every range/azimuth/elevation cell. Such extensive Doppler signal information will be useful in determining the presence of significant signal components.

6.2.2 Detection and thresholds

Processes of detection involve comparing the amplitude of a signal with a threshold. Thresholds themselves are based on the radar power budget, on the target radar cross section (RCS), and of other aspects of the radar's operation, and below we outline different methods of threshold setting.

The detection of a particular target by a scanning beam must occur during a short period when the beam direction matches that of the target, and when the delay of observation matches that of the signal scattered from the target, and its amplitude exceeds a pre-determined threshold. Detectability is then determined by the terms of the Radar Range Equation. The timing and margin of detection and the measured, usually ambiguous, Doppler shift are then used as basic information from which plot positions are determined and tracks can be estimated. Confirmation and update of that track are then suspended until the radar beam once again points in the direction of the target when detectability is tested anew.

For staring radar, detection occurs following an extended CPI when the presence, scattering properties and motion of the target, irradiated by the radar in the context of the EUNIT, yield EM fields across the receiving array that are compatible with one of the range of target models within the surveillance specification of the radar, and within at least one of the resolution cells provided by the radar. Because observation is continuous, the length of the CPI can be chosen appropriately for the required sensitivity, the range, the PRF and the expected target dynamics. The simplest target model is a Doppler-shifted sinusoid continuous over the CPI, but many others are possible.

The challenge associated with threshold setting is exacerbated by the presence of increasing numbers of objects in the airspace that move, the presence of moving aircraft components, and also the reducing size and RCS of many targets such as drones, which are of interest, and birds, which may be of interest depending on the site.

A further challenge is that the setting of the threshold and the fact of target detection are points at which conditional processing begins in the radar. Detection by means of the amplitude threshold leads conditionally to the estimation of the source's position and state of motion.

The derived data are separated from all others and submitted to the tracking procedure. The association and filtering processes used in tracking tend to be adaptive and strongly data-dependent, which makes them less suitable for highly parallel processors, whose capacity is dependent on being able to apply the same process across a wide field of stored signal data.

A prevalent method of threshold-setting is a CFAR algorithm, based on the state in neighbouring cells and designed to limit the rate at which false alarms are generated.

6.2.3 CFAR thresholds

The threshold for a resolution cell is calculated within each cell over time or in neighbouring resolution cells, based on what are assumed to be uncorrelated noise or stable clutter values. This reflects practice in BSRs in which reception makes maximum use of each single interrogation of a target and minimum demands on downstream processing. The CFAR algorithm must be conditioned by the dimensions of measurement, their degree of correlation, the threshold to noise ratio, the ratio of different signal components to the noise level, the provision of a radar map of fixed clutter, and the ability of the downstream process to isolate false detections from target trajectories.

The objective of CFAR processes has been to calculate an optimum threshold for each test cell, at which level the detection sensitivity specification of the radar will be met, while the rate at which non-target-related detections are generated will be kept within an appropriate bound that will not saturate the data processing capacity. Thresholds are calculated from leading and lagging reference cells, as illustrated in Figure 6.4.

More narrowly local thresholds may be set to account for terrain, structures or ground clutter, with the risk that multiple targets might desensitise the detection process.

The primary constraint under which the CFAR process operates is that the noise in the test cell can be estimated from a number of nearby training cells that are close enough to represent the test cell, but not so close that the target itself, its sidelobe returns or its Doppler spurs bias the noise estimate. Different and more complex

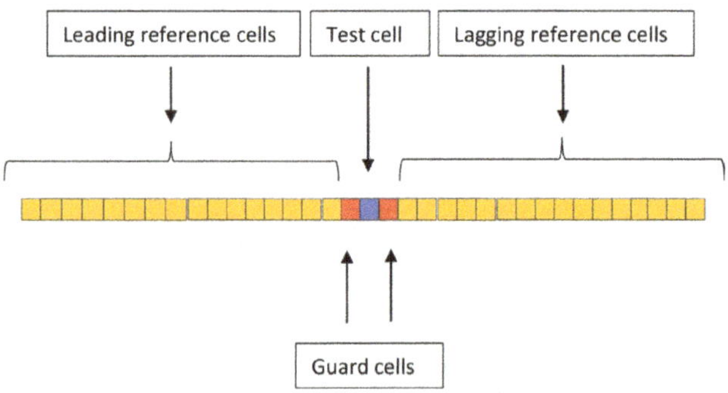

Figure 6.4 CFAR threshold setting: cells at successive ranges or Doppler indices

versions of CFAR processes may be required when ground clutter is inhomogeneous, when other targets may appear at close-in positions or, as could occur with HSR, where the range dimension may not be the primary source of noise training values.

For a BSR, the available dimensions are range (defined by resolution cells that are typically narrower in range than the beam-limited azimuth dimension) possibly plus low-resolution Doppler.

Variability in the threshold naturally has to be added to noise levels themselves in estimating the probability of detection for a target of known amplitude. Where the threshold is calculated on the basis of an average of a number of cell values, in principle the greater the number of cells to average, the lower the added noise. At the same time, the greater the number of cells, the greater the likelihood that the value is distorted by other targets or clutter.

A small number of range cells are used in the neighbourhood of the test cell with typically inboard and outboard guard cells. If 16 cells are used, the test cell and 2 guard cells would leave 4–6 for symmetric training on each side. A total of about 20 training cells, including Doppler neighbours, would yield added uncertainty of about 25% in amplitude if a multiple of the mean is used to set the threshold. If 32 cells are used, the training set is of order 100 cells and may add about 10% in the mean noise amplitude, equivalent to increasing the noise figure by about 1 dB. For these small numbers of training cells, a second target or clutter signal occurring among them at 20 dB above RMS noise may have a significant effect on the threshold, and therefore on the P(d). If the range cells used are in fact correlated because of finite range resolution, then this effect will be increased. Square root or logarithmic values may be used to set the threshold, which may be less susceptible to the presence of clutter or targets. CFAR processes for threshold determination are a continuing subject for research and optimisation.

For an HSR, extended Doppler spectra will be available in each resolution cell: 2 048, 4 096 points or more, and may yield a preferred source for CFAR reference.

This number may be sufficient to allow the exclusion only of near-zero-Doppler cells and the test cell from CFAR threshold training. By using a square-root or logarithmic encoding of the signal for estimation of the mean, it may be possible to set a threshold without exclusions in non-zero Doppler cells. However, signals in or near the zero-Doppler-index may be extreme in relative amplitude, and there may need to be separate thresholding treatment of those indices. This yields a training population of 500 to 2 000 cells, and a threshold training uncertainty of a few per cent of the noise level, leading to an increase in the noise figure of only 0.2–0.4 dB. The effect of similar secondary targets is then relatively small among this larger sample size.

Also for HSR, the Doppler spectrum for a given type of target, developed over successive, contiguous and longer CPIs, may be expected to contain more detail than for a radar that scans with a short time on target and a long scan interval. This means that greater care or different algorithms may be necessary for setting a threshold. As an example, the presence of a large airframe return might increase difficulties in detecting small micromotion components representing rotors or propellers.

For a given spatial resolution cell there may be multiple targets that include a wide range of radial speeds, increasing the need for thresholds calculated over different sets of Doppler or range indices.

6.2.4 Historical threshold setting

As an alternative approach for testing signal significance, rather than comparing a test cell with neighbouring cells at the time of measurement, the statistics of signals at representative positions over a previous time period are evaluated and used as the basis for setting the threshold.

This has the advantage that thresholds will be unaffected or minimally affected by the presence of neighbouring moving targets; nevertheless if the conditions of operation of the radar change, this method will be relatively slow to adjust. In particular, it will be slow to respond to changes in levels of air activity or of interference or jamming.

Considerable experience has been gained with this method of threshold setting and it has proven robust in several circumstances; however, it is not suitable to all cases, especially repetitive events.

6.2.5 VH data format

The EUNIT provides that target information will be accessible over time, but its processing may imply a large computing resource. Sufficient processing capacity is offered by a number of highly parallel technologies of continually increasing capacity, and in most cases reducing cost, each of which has a potential role to play, but also has characteristics that may limit how it can be applied. These include the fact that they demand large memory and data transfer resources, and may lose their effectiveness if memory access becomes conditional on point by point calculations in the course of processing.

The examples in Section 6.3 illustrate several combinations of signal forms that will be encountered by a surveillance HSR and must be accommodated by the radar reporting process. Those derived from the radar's known transmission, subject to the EUNIT, should be recognisable over time, provided both linear signal processing and sufficient processing capacity, and there will also be signals from unexpected and unknown sources that must be managed. Experience with combinations of different forms suggests a need for improvements over a single detection threshold leading to data-conditional processing.

A non-conditional process is needed to separate resolution cells between those that conform with noise and those that need further analysis before determining target reports.

A basic tool for characterising the amplitude distribution of a signal data set is the histogram: a conditionally determined format that counts amplitude instances between each successive pair of thresholds listed in the histogram abscissa. A histogram allows zero values, saturated values, normally distributed values, Rayleigh- or Ricean-distributed variables, etc., to be counted and assessed. It can be used to set thresholds, but in its basic form, it does not retain links between threshold test passes and the population from which they come.

We have referred in Chapters 3 and 5 to a form of digital decision-making which, rather than interposing tests within a running process that have data-dependent outcomes and resource demands, relies on decisions already made within the data flow: a VH (also referred to as a source-referenced histogram, SRH), in which each abscissa index refers not to a scalar aggregate but to an indexed list of data items (frequencies, phases, time, etc.).

Source referencing in a VH is designed to maintain those links, in which case the source addresses (e.g., the indices of a signal component in time or spectrum) themselves become available unconditionally as data values in one or more downstream processes.

An example input data set contains measured values assigned to each of a series of data addresses. This may be a set of scalar or complex values, on either the time or frequency dimension as abscissa.

To form a conventional, 'scalar' histogram is itself a conditional assignment process: for a variable value Y at address #A:

'If Y(#A) lies between thresholds M and N, increase X_{ord} (N) by 1'.

For a single threshold, that test itself is used to direct the downstream process. Multiple thresholds can be used to develop a histogram, whose distribution may also be used to direct the process, and it is these redirections that are inconsistent with highly parallel processing.

A VH, by contrast, treats the numerical data values as already-made, high-resolution value decisions. The VH process encodes each data value as a lower-resolution integer by means of an amplitude-compression table and without explicit tests. Each compressed value is then transposed as a histogram abscissa address and the relevant ordinate is increased in the usual way, but the links between that VH-addressed increment and the precise values it refers to are retained. The word 'vector' is used because the amplitude/ordinate indices identify a list of references to signal values matching that compressed amplitude.

The value encoding yields the effect of multiple thresholds; for example, encoding as an 8-bit unsigned integer acts as 256 thresholds.

A VH, applied to all components in the time or the frequency-domain, has the effect of a multiple threshold but not in the form of a conditional test. All the tests have already been applied during conversion, calculation and encoding; the VH then redistributes all elements according to their current, known content. The range of applicability of this data format is not yet clear, but it appears to support-decision making in highly parallel, memory-intensive systems.

Figure 6.5 illustrates the application of a VH. The modelled target information is the encoded spectrum (Figure 6.5B) of noise, clutter residues and a target airframe return with sidebands to be used in classification. Figure 6.5A illustrates a histogram showing the distribution of amplitudes in the signal, and Figure 6.5C shows a VH in which the amplitude has been compressed and the ordinates are colour-coded to reveal the direct link between the amplitude values and their spectrum indices. In the histogram (Figure 6.5A), the outstanding features – the target

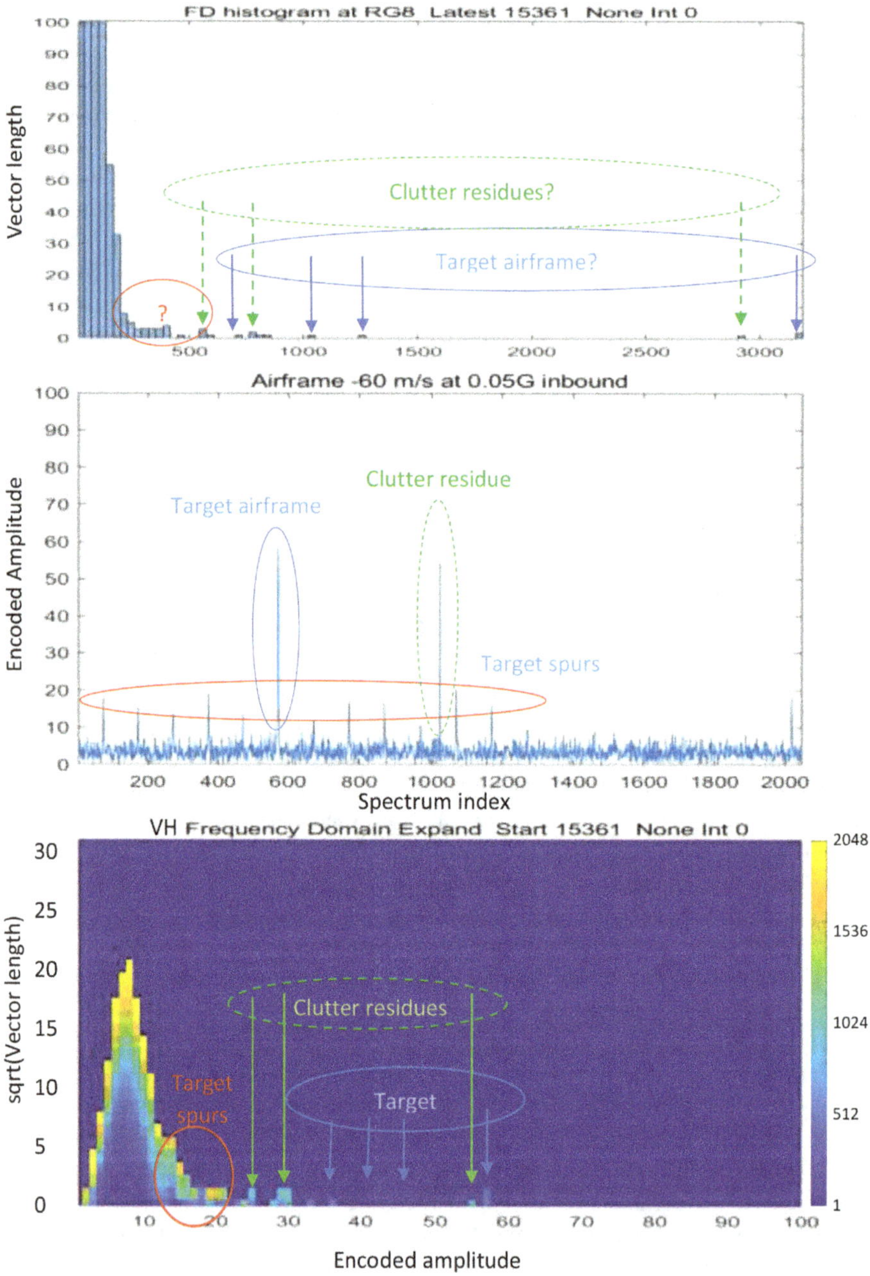

Figure 6.5 (A–C) Cell spectrum containing noise, clutter, target and spurs; histogram, spectrum and VH. The colour bar defines the time or frequency index encoding in the following VH formats.

(index 56 in the encoded amplitude) and the residual clutter (encoded amplitude 54) are undifferentiated.

For presentation, the Doppler colour coding shows the clutter peak (coded green) representing the zero- and near-zero-Doppler components. Three target components (coded blue) originate together in the negative Doppler segment (approaching target airframe), while the side spurs (propeller harmonics, circled in red) as a cluster are evenly (harmonically) spaced in frequency in the spectrum, both approaching and receding, in association with the airframe peak. The amplitude and phase of each is linked to its label, without any testing against threshold conditions.

The cell signal contents in time and frequency and their distributions are ordered by the VH and are available unconditionally. These contents may be used to discriminate targets, static and moving clutter, interference and malfunctions. In this example, high-amplitude, green-encoded values feed processes appropriate to near-static clutter; high-amplitude and locally colour-associated vales refer to targets and the effects of radial acceleration; points distributed separately but evenly relative to the target's airframe refer to micromotions (a propeller), and the low lobe illustrates the noise spectrum spread over several amplitude codes covering the whole spectrum width.

A VH offers a solution to the difficulty of incorporating multi-threshold target capture within a highly parallel and non-conditional cell discovery process.

The processing effort required to form a VH can be estimated as follows:

The starting point is a complex data time series or frequency spectrum. SRC and ATC configurations are both based on 2 048-point sample series.

The values are initially 32-bit floating-point numbers and are encoded as 8-bit integers by a look-up table using the exponent and the upper bits of the mantissa as input. This requires only memory access.

The VH itself is formed by transposing encoded 2 048 amplitudes as 8-bit addresses, incrementing each addressed ordinate as a 16-bit integer and storing the time, frequency or phase as 16-bit integers. This requires 2 048 lookup memory accesses, 2 048 arithmetic swaps, 2 048 additions and 2 048 output memory accesses, totalling 8 192 operations per VH per cell, or 16384 per cell, per CPI (2 seconds).

ATC may have up to 900 range gates, 200 azimuth beams and 30 elevation beams, totalling 2.7 million cells. If applied throughout the CVoR for each CPI, VH data operations will total about 20 billion per second, divided between the several sector processors.

With respect to memory volume, the VH may count up to 2 048 16-bit points in a single abscissa address. For full flexibility, each of 256 addresses may occupy up to 4 096 bytes, or a total of 1 MByte. A reasonable allocation of space with the address reduces this to 131 kBytes for TDVH and FDVH forms, and a total memory requirement of 755 GBytes for VH storage applied throughout the CVoR.

A VH can perform as a multi-threshold and Doppler-discriminating detector. It includes all the information in its resolution cell and domain, ordered in both amplitude and either time or frequency, as a direct basis for propagating information about each signal source type into appropriate processing channels:

a. Noise: cell status monitoring; interference monitoring
b. Clutter: cell clutter stability, suppression, repetition and correlation
c. Target airframe: radial speed, dynamics, scale
d. Micromotion: harmonics and reference to the airframe.

The following sections 6.3 to 6.5 illustrate what information arises from signal data and histograms, and explore the value of the VH form.

6.3 Cell discovery and analysis

The earlier outline of target detection was predicated on a process that aims to test in a single operation the difference between a target return and a form of random noise.

A typical BSR detection test occurs over a time period either inversely proportional to the bandwidth of the radar (order of microseconds) or close to the dwell time of the radar beam (order of milliseconds). At the point of detection, data that pass the detection test are deemed to represent a target of interest, and are separated from data that fail it and are effectively discarded. A tracker is then provided to estimate the real trajectory underlying the series of measurements.

This quick separation of signals of interest is a natural and necessary result of the beam-oriented radar method, and it has the value of ensuring that the demands on subsequent stages of processing and tracking are kept to a minimum. Its corollary is that extended target analysis must be exceptional.

This focus on speed of detection, followed by track association and filtering, can encourage a perception by designers that information reported by the radar is actually created by the downstream track processing algorithms, under the constraint of the detections with their delay times and directions determined within the radar.

In reality, every valid track report is derived from the inherent information content of the received signals themselves, as provided under the EUNIT, and extracted, filtered and formalised by the receiving process. The task of the processor is to find and separate that information from random or quasi-random disturbances or competing or distracting sources.

Staring radar is available as a solution when requirements include the determination of the nature or behaviour of the target in addition to detecting and tracking it. This may take the form of discrimination between differently moving objects not only in terms of their reported track but also of fine characteristics of the motion such as stability, acceleration, vibration, rotor motion or the number and characteristics of rotor blades, dismount events or the effects of damage.[a]

[a]A potential vulnerability of staring radar arises in the case of high-speed or rapidly manoeuvring targets, in which case their range or their Doppler shift may evolve during the coherent processing interval beyond the range or Doppler resolution of the radar, causing 'Range Walk' or 'Doppler Walk' and risking a loss in sensitivity. Within the HSR concepts, processing capacity can be scaled for possible subdivision of processing intervals in time, or concatenation of resolution cells in time or space, so that appropriate cell configurations can be achieved and sensitivity maintained. These subjects will be addressed further in Chapter 7.

Signal information is the radar's primary asset, especially as requirements focus on target attributes beyond presence and position. The information must be preserved by electronic and numerical linearity through the processing chain, and is placed at maximum risk by nonlinear processes, of which threshold detection is the prime example. At the point of detection, these inaccuracies are preserved and propagated, and at any non-linear step, such as saturation or binary compression, cross-modulation products are generated by the radar that bear no relation to individual contents of the volume of regard, violating the EUNIT. This chapter is to propose methods of 'target capture' that retain the great majority of received signal information. When the chosen process of capture is applied, it is critical that pre-detection information is propagated to contribute to further target analysis.

That information itself arises from irreversible physical processes, and once generated by EM scattering events it sets the context for future events. It can be compromised or lost, but recent, accurate signal information will have continuing significance in interpreting continuing observations.

Target discrimination depends on the information capacity and content of each resolution cell. These are set by the target within the limits of bandwidth, beamwidths, power budget and dwell time of the radar. The ability of the radar to process this information requires high speed, parallel computing, in addition to that required for multiple beamforming.

The process of cell discovery must enable an extended analysis of each cell's signal content and its evolution, while, as with CFAR, avoiding an ever-accumulating processing load. This will be at a cost in terms of memory capacity, as also discussed in Chapters 5 and 7.

The following sections focus on the nature of signal output by any HSR resolution cell, with respect to radar returns, noise, clutter and interference, in the time-domain and the frequency-domain, and their effects on sensitivity to the presence of targets of interest.

The immediate amplitude of signals is an important variable in rapidly determining their likely significance, but their real significance is best realised in terms of complex amplitude progressions, at the latest possible point in the receiving process. Indeed with high-resolution Doppler spectra, target information is often to be found in components at amplitudes lower than that of any single peak. A single threshold test will then result in their loss from the data record. Such 'Detection', which results in the discarding or degradation of already acquired, relevant signal data is undesirable, and we seek to avoid it in a staring radar, despite the increasing processing burden that that implies. While the relevant laws of physics remain stable, the availability of processing power is being transformed: it continues to grow in capacity and to reduce in size, energy use and price. This trend is unlikely to come to an end soon, and the historic incentive to minimise the volume of data to be treated by the radar appears counter-productive when more information and better classification of targets are increasingly required, and when staring radar offers the opportunity to acquire target information on a scale that can now be exploited cost-effectively.

6.3.1 Raw data throughput

For HSR, signal data processing requires high throughput, at the channel level (at the receiving array), for multiple beam forming and target analysis (at a converged processor), and for the final tracking process to select a reportable target and track information of value to the user. To acquire target detail implies that detection must be less than a once-for-all qualification for targets of interest near the front end, and more an opportunity to apportion a set of processing routes appropriate for signals of varying character.

The key functions of a staring radar require detailed target reports, consistent with the needs of the air traffic controller. These include, as prescribed in Air Traffic Management standards, that a surveillance report must occur within a latency period on the scale of seconds after the original radar return. This is consistent with the integration periods expected in staring radar practice and suggests a target of 2 seconds for the maximum latency.

With present and anticipated digital signal processing systems, this time constraint allows that large volumes of signal data may be acquired and computations carried out in determining whether a series of signals support a robust target report.

6.3.2 Signal acquisition

The premise of HSR is that the signal acquisition system observes and collects signal information from the whole CVoR over extended CPIs, storing it in memory for analysis. This has been illustrated in Figure 3.3. The entire VoR, in the form of complex amplitude sequences and regardless of application, is 'downloaded' from the airspace into memory, with the delay associated with the pulse transit time, following which its content can be assembled, searched and evaluated at the speed of memory access (in that sense, and after that minimum delay, much faster than the speed of light).

This approach provides that cell discovery, target capture and classification can be applied to extended target histories of radar-derived information. The single act of detection followed by efforts to position and track the target may be replaced by a more continuous process, using parallel and memory-intensive resources on a less conditional basis. It is consistent with well-understood principles of information theory, in which discontinuous or non-linear processing of sampled, information-bearing measurements, such as time series, including radio or radar signals, will introduce errors or losses in information quality.

6.3.3 Clutter and its suppression

Static scattering objects near the surface can be large and reflect 'clutter' signals with near-zero-Doppler shift. Resolution cells, especially those near the surface, typically contain clutter returns, which can be suppressed by Doppler filtering, but which often yield residuals that may still compete for amplitude with radar targets of interests (RToIs).

6.3.3.1 Clutter on the radar display

Radar clutter, to the user, consists of radar display paints or target reports that arise from objects, including weather features and birds, that scatter radar transmissions, as they must, but are not significant to operations. Paints consist either of visual bright spots or digitally defined icons.

On a BSR display used in Air Traffic Control, many 'paints' on the display may originate either from fixed clutter targets or from weather features in addition to targets of interest. The ability to exclude clutter from displays is a persistent theme of radar specification and development.

Clutter on the screen refers to the actual paints, plots or tracks presented to the user that do not represent targets of interest. Clutter, as referred to the objects themselves, consists of many different classes of terrain, structures, weather, surface disturbances and, in some cases, natural flying objects. Clutter signals are those received by the radar, due to scattering by clutter objects, which affect radar's function and tend to degrade performance in several ways.

First, clutter can cause false detections where the filtering and threshold setting process does not adequately suppress and exclude a clutter signal. Second, very high-amplitude clutter may desensitise the receiver, causing failures to detect significant targets. Third, where multiplicative phase or gain noise combines with a high-amplitude clutter signal to raise the noise level, especially at non-zero frequencies, it can cause either false detections or reduced sensitivity through the CFAR process. Clutter objects can also, in addition to the ground or ocean surface, result in reflections that contribute to multipath propagation.

For HSR, the effects of clutter can be measured by the radar, may be analysed and mapped, and, in the presence of real targets may be evaluated in terms of its effects on propagation in the CVoR, subject to a significant processing burden.

In normal radar operation, low-frequency quasi-static clutter signal components such as those seen in Figure 6.9A are isolated through the coherent integration and Doppler filtering process to yield the central, sharp peak indicated in Figure 6.9G (after high-pass suppression).

Methods used to reduce the effects of clutter include, in a first case, to establish a clutter map, against which thresholds at particular ranges and pointing directions can be adjusted according to known or updatable, geographically fixed sources of radar clutter. The second is to rely on clutter appearing only near the zero-Doppler frequency, which can then be filtered out prior to detection. The third is to use a CFAR detector such that the processing capacity of the radar will be presented with a manageable flow of detections prior to association and tracking. If static, these can be excluded from target reports.

The ability of radar to recognise and appropriately process clutter signals of different types is increasingly important as airspace becomes more congested, ground structures increase in size and RCS, and as the spectrum itself becomes more congested and bandwidth assignments under pressure. The treatment of clutter is a key aspect of the setting of detection thresholds.

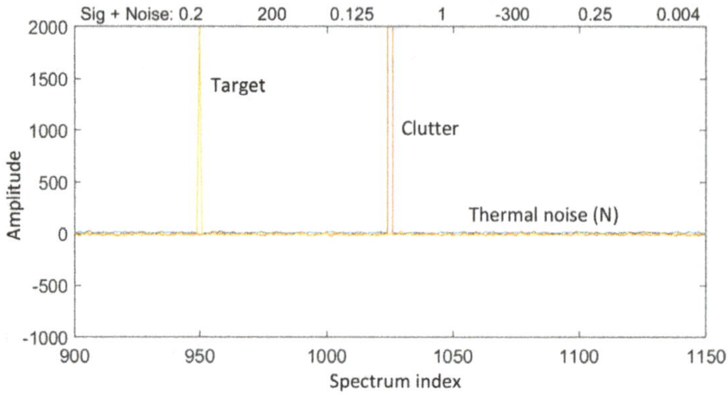

Figure 6.6 Target, clutter and additive noise

6.3.3.2 Clutter with additive noise (NXS)

Clutter related to terrain, buildings etc. creates returns with zero- or near-zero-Doppler shift. After Fourier transformation, noise added to the signal in each channel input is spread across the Doppler spectrum. A model of static clutter with additive noise is illustrated in Figure 6.6, after Fourier transformation. After Fourier transformation, the noise components are small and naturally uncorrelated and asymmetric. There is a significant, clutter-generated, DC offset and a single Doppler peak as would be associated with an approaching target.

In Figure 6.6, the noise is at a level representative of the mean receiver noise figure across the array.

6.3.3.3 Clutter with multiplicative noise (NCD)

After down-conversion and A/D conversion, and subject to typical, small levels of mixdown phase errors and timing jitter, the presence of the large clutter signal leads to an increase in noise levels. Figure 6.7 illustrates the increase in phase-derived noise (PN) that is added as their product.

The clutter remains, and the signal amplitude is disturbed slightly by the phase noise.

Close inspection of the noise at this stage shows that there is a symmetry between the 'approach' and 'recede' sides of the spectrum, which shows that the phase noise is not entirely random as is the additive thermal noise. Apparently random components of the I- and Q-phase noise spectrum components are seen to be matched for clutter-derived phase noise. In the figure, the symmetric Quadrature component is easier to observe than the antisymmetric in-phase component. Same-colour arrows here point to symmetric quadrature examples. Similar effects but different symmetry apply to 'gain noise'.

This provides an opportunity to suppress the jitter-derived noise. Provided that the time errors associated with jitter are small compared with the carrier period, it is

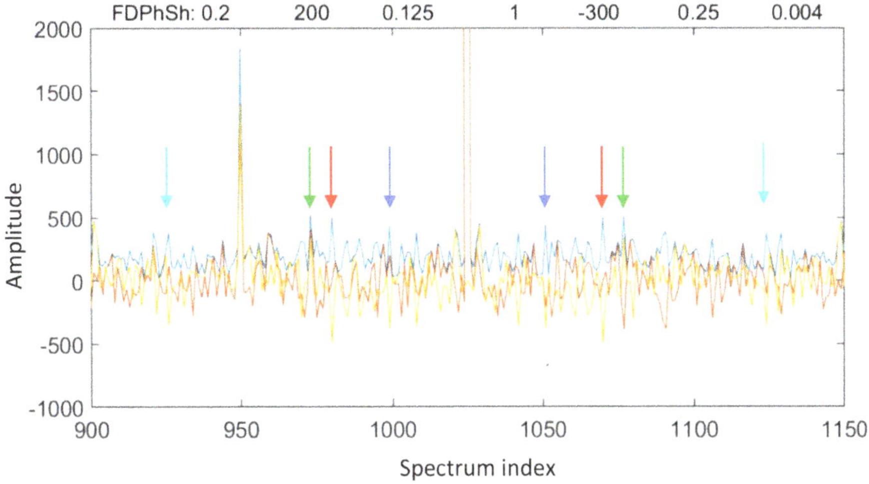

Figure 6.7 *I (red), Q (yellow) and modulus(blue) for clutter, signal and phasenoise-dominated noise*

found that a simple arithmetic process allows the signal-to-noise ratio (SNR) to be recovered. The effect is illustrated in Figure 6.8.

Clutter has been observed at over +60 dB with reference to $1m^2$ and can itself result in desensitisation due to saturation of the receiver, particularly at ranges significantly less than the maximum.

Phase noise can be expressed in terms of decibels relative to the 'carrier' signal (in this case the static clutter), per Hertz, that is, in dBc/Hz.

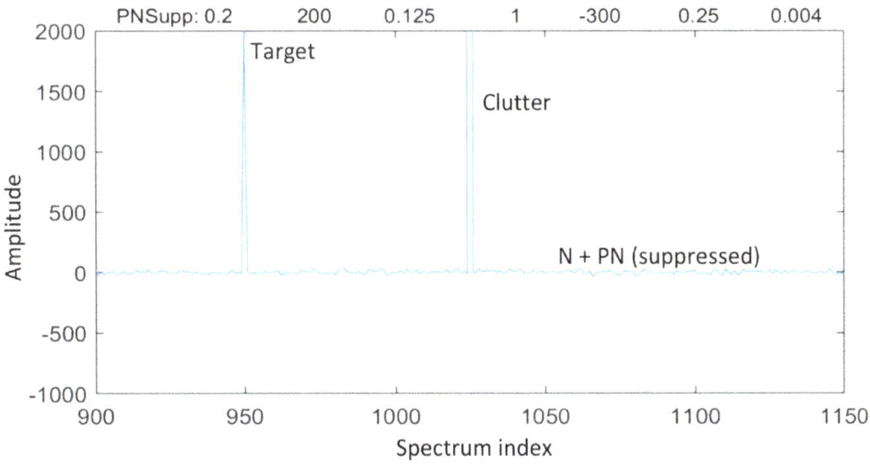

Figure 6.8 *Target and clutter after phase noise suppression*

For phase noise of −80 dBc per FD bin, as an example, a target of +60 dBsm is translated to a resultant noise level of −20 dBsm, or close to what may be a preferred minimum SRC target level, and threatening a threshold increase to 0 dBsm. Reduced illumination of such clutter might be required, reducing sensitivity near the clutter location. HSR dynamic range promises in the region of 130 dB, but where targets are close to large clutter the effective dynamic range can be reduced to nearer 80 dB (close to the phase noise figure).

The symmetries in phase noise, which are random phase modulations of a measured carrier (the clutter complex amplitude), are different from random additive noise and result in these different spectral characteristics. These allow a degree of compensation or recovery of performance in the presence of very high cross-section-built clutter.

The item of intense, static clutter is measured and (provided that it does not itself saturate the receiver channels), its phase, which is related to that of the phase noise components, can be used to adjust and then cancel the PN components. The resultant is illustrated in Figure 6.8. In this simulation, which is comparable with observation, the noise is returned to close to the thermal level and the signal peak level is restored.

A similar effect can be caused by errors in the frequency or A/D conversion gain of a channel ('Gain noise'). Gain noise refers to variations in the conversion of EM signals to digital values. For a 16-bit converter with a voltage reference accurate to 10 parts per million, errors should be below −100 dBc. An increase to 100 ppm will increase this to −80 dBc. Some aspects of phase noise may be common to all array channels, but in many HSR designs gain errors will be averaged over the array and so less significant. We have considered primarily phase effects with respect to multiplicative noise.

6.3.3.4 Clutter from fixed but moving targets

Targets under surveillance should be resolved either in range, in elevation, in azimuth or in their complex Doppler spectra. This is the fundamental function of the extraction of target information from received signals, and the requirements for adequate resolution should be user-defined quantities.

In practice, the range of objects that meet the definition of targets or clutter is evolving with time and may challenge the methods available to a particular radar configuration. Examples are large wind turbines (addressed in Section 6.4.5). Radar that has the freedom to shift, extend or reduce its CPIs and therefore its dwell times may be well-adapted to performing multiple functions in the presence of targets of new or different kinds.

Further challenges exist such as small, unmanned aircraft (addressed in Section 6.5.4), whose RCS competes with those of birds, bats, etc.

6.3.4 Target discrimination

6.3.4.1 Single threshold detection

The use of a single threshold to initiate plot association and track filtering is vulnerable to the exclusion of information that fails to meet that threshold criterion – that is, for dynamic targets that yield lower-amplitude Doppler returns than expected, for target fading and for low-amplitude signal components. Further, a form that quickly recognises time- or frequency-specific characteristics (such as repetitive signals in the time-domain or harmonics in the spectrum), but may vary downwards in amplitude, will have substantial benefits in realisable performance.

6.3.4.2 Multiple thresholds

An alternative method is to begin cell analysis with a check not on whether a single threshold is exceeded, but by evaluating the distribution of signal properties as a whole while retaining their links to specific time- and frequency-domain signal information, and without adapting a threshold to transitory conditions in defence of processing resources.

We have outlined the meaning and use of a VH. The VH can be seen as an alternative to the conditional process in which the post-threshold sequence for a detected target is determined by specific signal data test results. With millions of resolution cells to process, each VH can direct its cell towards one of several different but fully populated processing routes depending on the cell position and signal content.

6.3.4.3 Cell discovery with noise, clutter and targets

A staring radar maintains observation of all the resolution cells within the VoR, using a large resource of data memory and processing capacity. According to this approach, and since surveillance requirements increasingly include a need for classification and prioritisation for different types of observable targets that naturally benefits from staring persistence, the key function is to determine which resolution cells contain information of non-random significance, which processes will be appropriate to evaluate their content, and which extracted targets should be reported and tracked. This is the process of 'Discovery', to be followed by evaluation of continuing information, which contrasts with that of 'Detection' followed by tracking of intermittent updates.

Ultimately it is necessary to choose targets of interest and initiate track filtering and reporting. For a target in the presence of thermal noise and static clutter, the VH responds to a spectrum form similar to that in Figure 6.5B, yielding the VH in Figure 6.5C.

Comparing this with Figure 6.5A, the quantitative labelling (colour-coded for visualisation) of the higher-amplitude encoded elements allows for the target and clutter to be distinguished directly, and for the target-associated sidebands to be collected and correlated.

The VH for each resolution cell avoids the setting and application of one or more adaptive thresholds and is expected to be compatible with the computing resources identified in Chapter 7 (SRC and ATC).

The cell illustrated in Figure 6.9A–H yields RMS noise of 8.66 amplitude units including clutter and 0.96 after clutter suppression.

In the presence of substantial near-static clutter, the time-domain histogram is double-peaked; the lower peak includes the noise, but the very low-frequency 'wobbles' arise from near-static clutter and can form a higher-amplitude lobe. The noise component yields close to a Rayleigh distribution, increasing linearly and decaying exponentially with amplitude.

In the plots in Figure 6.9, the left column shows signals or signal amplitudes in the time- and frequency-domains; the right column shows distributions as histograms for the resolution cell selected. Figure 6.10 illustrates in-phase and quadrature components of noise and clutter, again partially suppressed. Figure 6.11 is formatted similarly to Figure 6.9.

In the frequency-domain (Figure 6.11A and C), there are several added components at distinct frequencies. The modelled aircraft in this case is single-engined and propeller-driven. The largest component at (459, 3 148) represents the Doppler-shifted signal scattered by the airframe. Distinct spurs are seen with amplitudes near a few hundred scale points. These will be discussed in Section 6.5.5; they are modelled as a 2-bladed propeller, but are known to be realistic. The component at bin 1 025 is residual static clutter. Bins 1 013–1 035 are also attenuated with the clutter. The airframe in this example is clearly distinguishable, but the associated spectrum peaks are reduced by about 20 dB and may or may not exceed a threshold.

By setting a single threshold at a value of 1 000 in the figure, the airframe will be comfortably detected, but the additional peaks will be ignored and any associated information lost. A threshold value of 100 will yield 31 threshold crossings. It is not desirable to initiate independent detections for these, some of which are at marginal SNRs, risking greatly degrading the false-report performance of the radar, but to enable associated detections may be valuable. In this case, all the subsidiary peaks are related to the primary airframe return; they shift with the airframe Doppler return as a reference. They represent modulation of the target scatter by the aircraft's propeller, and provide a powerful basis for target analysis and classification, but they risk uncertainty for threshold-conditional processes. A VH offers an improved basis for cell discovery, as indicated in Section 6.3.6.

6.3.5 Channel and system malfunctions

6.3.5.1 Excess or degraded, persistent, random channel noise

Signals from each channel, taking account of calibration results, allow underlying noise and continuing changes in the performance of each channel to be monitored. Comparison with adjacent channels or past conditions may indicate a channel failure. Channel saturation may also be detected, and channel instability may be determined. An appropriate response may be to exclude specific receiver channels from beamforming, whose effects in small numbers can be suppressed. In the absence of interference, the level of the noise

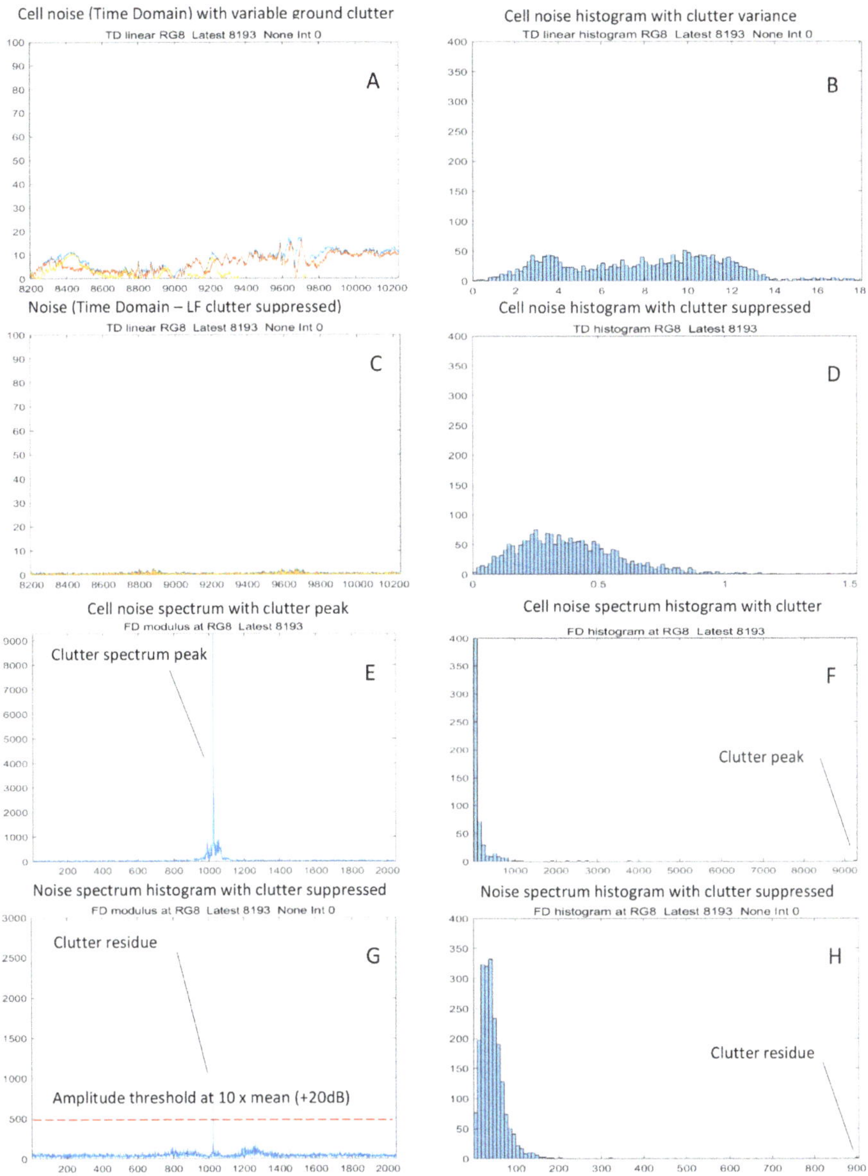

Figure 6.9 *Panels (A–H) illustrate stages of processing with the effects of noise and clutter. Note vertical scales and fixed histogram cell count.*

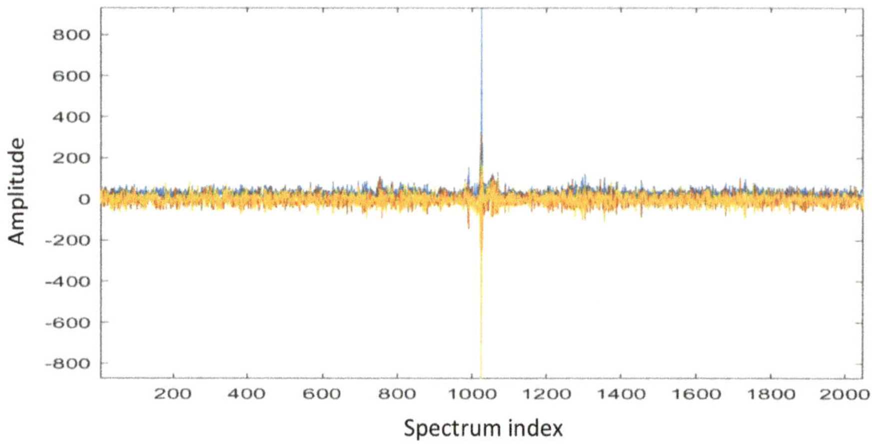

Figure 6.10 *Frequency-domain modulus, in-phase and quadrature components of noise with a partially suppressed, zero-Doppler clutter component*

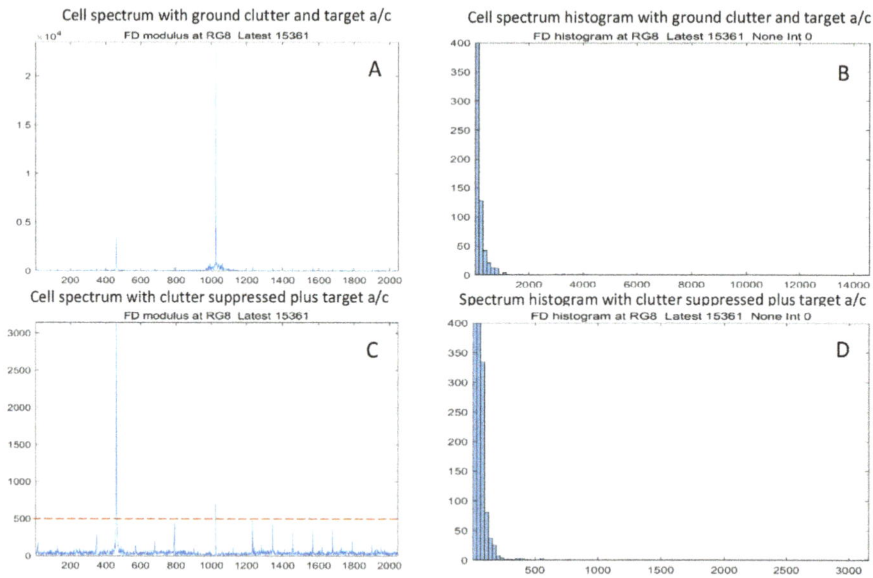

Figure 6.11 *A–D illustrate the effects in the frequency-domain of introducing a target – a single-engined, propeller-driven model. In the time-domain, the target is not discernible.*

floor should be close to that expected from the front-end design. Different forms of interference will be considered in Section 6.4.

Where high levels of fixed clutter are observed these will affect noise levels through the multiplicative effect of phase noise. Where this effect is recognised it may provide a basis for the measurement of actual jitter and phase noise levels. Where the effects of high clutter and phase or gain noise are discovered, the cell in question then becomes a candidate for processing to recover signals in the presence of the particular effects of phase noise, as described in Section 6.3.3.3.

6.3.5.2 Specific, fixed spectrum peaks

Real targets yield Doppler components related to their trajectory, and fixed clutter tends to generate strongly zero-oriented Doppler components. Fixed spectrum features that do not conform with these patterns may indicate distinct interference, or else internal system malfunctions, such as unintended resonances or modulation of the transmit signal.

These can be made available in one or more RAED channel-wise components (in the case of array malfunctions) or beam-wise components (in the case of interference) and passed to the central processor, where the distribution of VH output findings should clearly identify the source.

6.3.6 Cell classifications

Tools such as a VH allow measurements quantifying the presence of noise, clutter, phase noise, repetitive clutter, translating targets and dynamic targets, and allow us to assign each cell under categories such as the following list:

1. Noise plus static clutter within specification (NSC)
2. Noise out of specification (NXS)
3. Noise plus static clutter-related degradation (NCD)
4. Noise plus repetitive clutter degradation (NRD)
5. NSC plus Newtonian targets (CST)
6. NCD plus Newtonian targets (CDT) and multipath (CMT)
7. NRD plus Newtonian, dynamic targets (RDT)

6.4 Measurement of interference

Externally generated radio signals received at the array are subject to the EUNIT, but may occur within the receiver passband. By definition, they will not be consistent with the irradiation source from which target returns originate. The HSR receiver has the task of identifying interference and suppressing it as far as possible. Cells affected by interference would be classified NXS.

The most destructive forms of interference result in saturation of the HSR receivers. It is simple to infer from an amplitude histogram at the channel-wise view within the RAED structure but difficult to suppress without loss of sensitivity.

6.4.1 Radio interference (NXS) – scalar histograms

Radio interference may occur because of malfunctions in an external transmitter or failures in spectrum management. In cases where the interfering source is localised in a beam, the signal is persistent compared with the transmitted radar waveform, or where it causes the observed noise level to increase substantially, then it may be identified. A number of strategies may be used to avoid false alarms while minimising loss of sensitivity. Where constant, low-level interference is caused, for example, by spurious radiation from a radio station, a CFAR threshold will be raised or else a local beam null can be formed, reducing sensitivity only in the direction of the emitter. Radio interference may be modelled in terms of a static FMCW station, radar interference, active jamming within the bandwidth of the HSR, and repetitive and persistent clutter signals.

In Figure 6.12A, partially suppressed but variable clutter is seen in low-frequency fluctuations in the time-domain, superposed on the noise. These are clutter-derived and shift and spread the typical Rayleigh distribution peak (Figure 6.12B). In Figure 6.12C, the 'noise' becomes dominated by superposed complex CW broadcast interference. The centre of the amplitude distribution (Figure 6.12D) is shifted by the interference, but it has similar amplitude variance and becomes closer to a Ricean distribution. With a steadily approaching target present, the overall noise and distribution are similar (Figure 6.12E and F).

In the frequency-domain view, in the absence of interference (Figure 6.12G), the clutter is focused in the near-zero-Doppler frequency bins. Noise is distributed over the whole spectrum width, focused in the lowest bins of the histogram (Figure 6.12H), with the single clutter-derived peak near 2 000.

In Figure 6.12I, clutter and noise are superposed on higher-amplitude complex FD noise from modelled FMCW interference, yielding a broader FD amplitude distribution (Figure 6.12J). In Figure 6.12K a target Doppler component is clearly visible out of the noise in the frequency-domain, but its airframe return is comparable with the clutter residue. The distribution (Figure 6.12L) has upper points referring to the clutter and the target residue that are not distinguished, and both would be likely to cross any appropriate single threshold.

The data illustrated in Figure 6.12A–L are oriented towards a threshold decision that in this case, if set at a level able to yield a detection from the target signal in the context of additive thermal noise (say SNR = RMS + 20 dB), would detect the target and the clutter residue but with very many false alarms in the absence of effective suppression. An effective detection threshold has to be set based on the variable neighbouring noise and clutter, using a form of CFAR process. In this case, the threshold is set at 10× the mean noise level, yielding uncertain detections, and no micromotion indications.

A more flexible and less coarsely conditional cell discovery and interpretation process is needed and may be offered by a VH, in which each instance is referenced to its original source, including harmonic components near the noise level.

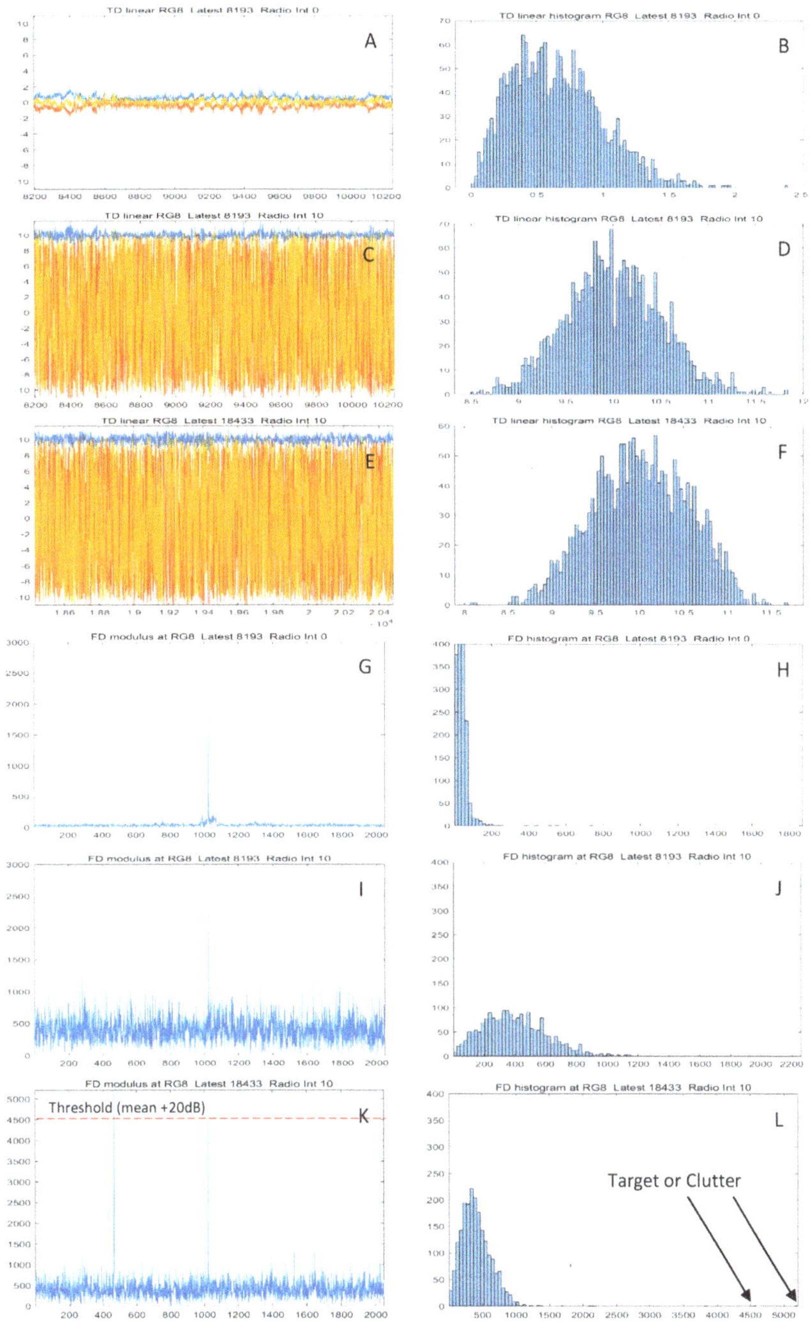

Figure 6.12 *(A–L) Time-/frequency-domain cell outputs and histograms without/with radio interference and targets. Clutter has been partially suppressed.*

6.4.2 Radio interference using VH format

Figure 6.13A–L illustrates outputs of the amplitude-encoded, VH approach, in which the same time- and frequency-domain data are explored, potentially yielding information in more dimensions, with a non-conditional, multiple-threshold equivalent.

This approach aims to resolve the facts that time- and frequency-indexed signal data contain many items of information relevant (via the EUNIT) to each observed target, but that a complex conditional thresholding process results in downstream processes that are difficult to account and yield, in general, only a few bits of information until a tracker assembles them into a conditioned sequence.

VH formatting, described in Section 6.2.5, as an alternative to a single threshold decision, allows data to be ordered by amplitude and referenced in time and frequency, as illustrated in Figure 6.13.

In Figure 6.13A the encoded (compressed) version of the amplitude, in the absence of radio interference, is shown as a function of time. In Figure 6.13B the abscissa is the amplitude value (as in a histogram), the ordinate is the number of occurrences, and the colour (a value 1–2 048 in this case) denotes the time sequence of the samples. The mean encoded amplitude is close to 10 here. The encoding used is the square root of amplitude expressed as an unsigned 8-bit number (1–256).

Figure 6.13C illustrates the values in the presence of radio interference, and the VH in Figure 6.13D shows the shifted centre of the amplitude distribution.

Figure 6.13E shows the amplitude as a function of time in the presence of a target (here the target was beginning to depart the particular range gate, and in Figure 6.13F the VH shows the later times (colour-coded orange > yellow) biased towards the lower amplitudes.

Figure 6.13G shows the encoded clutter residue and noise in the frequency-domain, and Figure 6.13H, the VH peaks near 5 with a few higher-amplitude components, encoded green, denoting the near-zero-Doppler frequencies of the clutter.

Figure 6.13I shows the encoded FD signal with radio interference and the clutter peak, and Figure 6.13J the VH distribution of amplitudes and frequencies with radio interference, showing a single green clutter-residue outlier.

Figure 6.13K shows the encoded signal with interference, clutter residue and the target Doppler. In Figure 6.13L the VH illustrates the target, encoded pale blue (representing its approaching Doppler frequency), and so resolved from the clutter, encoded green. The VH also resolves spur harmonics in frequency.

VH data formatting again provides a way of ordering a full set of signal data in relation to its amplitude, as an alternative to selection according to a single threshold condition. Distributions of amplitudes and frequencies can then be used to allocate between process resources, and then minimise the noise margins required for target capture. Here the signals more than 6 dB above the mean encoded noise contain multiple harmonically related bin indices that discriminate in favour of a propeller-driven target.

Figure 6.13C and D contains the information that CW signals are present at this resolution cell. To derive the VH, 2 048 values are encoded, 2 048 assignments arise from the value/address exchange and 2 048 increments are performed: a total

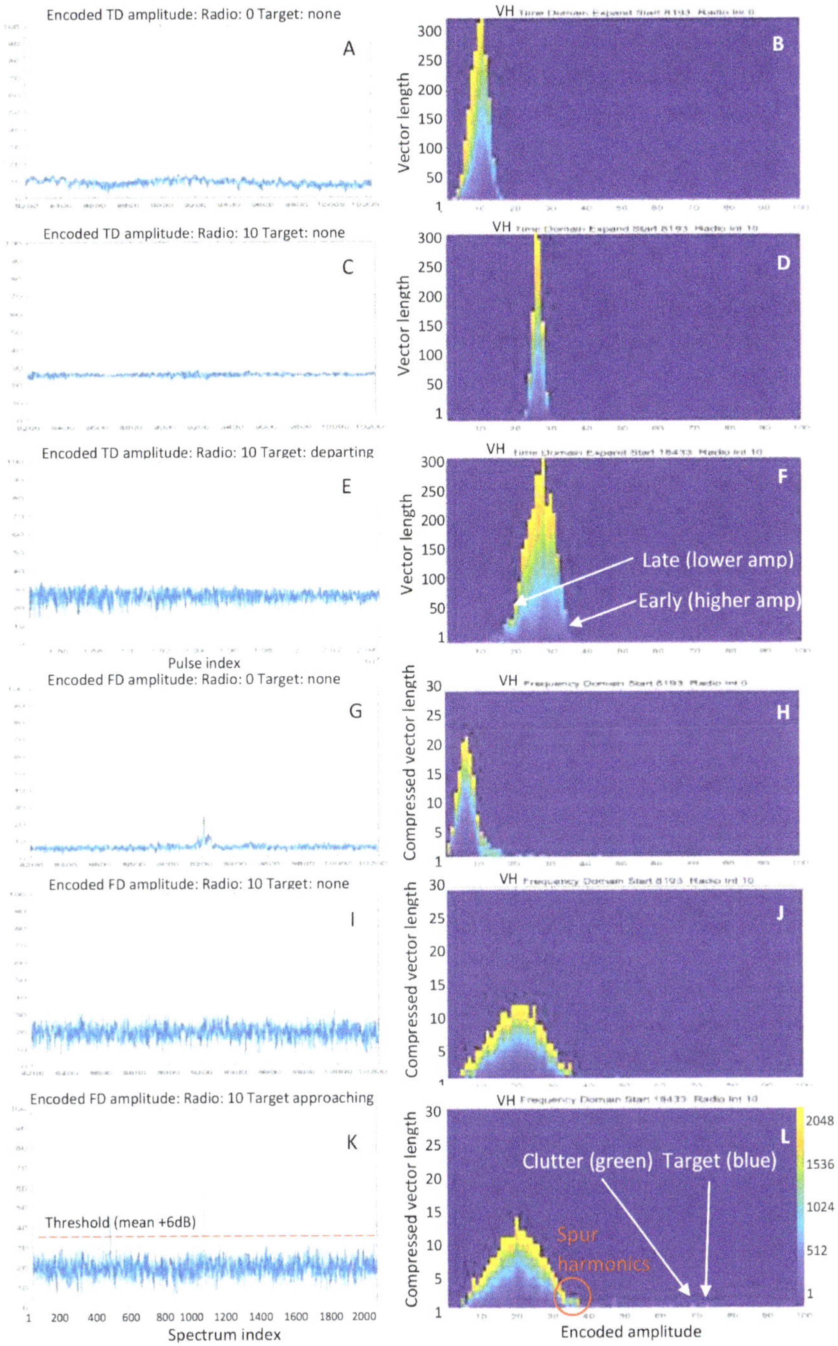

Figure 6.13 (A–L) Amplitude encoded cell outputs and VHs without/with radio interference and target, in the time- and frequency-domain

of 4 096 operations. As for the histogram, 6 144 arithmetic operations yield the mean and the RMS width of the distribution. The encoded TD VH version including a target refers back to the trend of amplitude with time (late, yellow-coded bins are skewed to the low-amplitude side of the distribution), but does not reveal directly whether this arises from the departure of a target or variance in the interferer.

In the frequency-domain, the target airframe is resolved from clutter, and the target side spurs are grouped; their frequency addresses will indicate the propeller rotation rate, but their spread will be masked here by the radio interference spectrum. The spectrum of interference will be characteristic of the type of signal, and its suppression will be discussed in Chapters 7 and 10.

The use of a VH allows a considerable degree of signal source discrimination and target capture within the parallel processing stage, rather than depending on explicitly conditional or threshold-dependent process segmentation.

6.4.3 Radar interference (NXS) – scalar histograms

Signals including repeated, high-amplitude pulses at regular intervals (0.1–10 milliseconds), with scan groups spaced at longer intervals (seconds) model external radar interference. Provided that frequency management is effective, external interference should be at a low level after filtering at the front end, and at asynchronous intervals; nevertheless, spurious emissions may be present in the operating band, and interference may still be at an appreciable level.

Figure 6.14A–L TD noise without radar interference (A) with interference (C), and with a target present (E), with associated histograms (B,D,F), followed by FD noise with residual clutter peak (G), spectra of the interference (I) and with the target (K), and associated histograms (H,J,L).

The time scale is a full CPI in Figure 6.14A, C and E, showing the pulse amplitude, in-phase and quadrature components.

In Figure 6.14B, the pure noise is close to a Rayleigh distribution, tending linearly up from (0.0) and reducing exponentially as the amplitude increases, with the peak near 0.5 and slightly flattened by the clutter residue. With radar interference the time domain amplitude is shown in Figure 6.14C, at extended time scale for clarity. Eight pulse peaks are seen, representing a Gaussian scanning beam profile. In the histogram in Figure 6.14D the noise is concentrated in the lowest bins, with undifferentiated amplitude peaks.

The target is present in Figure 6.14E and F but makes no visible impact on the plot or the histogram.

Figure 6.14G and H repeats the undisturbed noise and clutter of Figure 6.13G and H. The spectrum illustrates the clutter peak near-zero Doppler, and the FD noise dominates the low histogram bins.

Figure 6.14I has a series of high-amplitude but spread peaks, centred near harmonics of their beat frequency with the HSR PRF, and the frequency-domain distribution (Figure 6.14J) is extended in amplitude to near 500, with the single peak at zero Doppler remaining near 2 000.

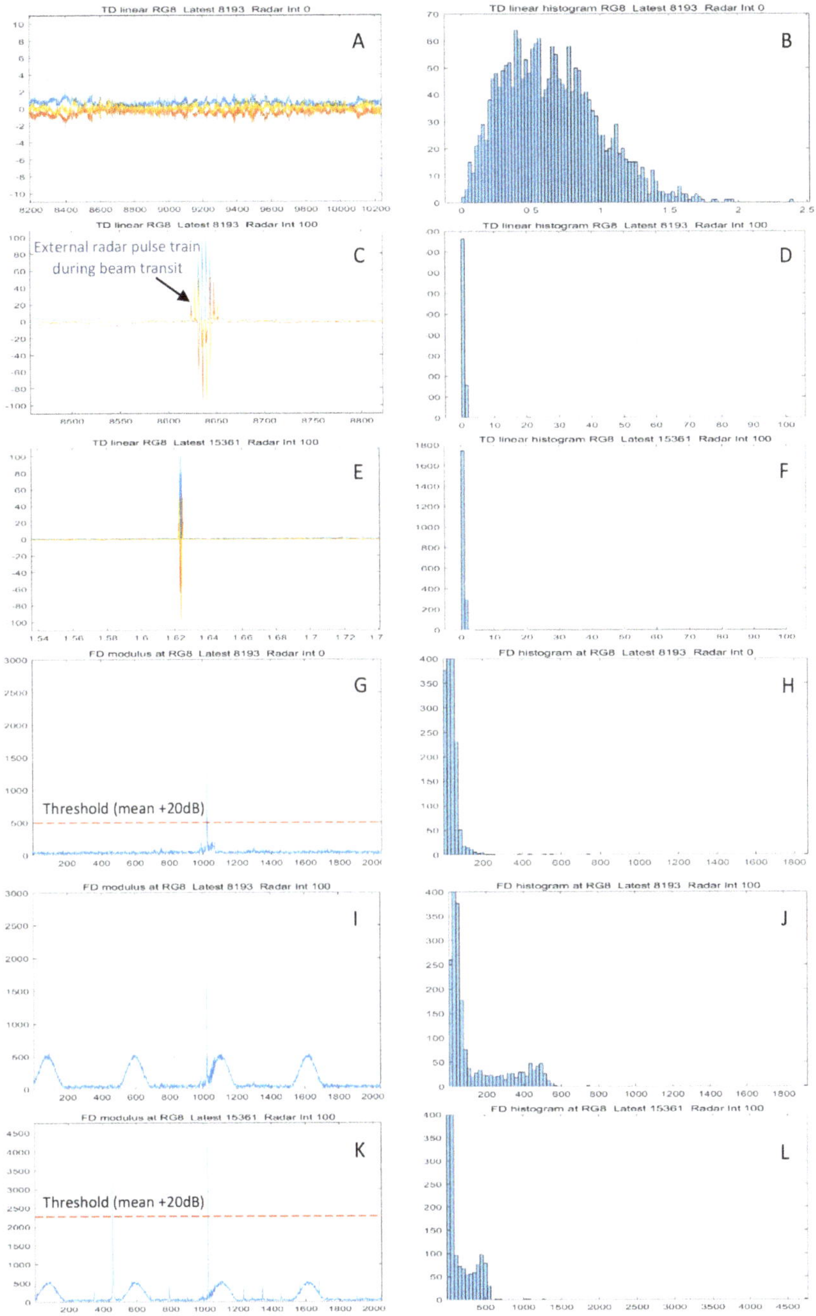

Figure 6.14 *(A–L) The effect of interference from an external radar transmitter, when the interfering radar PRF is close to that of the HSR model, yielding harmonics of the beat frequency*

With a target present, the cell appears as shown in Figure 6.14K and L. The target in the spectrum plot can be resolved near amplitude 3 000 from the radar spectrum components, but at 3 000 in the histogram it is not distinguished from the residual clutter signal near amplitude 4 500, and its harmonic spurs are not detected.

6.4.4 *Radar interference using VH format*

Figure 6.15A shows a plot of the encoded amplitude of noise and residual clutter, and Figure 6.15B shows the associated time domain VH. The noise amplitude distribution is centred near 10.

Figure 6.15C shows the superposed radar pulse, and the VH in Figure 6.15D and Fshows that the signal distribution is similar with the addition of the radar pulse. Figure 6.15E shows that the presence of a target may have a perceptible effect on the apparent noise distribution (Figure 6.15F), without reaching a threshold such as RMS noise + 20 dB.

It is important to reiterate that in this scheme the 8-bit values used to encode amplitudes to address the VH provide multiple fixed thresholds, rather than being adapted to the observed signal levels by a CFAR process.

Figure 6.15G and H once again shows the undisturbed FD signal and distribution.

Figure 6.15I shows the spectrum including noise, clutter residue and the effects of radar interference, and the VH is at Figure 6.15J.

The most significant displays are in Figure 6.15K and L; also reproduced at a larger scale in Figure 6.16A and B, the harmonic spurs can be captured.

Coloured points in the VH denote features in the spectrum that exceed a known set of thresholds and simultaneously point to data features such as complex spectrum amplitudes and spectrum offsets. The VH content is used to apportion resolution cells among different algorithms, in this case for target capture with suppression of radar interference.

Figure 6.16A and B further illustrates the VH expression of clutter and target in the presence of radar interference.

In Figure 6.16A, grouped separations in frequency (colour coded for visualisation) between radar lobes indicate the PRF of the radar, and between target spurs will indicate the presence of and rotation rate of a target's propeller (or rotor in the case of a helicopter or multirotor drone). Over time, the content of the VH for a resolution cell can be used to set process resources.

6.4.5 *Wind turbine interference (NRD)*

A key disturbance for air surveillance radar is continuing series of returns from wind turbine generators ('WTGs'). These are not targets of interest to the radar user, but constitute high cross-section targets whose blades generate repeated 'flashes', as illustrated in Figure 6.17. In combination with the scanning radar method they are unpredictable in timing, and in amplitude since each turbine (and there may be hundreds in a wind farm) moves at varying speeds and can face in different directions with varying blade pitch. They move with tip speeds up to over 100 m/s, and their returns tend to obscure those of aircraft. Without mitigation, they and their

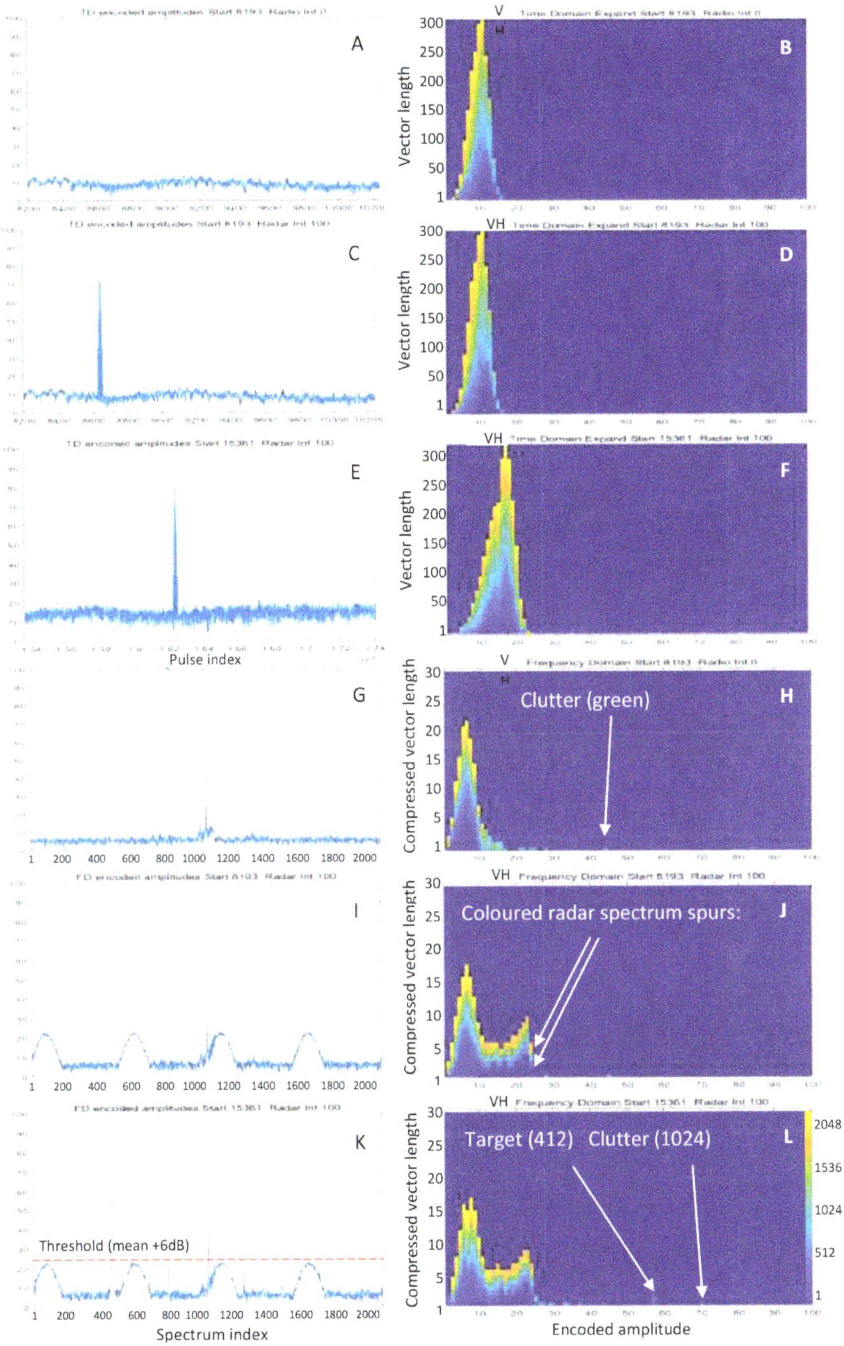

Figure 6.15 *(A–L) Encoded signals and spectra with radar interference, residual clutter and target*

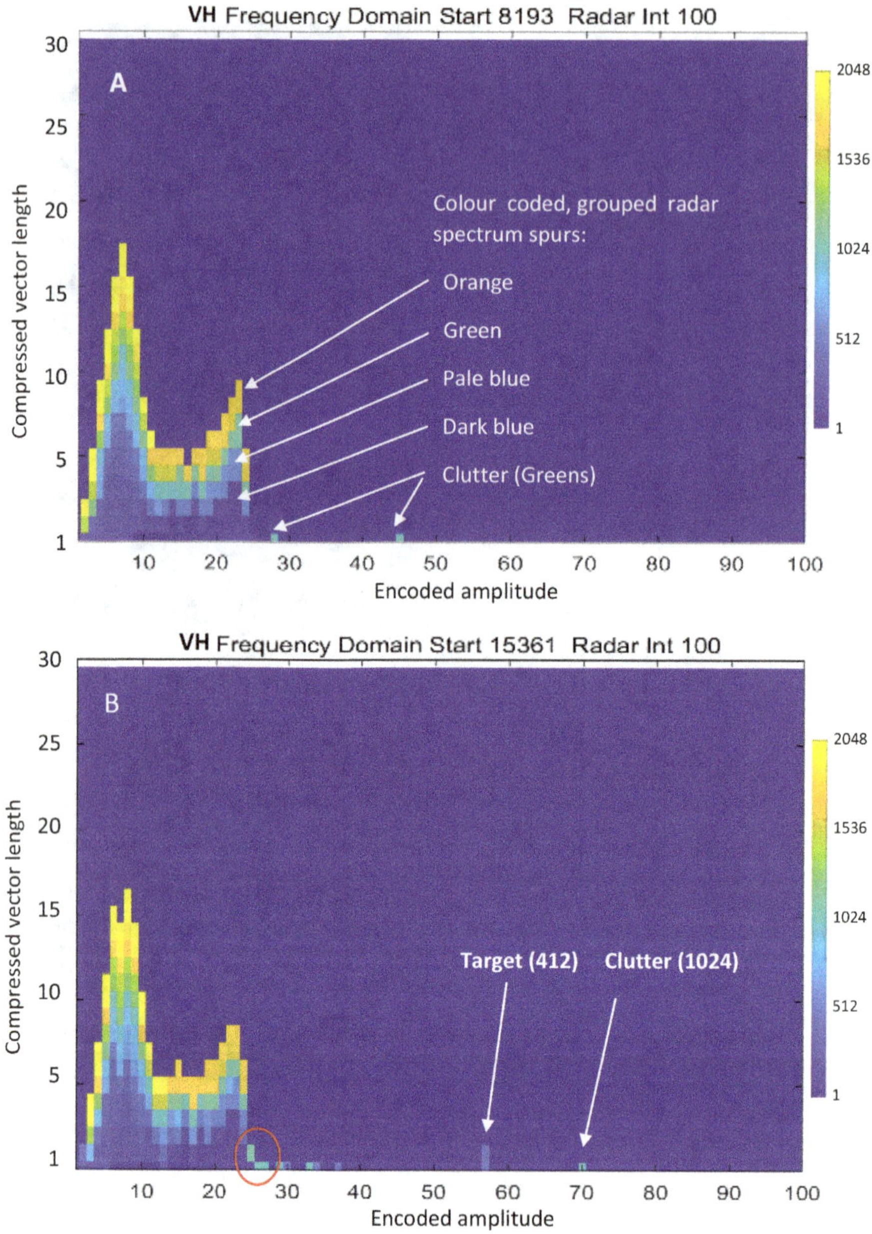

Figure 6.16 *(A and B) VHs. Resolution of clutter and target in the presence of radar interference.*

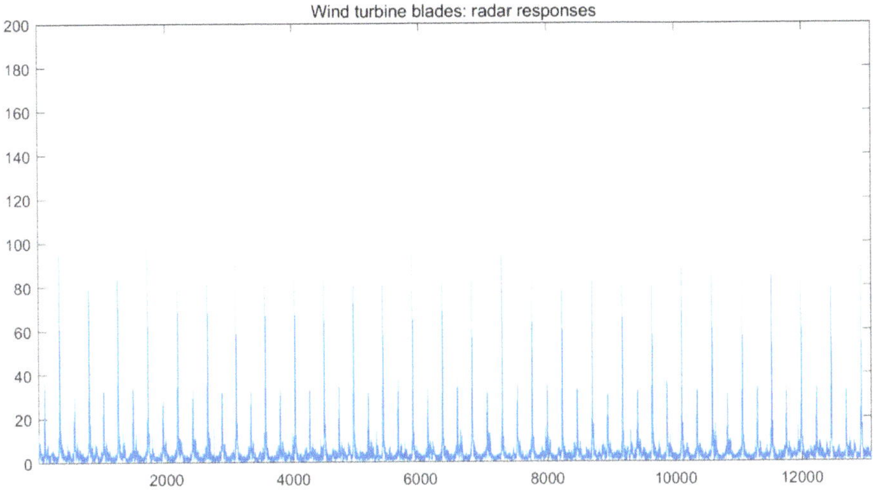

Figure 6.17 Time-domain model of a wind turbine blade flash sequence as for HSR

proliferation are therefore incompatible with the operation of both Air Traffic Control and Air Defence radars.

Many approaches have been tested to overcome this incompatibility, and some have had some success. The earliest successful method was that of terrain shielding, in which radar is sited so that the wind farm is hidden behind terrain, masking returns from the turbines. Other successful efforts have used high-resolution radar to minimise the areas obscured by turbines, but, as turbines increase in size and expand in quantity, generally applicable, long-range radar solutions are needed.

The growth of WTGs as a strategic energy resource is a threat of increasing importance to radar surveillance, and this section addresses the immediate form of the signals that constitute this form of interference.

Holographic Staring Radar has been tested and used successfully for discrimination between the radar returns scattered by aircraft and those returned by turbines, and its potential for suppressing WTG interference in the future is discussed in greater detail in Chapter 8.

WTGs generate steadily (but not precisely) repetitive waveforms, modelled in Figure 6.17, and their 3-bladed rotors yield returns in the form of regular reflected flashes as the turbine blades rotate through the normal to the direction to the radar, but which vary in response to changes in wind direction. For each turbine, these are correlated over seconds and provide persistent alerts to a surveillance radar that this form of interference is present.

For a scanning radar, the reflected flashes cannot be synchronised with the beam. For any turbine, they may not occur for several rotations of the radar. Nor are they coherent, presenting no sampled sequence capable of analysis by a scanning sensor. As a result, only blanking in one of its forms is effective in removing these returns, and as a result detection performance can be compromised.

For a staring radar, every flash of every turbine blade in the field of view of each resolution cell is observed. Each can be analysed and processed in real-time to extract sufficient signal information to separate aircraft returns from the turbine clutter.

During time-domain processing, the sequence of turbine flashes can be measured, and their effects suppressed. This subject will be explored in detail in Chapter 8, but since each turbine is a static feature, with known behaviours, and each one is associated primarily with a single resolution cell, for each such cell continuing measurements are used to establish the status and operation of that turbine.

Turbine status can be established primarily in the time-domain and contributes to the subsequent process leading to suppression of its radar effects.

This is one of the radar functions that directly benefit not just from the continuity but the more extended history of associated signals and is discussed further in Chapter 8.

Observed from a distance, in a single resolution cell, each blade of a turbine (almost universally three-bladed) presents its leading edge (with approaching Doppler) and trailing edge (with receding Doppler) to the radar, yielding three sharp flashes at very nearly regular intervals. The blades are mechanically very similar but are not identical. The model emulates the amplitudes of each cycle of three flashes, subject to noise and modelled curvature variances. The approaching edges, as for turbines in the field, yield sharper, more intense flashes than the trailing edges.

The autocorrelation function of such flashes is illustrated in Figure 6.18, where the sharper peaks represent leading edges. The trailing edge of an aerodynamic rotor

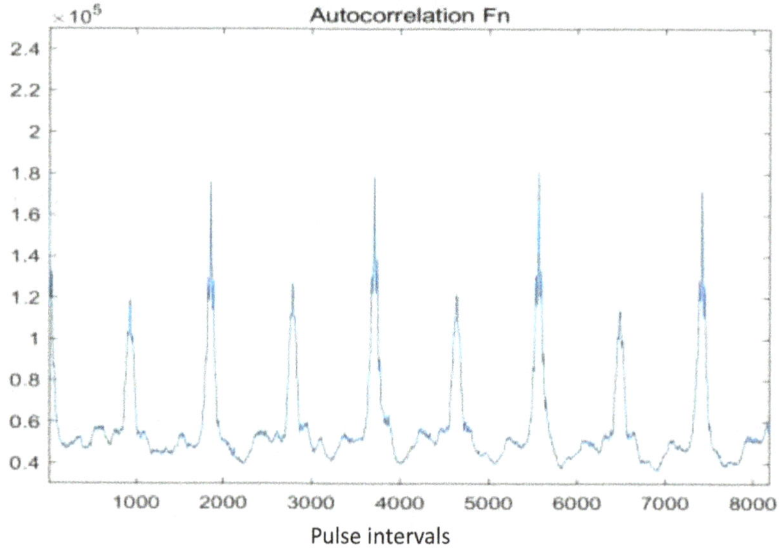

Figure 6.18 Autocorrelation function of data in Figure 6.17

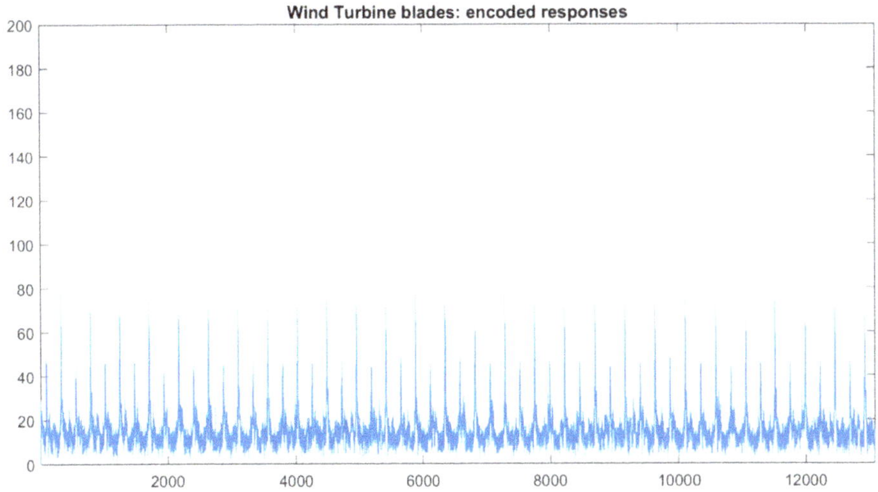

Figure 6.19 *VH-encoded version of the flash amplitudes in Figure 6.17*

blade is typically more curved and is sharper than the leading edge. The slower trailing peak is also represented in the autocorrelation.

These figures are based on a large wind turbine model. In this case, each rotation of the turbine takes about 5 600 pulses or 4 seconds.

Amplitude compression for VH formation yields the encoded values shown in Figure 6.19, and the VH is shown in Figure 6.20. The peak flashes, in addition to occupying the upper set of histogram points (circled), include as the colour coding

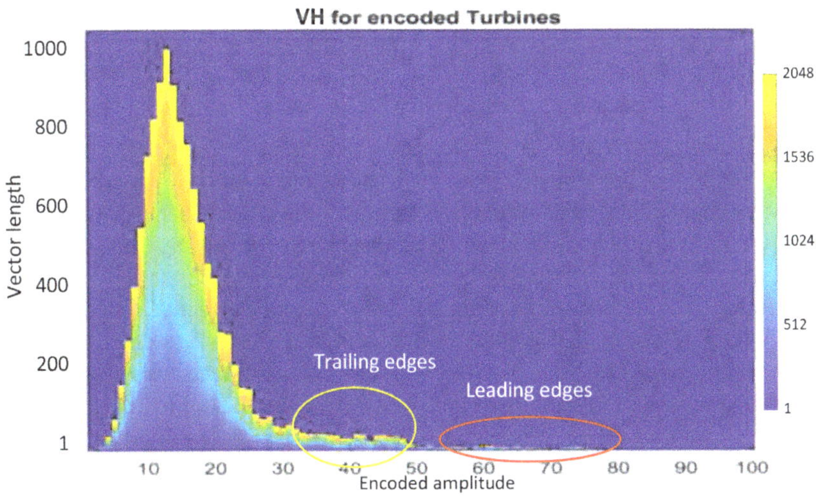

Figure 6.20 *Time-domain VH derived from the encoded signal flash sequence in Figure 6.19*

Table 6.1 Time indices of peaks captured in the VH

Timing (pulse indices) of peak samples in a wind turbine flash sequence							
8226	790	2638	1252	4026	10104	3101	328
1714	2176	3564	4489	4490	12939	5884	3563
4953	6819	10575	7288	8225	5418	7757	11520
12467	6351	8695	8694	11047	9163	11994	9633

of each point the time (measured at the PRF) of the flash. This compact dataset contains all the information necessary to register the turbine suppression process to the observation of each blade.

The VH in Figure 6.20 shows 100 threshold values in parallel in preparation for downstream processing. In the case of wind turbine suppression, the time-domain VH provides the basic information to speed the process. In the VH the upper values (circled) contain the time addresses of repetitive peaks. The content of this cell, which can be recognised as the location of or close to a known turbine, is passed to the process and memory segment provided for suppression.

The red-circled instances contain a series of numbers related to the timing of each of the amplitude peaks shown in Figure 6.19 and Table 6.1.

These apparently random numbers in fact represent a sequence of turbine blade flashes varying from 1.26- to 1.28-second intervals. This process provides for all the different signal amplitudes in the series and illustrates the flexibility of multiple-fixed-threshold signal discovery.

Note that this does not itself achieve turbine clutter suppression; it acquires the time sequence that will be necessary for the suppression algorithm itself, which exploits the repetitive form of the turbine flash returns. Turbine clutter processing will be addressed further in Chapter 8.

6.4.6 Intentional interference

The potential for electronic countermeasures (ECM) may be a key feature of military surveillance and in that case must be detected and countered. Jamming sources aim to cause interference and potentially saturation in subject radar receivers.

There are several forms of ECM. The most extreme is 'barrage' ECM in which the intent is to saturate the victim receiver and prevent its function in detection. A less-extreme form of countermeasure, but one that can be effective at longer range, is that where, although the receiver is not saturated, the victim detects, possibly via its antenna sidelobes, false targets that are not actual threats but may exceed its capacity to process and recognise them. In a further, more sophisticated form, signals are imposed, again via sidelobes, that appear to the victim to represent targets in consistent but fictitious trajectories.

Other than saturation, the interactions of ECM with scanning radars are very complex, often exploiting either the relatively low-dynamic range of a single-receiver radar, or confusing its function via its sidelobes. Counter-counter measures tend to involve agility of various kinds including frequency agility, scan agility and waveform agility.

Under ECM conditions, the first requirement for detection is to identify its presence in any of these forms. For barrage jamming, persistent triggering of thresholds throughout the range window, the Doppler spectrum and across the beam set is a strong indicator. More sophisticated jammers at a longer range may raise effective noise levels through radar sidelobes, and this can be detected through an associated general raising of thresholds.

Spoof target countermeasures may be a greater future threat. Where a narrow azimuth beam is scanned in direction, they may be introduced through sidelobes, and their trajectories constructed to deceive the tracker.

For a staring radar, all channels in whose field of view the source sits receive the same signal, phase-shifted according to the position of the channel antenna in the array. As with a scanning radar, receiver saturation is difficult to counter except by frequency agility or direct blanking during the imposition of countermeasures.

Unlike a radar distracted via its antenna sidelobes, in which case the direction of arrival is unknown, the azimuth direction from an HSR to the source is unambiguous. Although sidelobes exist, the primary direction of transmission is always under observation. Spoof trajectories will therefore always be found in their actual direction of transmission.

The properties of HSR lead to potential ECM recovery opportunities, outlined in Chapter 7.

Staring radar may have constraints in the available degree of frequency agility, but it has alternate degrees of freedom in achieving resilience so that the probability of detection may be recoverable. We shall not address the intentional jamming itself further, but we shall discuss the suppression of other forms of interference that are expected to be relevant, in Chapters 7 and 10.

VHs operating at each of a set of HSR resolution cells near the direction of an interfering source will acquire characteristic signal data that will provide a basis for counter-countermeasures; first that signals are arriving that cannot represent actual scattering from targets irradiated by its transmitter.

6.5 Target capture for staring radar

In each of the cases described, a single threshold that is adapted to current signal conditions or to an unknown history yields uncertainty in the probability of detection of a target, particularly under conditions with clutter or interference present, and it appears that, provided adequate memory and processing capacity, an HSR using an approach implementing multiple thresholds can be more responsive to different signal conditions, through which targets can qualify as RToIs.

6.5.1 Nominal conditions (noise, clutter and target features)

To make use of the parallelism of HSR and the nature of the CVoR, a possible target search method accesses all azimuth directions within each beam sector (32- ×

1.8-degree beams for ATC) in parallel, with subsets of successive range indices, at a series of elevation values beginning with the lowest. The number of range indices is determined by the number of parallel processors available within each sector processor; 512 is assumed here, with 16 azimuth and 32 elevation indices. Frequency-domain VHs, which should be the vehicles for capturing RToIs in Newtonian motion, should be available less than 10 milliseconds after the end of signal acquisition for the CPI, from which very low-frequency clutter is excluded.

Parallel patches of range/azimuth cells are set to overlap so that sidelobes in each dimension are included. In the model, patches of 512 FDVHs are called, and sets of high, contiguous VH features are extracted from each and sent to a CPU.

Matching Doppler indices, range, and the peak amplitude for each are identified and captured as targets within the patch. Certain conditions render such detection more complex, including multipath conditions, target dynamics, target micromotion, etc.

6.5.2 *Surface multipath conditions*

Surface multipath propagation is illustrated in Figure 3.4 and can cause loss of a target return.

Paths A and C in the figure can cause a failure in irradiation of the target, and in that case, provided that the CVoR is safeguarded close to the radar there is no recourse other than to wait for the effect to evolve towards the constructive interference case.

Interfering multipath occurring on the return path from target to a receiver (Paths B and D) will result in variation of the signal amplitude at the appropriate Doppler frequency in the vertical axis on the array: a direct indication that multipath propagation is occurring. Target fading will occur with changes in range or altitude, as the different path lengths and their interference effects change.

Overall, surface multipath results in an increase of received signal power, varying sinusoidally between 0 and 16 times the nominal intensity, but destructive interference leads to missed detections in the scanning case. In the staring case, the ability to store and refer to extended signal histories will allow the target's vertical position to be measured over a period that depends on the trajectory and the geography. In outline, the higher the target, the more rapidly the interference patterns transits the array. This will be dealt with in greater detail in Chapter 7.

An extended time-domain VH will provide the basis for identifying these conditions, but a full method of interpretation and recovery will need access to channel-wise Doppler transforms.

6.5.3 *Target capture with dynamics*

When radial acceleration occurs, signal energy is conserved in the scattered signals, but, after the application of a Fourier transform, it is spread in sequence across a contiguous band of Doppler frequencies, in a pattern that varies with the actual acceleration profile. This is not a case of decoherence, but the application of a single stage of Fourier analysis does not result in optimal detection via a pre-set threshold. This case is introduced in Chapter 3, section 3.11.

Signal energy can be recovered in a number of ways, and most simply, but not computationally cheaply, via a direct, coherent but phase-compensated frequency-domain transformation provided that the correct, wide range of dynamic cases is considered.

A less costly form of filter allows the moduli of a noisy, spread-Doppler spectrum (as in Figure 3.6) to be aggregated by means of graduated scalar filters, with the complex Fourier transform as its starting point.

This yields full detection sensitivity and also provides a measure of the rate of acceleration. For a total of 45 000 unconditional scalar operations per CPI, in addition to 22 500 complex operations for the 2 048-point FFT, per resolution cell, sensitivity can be sustained for accelerations up to 5G.

This result illustrates the power of evaluating trajectories analytically. An agile aircraft moving at an initial speed of recession of 100 m/s, and then accelerating at 5G towards the radar will reverse its direction of travel in the course of 4 seconds. A tracker operating with a radar with a 4-second update rate will find it difficult to associate the beginning and end-points of this time, whereas the HSR will estimate the acceleration accurately and provide the tracker with fully adequate input to follow the curve.

In the case of constant acceleration only, the spectrum is spread evenly between Doppler bins. Where the radial acceleration rate varies through the CPI, signal energy will preferentially occupy Doppler indices representing speeds at which acceleration has been slower.

To retain sensitivity and extend accuracy is the fundamental objective of HSR. Chapters 7 and 8 will address this effect in the discussion of possible vulnerabilities of HSR and of ways of ordering target information efficiently over time. Figure 7.8 illustrates the effect on the Doppler spectrum in the case of constant acceleration, with the application of a VH, which offers a more versatile way of ordering signal information than the non-coherent filter described earlier.

6.5.4 Target signatures and micromotion

To illustrate the capacity of such a system for target analysis and recognition, we may consider not only the Doppler sequence, as illustrative of target dynamics, but also the sequence of complex amplitudes.

Varying amplitudes will result from variable rates of acceleration, but they may also, independently, represent information about the slow fading of the target as it moves into different viewing directions from the radar, or from the effects of multipath interference. Each of these effects can be interpreted provided that each occurs in isolation, but will be challenging to separate directly if coincident.

Where the target cross section varies with time (e.g., where a low fade occurs during a CPI) but the radial speed is constant, the spectrum will be modified as in the case of amplitude modulation, with symmetric, close-in sidebands or a symmetric broadening of the peak, whose components represent the scattering nodes.

In the case of multipath interference, as we shall see in Chapter 7, the amplitude of a target approaching radially, without acceleration, should be expected to vary sinusoidally but slowly with distance, in the vertical array dimension.

To determine which of these effects is being observed, assuming it is only one, is not possible for discontinuous observations where many different interpretations are possible for any Doppler sequence. Provided that the phase sequence is continuous, then in principle this information is available from a staring radar, and will be accessible via a VH and concatenated CPIs.

We have frequently referred to frequency-domain spectra in terms of Doppler shifts; however, these spectra may include both signals determined by the formal Doppler effect, associated with a target's whole-body motion, and those originating with other signal modulations that are related to non-translational motions. We have met such signals earlier in the form of 'micromotion'.

In Figures 6.5B and 6.11C, the spectral signature of a single-engined aircraft was modelled, containing multiple high-speed components at an amplitude level lower than that of the airframe. These are clearly harmonically related at bins 348, 570, 681, 791, 1 235, 1 346, 1 457, 1 568 and 1 679 (separated by multiples of 111 bins, +/−1). They represent the scattering amplitude modulation effects of the aircraft's propeller: they can be used to determine the rate of rotation, the number of propeller blades and also, from the extent of their spread, the length of the aircraft's propeller blades. In this case, the propeller blade length is modelled at 0.85 metres.

These internal features of the target are of value in discriminating between different target types, and potentially in target behaviour analysis. However, they tend to yield returns that are smaller in amplitude than that scattered by the airframe. An ability to explore signals for components beneath the level at which a threshold has been crossed by the airframe return is essential in such spectral target analysis, and in some cases may require an extension of the CPI to maintain the necessary probability of detection.

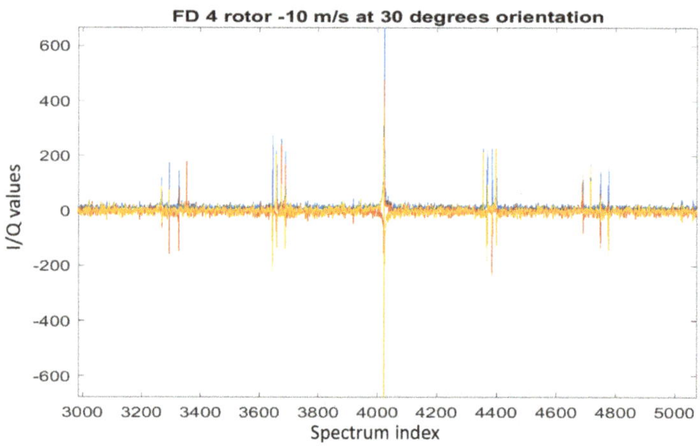

Figure 6.21 UAV model with four rotors, generating four harmonic components

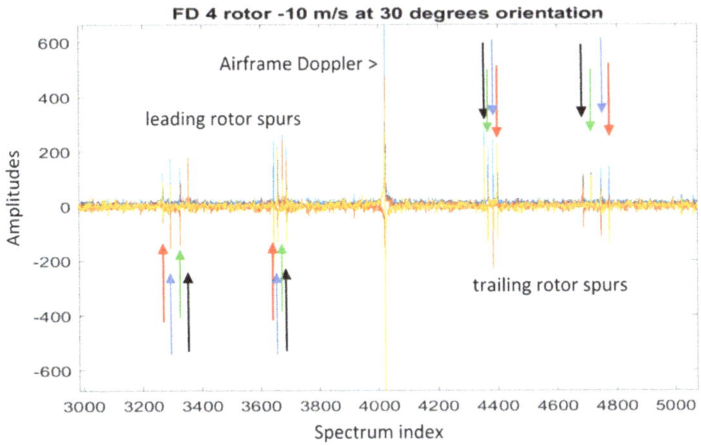

Figure 6.22 *Four-rotor drone as in Figure 6.21: 4 sets of 4 harmonics are indicated by coloured arrows and indicate number and speed of rotors*

Micromotion may be detected subject to independent detection of other target features, and in the case of VH-based assessment, the combined signature becomes available. Historically, radar targets have been defined in terms of the intensity of steady signals scattered back by their airframe in the direction of the illuminating radar: however, these effects are a function of the type and trajectory of the target, and its attitude and direction of motion relative to the radar. Figure 6.21 follows Figure 3.7; modelling a 4-rotor drone.

All of these aspects of radar signal reception contribute to an increasing pressure to go beyond single-threshold detection, to measure features of the target that affect the scattered signal but at a level sometimes well below that of the primary scatter from the airframe. These features may take the form of jet engine modulation, propeller or rotor scattering, surface vibration, propeller or rotor tip speeds, accelerative motion, imageable features, etc.

Figures 6.22 and 6.23, illustrate respectively, colour-matched harmonics representing UAV rotors, and successive returns expected to represent manoeuvring birds.

The detailed content of these radar signatures is available in the pre-detection radar returns from these different sources, and the use of a VH can support rapid identification of birds in flight near airports, aircraft or wind turbines, as well as noise, clutter, phase noise, radio and radar interference, targets of interest and targets that do not qualify for reporting but may degrade probabilities of detection.

6.5.5 Target tracking

The form of cell analysis we have discussed has the potential to provide continuing cell measurements, such that the function of tracking, in which a target's trajectory is interpolated or extrapolated based on Kalman filtering or a similar process, is effectively subsumed into the analytical process.

A role in staring radar similar to that of the tracker in a scanning radar is that of assigning a target to a combination of cells as in the case of Range Walk. At each point of reporting, it will be necessary to form associations between past reports and new reports, and to update the history and classification of a target. For staring radar this is the point at which measurements are reported, discarded, or used to inform future measurements.

The distinction between staring and scanning is expressed clearly in the different constraints on deviations from constant velocity for tracking to function.

In the scanning case, for plot association to continue, the deviation in a range from a linear track should be comparable with the radar range resolution, within the scan interval. In that case

$$\begin{aligned} \nabla R' &= \frac{1}{2}.\mathbf{a}.\nabla t_{scan^2} < R_{res};\ \text{for } R_{res} \sim 150\text{m},\ \nabla t_{scan} = 4\ \text{seconds,} \\ &\text{so that } \mathbf{a} <\sim 20\ \text{m/s}^2,\ \text{or} <\sim 2\text{G}. \end{aligned} \tag{6.1}$$

In the staring case, to achieve the expected coherent gain in the CPI, the degree of ambiguity must be maintained, and range deviations due to acceleration must be less than a quarter-wavelength within the sampling interval (for a pulsed radar, the PRI). As indicated in Chapter 3, the allowable acceleration for maintenance of coherent gain is greater for both SRC and ATC, while the scanning case allows for accelerations up to a few G.

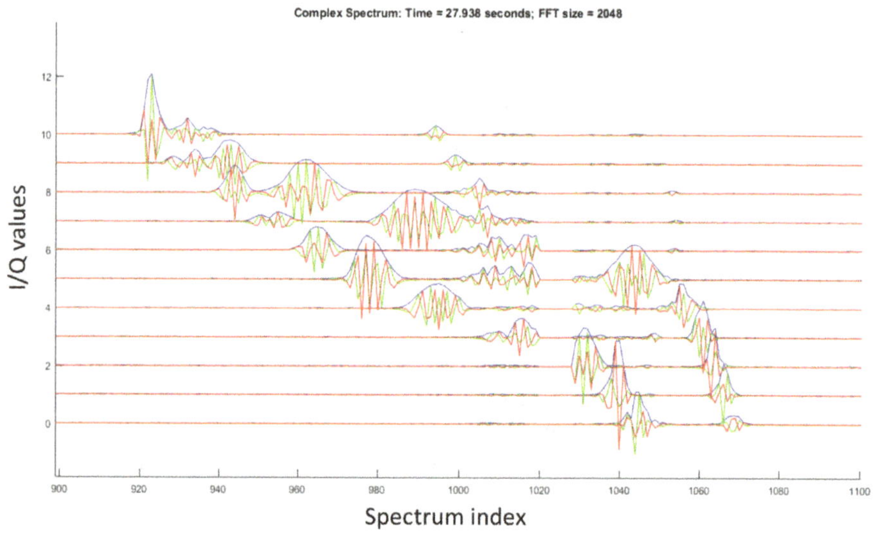

Figure 6.23 Doppler spreads lacking harmonic components represent active bird signatures

6.6 Conclusions on cell discovery and target capture

Chapter 3 provided the basis for acquiring information about objects within a CVoR, using known, persistent electromagnetic irradiation and an aperture providing continual reception, consistent with the EM Uniqueness Theorem.

On that basis, this chapter has described the range of signals that can be expected at the aperture and its constituent receiving array, and methods by which the presence of targets may be captured and other influences including clutter, noise, phase noise, multipath and interference may influence the process.

Staring radar is process-intensive, and the necessary capacity will only be manageable provided that it can use extensive parallel computing capacity efficiently. It will then support surveillance requirements that include classification and discrimination between different targets and behaviours. The ATC example is used to investigate a high level of processing demand.

HSR is likely to use highly parallel processing to accommodate the large volume of signal information it receives, which needs to avoid signal-dependent analysis to the maximum extent. This effectively means avoiding adaptive, single thresholds derived by CFAR processes in real-time, and we have described one method with the effect of multiple fixed thresholds. This is expected to allow each resolution cell to be 'discovered' as containing one or more signal components that may represent targets of interest to the surveillance user, which may then be captured for further analysis and the generation of reports.

The method of VHs is one way of allowing the information encoded in these signal components to be kept together and allows cell discovery to go beyond a process of single thresholding in selecting 'signals of interest', and to provide the basis for directing the downstream analysis of these components. The use of this format suggests a more closely linear and flexible way of extracting and exploiting the information embedded in the scattered target returns.

Chapter 7
Vulnerabilities and resilience

This book relies on the robustness of the electromagnetic uniqueness theorem. It presents opportunities that might stress practicality; in particular most functions of a staring surveillance radar will depend on substantial computing resources that require the capacity and speed of several still-advancing technologies. To assure surveillance functions under challenging target conditions that will test the resilience of these theories and methods, we shall consider what methods can work, what capacity will be needed for the performance of staring radar in reality and whether that is likely to be affordable.

As we have said, beam scanning radar exploits a high intensity of illumination and a short period of interrogation for each target, thereby minimising the complexity of processing of received signals prior to reporting a target.

With longer dwell times in which each target is irradiated, at a high rate determined by the PRF, and coherently (i.e., the complex amplitude of the persistent irradiation is under accurate control), we know that more and more precise, as well as enriched information about the target should become available, but we have to show whether it can be extracted and reported within reasonable resources and an acceptable time.

This chapter will now examine whether a long dwell time may allow changes to occur in the target return that might introduce risks or disadvantages for its use in particular circumstances. In some cases, we shall show that complexities associated with staring indeed require more computation, but that this is because the nature and precision of the information is greater and hence yields better functionality. Below we consider the nature and significance of a number of these circumstances, and whether they can be overcome or exploited.

Several possible downsides will be further discussed: loss of transmission gain, decoherence, multipath propagation, target walk and dynamics, interference and the cost in terms of data processing.

7.1 Transmission

7.1.1 Reduced gain on transmission

We have seen in Chapter 1 that the staring approach does not offer the high radiative transmit gain offered by a parabolic reflector or a phased array antenna. The staring concept implies that all the power of the transmitter for every pulse is spread over the whole field of regard of the radar, so that by comparison with a BSR, the illuminating intensity at a target is reduced. For example, in comparison with a 1.8-degree azimuth transmitter beamwidth, the transmitter gain of a 360-degree HSR is reduced by 23 dB, a loss of sensitivity that will adversely impact detection range unless it is recovered otherwise.

7.1.2 Recovery of sensitivity

To recover this loss completely requires coherent integration over a similar period to the rotation interval of the BSR. Successive pulse returns are coherently integrated, and the presence of targets travelling radially at different speeds and Doppler shifts is accommodated by Fourier transformation.

Sensitivity is then recovered by the same ratio – in the ATC example, 23 dB for an HSR with 1.8-degree azimuth resolution on reception. This would equate to integrating 2 000 consecutive pulses in place of 10 pulses integrable in the scanning case. The reliability and precision of target information is then again bounded by the receiving array geometry, the bandwidth and power budget and by the choice of the coherent processing interval in each resolution cell.

The key dependence of staring radar in achieving sensitivity and reliability is whether this recovery is or can be made truly robust under adverse target and environmental conditions.

7.2 Decoherence during the CPI

The gain available from extended coherent integration for a radar sensor may be understood to be limited due to 'decoherence' over the period, in which the phase evolution of a return signal departs from the linear progression represented by a constant Doppler frequency.

We have referred to the issue of decoherence in Chapter 3, noting first that here the term does not refer to quantum decoherence, which is a fundamental and well-attested phenomenon in subatomic particle interactions. Second, decoherence is not simply any departure from a monochromatic Doppler shift.

Relevant forms of decoherence are noted in terms of Doppler ambiguities, target dynamics, other motion disturbances and the effects of phase noise within the radar or vibration within the target.

7.2.1 Doppler ambiguities

In fact, provided that the motion under observation is inertial and the resulting Doppler shift falls within Nyquist-Shannon limits, instead of degrading the coherent gain of Fourier transformation, the process can yield a much more detailed analysis of the radial motion than merely a single Doppler shift measurement. Nyquist's theorem here applies as described in Chapter 3.

In its basic form, Nyquist's theorem requires that the phase of a complex sinusoid not change, between samples, by more than π. For a sampling frequency $\mathbf{F_S}$ this limits the frequency to between $-\mathbf{F_S/2}$ and $+\mathbf{F_S/2}$. Thus, phase may only change, between successive samples, up to between π and $-\pi$.

This rule can be extended to the case of signals at any frequency but whose phase change between samples (the true Doppler shift) lies within a range $(\mathbf{n} - \mathbf{1}) . \pi$ to $(\mathbf{n} + \mathbf{1}) . \pi$, where $\mathbf{n}$ is a constant positive or negative integer.

This expresses Nyquist's theorem for non-baseband signals and includes the possible ambiguity of the measured frequency. In that case, the observed frequency ($\mathbf{F_{obs}}$) is related to the true Doppler frequency ($\mathbf{F_D}$) by:

$$\mathbf{F_{obs}} = \mathbf{F_D} + \mathbf{or} - \mathbf{n} . \mathbf{F_S} \qquad (7.1) \text{ and } (3.9)$$

Doppler shifts beyond $+/- \mathbf{F_S/2}$ can be processed linearly provided that this more flexible rule is observed.

The actual radial target speed is then:

$$\mathbf{V_r} = (\mathbf{F_{obs}} - / + (\mathbf{n.F_S})).\mathbf{V_C}/\mathbf{2F_C}, \qquad (7.2) \text{ and } (3.10)$$

where $\mathbf{V_c}$ is the speed of light and $\mathbf{F_c}$ is the carrier frequency.

The value of **n** can then be determined by range walk measurements (see below and Section 7.4.2).

At S band for SRC the first ambiguity translates to about 200 m/s assuming a fixed PRF of 8 kHz. Doppler ambiguity is then unlikely for SRC targets. This limit clearly reduces at higher frequencies. For longer range radars, with lower PRFs to achieve high unambiguous range, the maximum unambiguous Doppler frequency is reduced and Doppler ambiguities are probable.

The ATC concept, again, operates in L band, and to achieve extended unambiguous range $\mathbf{F_S}$ (equal to the PRF) may be in the region of 1 kHz. That leads to a range of unambiguous speeds of about +/−60 m/s (or +/−120 kt), with ambiguities arising at intervals of 120 m/s. For air traffic surveillance, this means that the HSR receiver must be capable of functioning in the presence of a number of ambiguities in measurement. For military air defence or traffic control, speeds over 360 m/s (700 knots) will be encountered, and instrumented range of at least 111 km (60 nautical miles) will be necessary. For targets moving radially at 360 m/s at L band, each discovered cell must consider typically 5 possible values for the Doppler shift of a target, as illustrated in Section 7.4.1. At S band it is 11, at C band it is 20 and at X band it is 38 possible values. In these systems there is a natural preference for operation at lower frequencies, but it is likely that ambiguous Doppler shifts will result.

This will require disambiguation to fully exploit the Doppler information, and for conventional radar waveforms, appropriate methods are available to resolve such ambiguities. The most straightforward method is to compare the various ambiguous Doppler values with the trend of range measurements for a tracked target, selecting the value that best matches the measured **dR/dt** (i.e., the range rate).

We shall see in Section 7.4.2 that methods designed to accommodate the effect of range walk under high radial speeds will also allow ambiguous Doppler solutions to be resolved directly.

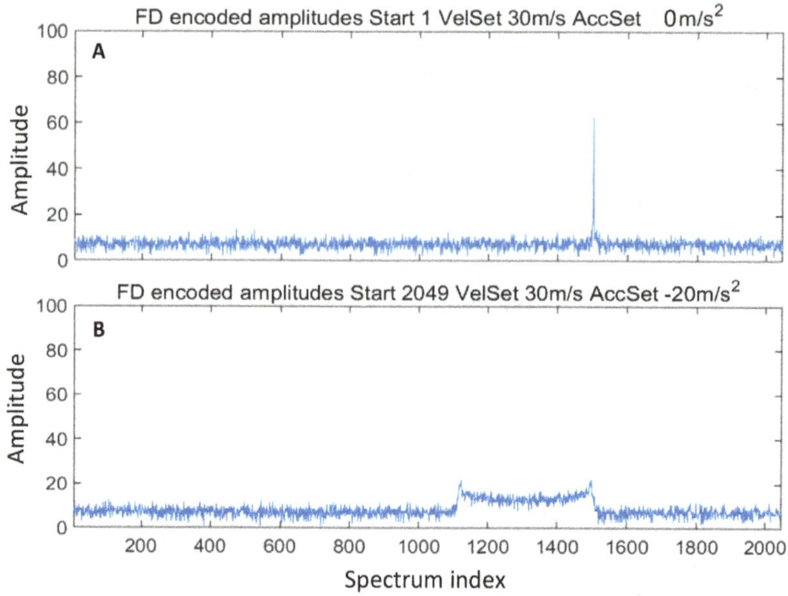

Figure 7.1 (A,B) The effect of dynamics on signal amplitudes for different levels of radial acceleration

7.2.2 Target dynamics

The extended Nyquist criterion places a high limit on target acceleration or deviations from linear motion. However, for simple Doppler processing over a CPI of 2 seconds (as ATC), radial acceleration should be less than 6G.

Subject to this rule and provided adequately accurate temporal sampling of the waveform, departures from a single frequency arise systematically and linearly from the actual motions of the target. Any arising gain reduction reflects the mathematical restriction that Fourier transformation itself imposes on target motions to yield coherent gain. Phase changes that occur as a result of tangential motion or radial accelerative, inertial motion are encoded in the scattered signal. The associated information is captured by the staring radar, but is spread into multiple branches by the fast Fourier transform (FFT) process (see Figure 7.1A,B). This reduces the peak signal amplitude, which in a BSR would be tested by a single detection threshold, however that is derived.

Random signals yield a similar modular spectrum, but for inertial targets the multiple Fourier components are correlated in phase, and signal energy can then be recovered.

The effect of the Fourier transform is achieved by multiplying the signal with a harmonic series of regular-frequency complex sinusoids, with the effect described above in terms of spectral spreading under acceleration. To refocus spectrally spread but coherent signal energy in a single FFT bin a more flexible (and more computationally demanding) process may be applied in which the signal is modified by each of a set of second- or higher-order, Nyquist-compliant and inertially consistent, sequential phase variations before the transform is applied. As with the FFT, the maximum amplitude is obtained where a phase sequence (including the optimum Fourier component) matches the actual radial acceleration of the target.

Alternatively, the output of the 'native' Fourier transform may be formatted as a vector histogram, the form of multiple thresholding described in Chapter 6 and in Section 7.4.1. This enables target capture that is less susceptible to these forms of decoherence and provides estimates of the acceleration from which the actual dynamics can be recovered. After application of coherent integration and suitable filtering and formatting, as described above, the resulting signal to noise ratio (SNR) is recovered; meanwhile the precision of the information obtained is greater by the same ratio due to continuous illumination and reception.

For the shorter-range SRC concept, a target translating radially at constant speed and interrogated over 0.5 seconds, and assuming a PRF of 8 kHz and a 4 096-point FFT, the received signal energy will be concentrated in only one of 4 096 Doppler bins, with frequency resolution of 2 Hz and a maximum unambiguous Doppler frequency of 4 kHz. At this high PRF and short CPI, acceleration will be of less significance than for ATC, but will obey the same rule. The 4 096 point FFT, applied in this way, also results in up to 36 dB of compensatory SNR gain.

7.2.3 *Motion disturbances*

Target information contained in the Doppler spectrum is not limited to radial speed. We have seen that rotor motions generate modulation sidelobes; we can measure radial acceleration directly; we shall see a range of manoeuvres solved in Chapter 8 and Appendix 1, in Chapter 9 that tangential motion can be estimated and in Chapter 10 that using a network of staring radars it is possible to derive the vector velocity of a target within a single CPI. In addition, aspects of target shape and motion can be found in the Doppler domain, including position derivatives, characteristics of rotating parts such as rotors and propellers, and events such as dismounts, impacts and vibrations.

Each of these aspects of target motion might be seen as threatening the coherence of target radar returns; however, we find that compliance with the EUNIT and extended Nyquist offers the ability instead to extract additional target information, provided, of course, that appropriate linear signal processing is used.

7.2.4 *The effects of phase noise*

Phase noise, as discussed in Chapter 6, places an upper limit on the dynamic range that can be achieved in the Doppler domain between targets in the same spatial resolution cell.

In very general terms, targets with very large RCS values tend to be clutter, characterised by zero- and low-Doppler frequencies, and resolvable from small aircraft targets that move at higher speeds and consequently generate higher Doppler shifts. However, phase noise is multiplicative with the amplitude of any target. A large target at zero Doppler shift whose measured return is affected by phase noise distributes proportionate quasi-random signals to all the upper- and lower-Doppler frequency indices because of the random effects of phase noise.

Phase noise is measured in terms of a fraction of the 'carrier' signal per Hertz of Doppler bandwidth (**dBc/Hz**), and the components at any Doppler frequency are the linear sum of the phase noise contributions from all targets at the relevant Doppler offset. This is typically, but not always, dominated by zero-Doppler clutter since this usually contributes the largest echo signal received. The level of phase noise takes values measured in minus tens of dBc/Hz and reduces with Doppler frequency offset from zero.

The use of low-jitter clocks and clock distribution, plus low-phase noise oscillators both on reception and transmission, is an essential design aim for staring radar [1].

However, the term 'quasi-random signals' is used here with intent, because the distribution of such phase noise-derived signals is distinct from that of additive thermal noise, and may be amenable to recovery, as described in Chapter 6.

7.3 Multipath propagation

We have said in Chapter 3 that multipath effects are a consequence of the electromagnetic uniqueness theorem, which describes measurements of electromagnetic fields in the presence of moving electric charges. The EUNIT refers to the effects of retarded potentials, resulting in linear, diffracted or reflected wave propagation paths. For any target in the Volume of Regard of a radar sensor, there will be paths for transmission and reception that, rather than passing directly from transmitter to target and from target to receiver, may include reflection in or diffraction around interposed surfaces, including the ground or built structures. Such reflected paths are termed multipath and are typically treated as a form of interference and false reports. The concept of staring radar experiences these effects under twice as many conditions as occur for scanning radar but provides options either to mitigate their effects or to exploit them in a useful way.

Here we shall address cases where the multipath propagation is adequately described in terms of surface reflections. In Chapter 9, the case of multipath by reflection in the azimuth plane from clutter such as large buildings (rather than by diffraction, for which effects are of a lower order) will also be considered in detail.

7.3.1 Surface multipath propagation for HSR

As outlined in Chapters 3 and 6, surface multipath propagation (SMP) occurs, in the same way as for BSR, in two distinct ways. The first is that the signal transmitted from the radar to a target arrives by two paths that will overlap; one travelling directly from transmitter to target, the other after reflection in the ground or sea surface. The reflection coefficient of the surface will determine, in part, the phase of the reflected component, its magnitude and polarisation. This increases the total irradiating intensity, but the surface wave may also interfere with the wave propagating directly. Interference will occur if the two path lengths differ by less than the effective pulse length of the radar transmission; it may be constructive in some geometric cases, increasing the field intensity at the target, but for other target positions it will be destructive, leaving the target without irradiation. If the interference is destructive the situation is not recoverable on the receive side. The effects of transmit multipath can be limited either by treatment of the ground surface near the radar, e.g., by introducing a reflecting fence near but below the level of the HSR transmitter, or by ensuring that transmitter sidelobes directed at elevations below the horizon are minimised compared to those transmitted at the same elevation above the surface. This can be achieved to a substantial extent by careful choice of phase and amplitude adjustment in the transmitter, but is more challenging the closer the target is to the ground or horizon. Multipath effects on reception, discussed below, may also be addressed by ground surface treatment and may affect the choice of positioning of transmitter and receiver.

The second source of SMP occurs after the signal is scattered by the target, by reflection in the ground surface below the path from target to receiving array. In the absence of SMP the signal received at the array is effectively a planar complex sinusoid (i.e., it is described by a superposition of sine and cosine functions of time and distance), yielding a constant signal amplitude across the array elements at the range and Doppler frequency associated with the target, and at relative phases depending only on its direction of arrival. The presence of surface multipath again results in interference, either increasing or decreasing the intensity at any array element, but in all cases causing the power received at the array aperture to vary with position: a horizontal reflecting surface causes the power to vary with vertical position, while a vertical reflector results in differences in the horizontal direction. Here we consider horizontal reflecting surfaces.

This effect of multipath on reception, in which the received amplitude varies over the HSR array, can therefore, in principle, be detected directly, provided that channel signals are still available at the point of target capture.

We repeat Figure 3.4 (Chapter 3) here as Figure 7.2 to illustrate the propagation paths available where a surface reflection occurs. Note that for the SRC and ATC concepts the transmitter is vertically separated from the receiving array, leading to independent (but usually neighbouring) paths.

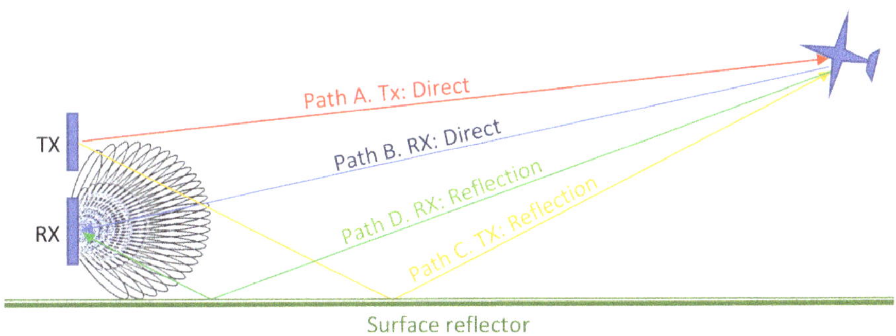

Figure 7.2 Multiple propagation paths between radar and target

In Figure 7.2, Path A is the transmit-direct path and B is the receive-direct path. C is the transmit-multipath and D is the receive-multipath. Interference occurs at the target between A and C. Interference at the receiver occurs when Path D interferes with Path B. For surface multipath, the path differences associated with direct and reflected waves for targets at low-elevation angles and radar near the surface are typically smaller than the effective pulse lengths used by surveillance radar. This has the result that signals received via the reflected path are not resolved in delay from the direct path and interference results. The four routes AB, AD, CB and CD are coherent but are received at different phases; they will interfere, and a single-CPI Doppler measurement may not resolve them. Observation over successive CPIs may be sufficient but will be slow. If the radar is located high above the surface the different paths may be resolved in delay, leading to separate signal receptions and potentially to targets at different ranges and elevations.

7.3.2 HSR elevation measurement with surface multipath

As described above, in the absence of SMP, the received signal varies only in phase with height on the array, the rate of change being proportional to the elevation angle and inversely proportional to the wavelength. This allows the target elevation to be determined by directly measuring the rate of change of phase with height on the array, or indirectly by applying a spatial Fourier transform to the channel signals.

The arrival of a second 'plane' wave via a surface reflector, unless the two arrivals are separate in their time delays, leads to interference between them. There will be a sinusoidal variation in signal intensity and phase along the vertical array columns. (This is known to give rise to ambiguity in true direction and, for homing missiles, can cause the missile to dive into the ground or sea as it seeks along the incorrect path.)

As an example, a target return, of sufficient SNR, may arrive at an elevation of ε_{Tg} radians above the horizon, at a carrier frequency of 2.7 GHz and wavelength $\lambda_c = 0.11$ m (e.g., a SRC implementation) above flat ground with an amplitude reflection coefficient $\rho = -0.75$ (with inversion of the reflected E field).

For a HSR receiver, since reception at different rows of array elements can be compared, methods of measuring and accounting for surface multipath will be available.

This case yields a signal intensity (power) that varies approximately sinusoidally in the vertical dimension with a wavelength of variation λ_v given by:

$$\lambda_V = \lambda_C / 2\epsilon_{Tg} \text{ metres} \tag{7.3}$$

Here $\varepsilon_{Tg} = \operatorname{asin}(H_{Tg}/R_{Tg})$, R_{Tg} is the slant range and H_{Tg} is the target height above the surface.

A unit (1 V/m) incident field is reduced by reflection along the horizon ($\varepsilon_{Tg} = 0$) to 0.75 of the incident amplitude and is inverted. Interference results in a combined field of 0.25 V/m, and the combined wave intensity is reduced by 12 dB.

For SRC and a target 4° above the horizon the first maximum will be approximately 0.4 m above the surface, and maxima are repeated at 0.8 m intervals in

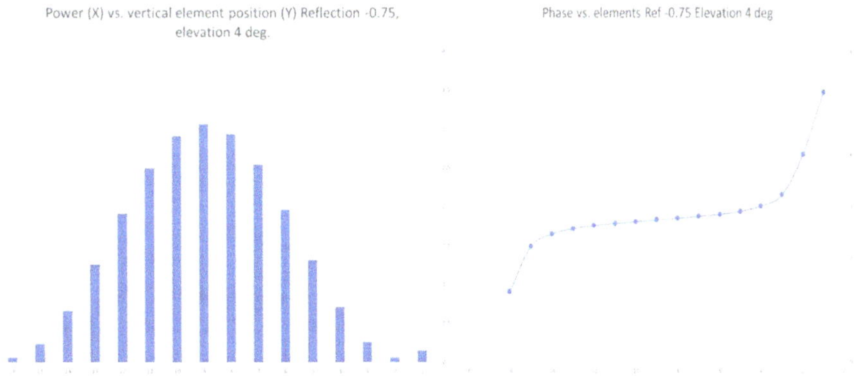

Figure 7.3 *Power and phase distributions vertically on the SRC array near a constructive maximum*

height above the surface. In this case the SRC array accommodates a single vertical interference cycle, and the elevation can be determined directly from the vertical correlation length ($\boldsymbol{\lambda}_{\mathbf{cv}}$). The SRC array concept is close to 1-m tall and will allow a direct inference of target elevation above 4° in the presence of reflections from flat ground. Figure 7.3 illustrates a maximum power distribution and the associated element phases.

In all cases where the vertical wavelength of the power distribution is less than the array height, the mean power is greater than that in the absence of multipath and the target elevation can be measured from that length, given the carrier wavelength. Note that the phase varies non-linearly up the array, making a direction-of-arrival measurement by FFT problematic.

Below 4° for SRC the vertical correlation length of the power distribution on the SRC array will be greater than the array height, so that a complete sinusoid is not available, but a parabolic fit can provide an estimate of elevation over successive CPIs.

In that situation, it is possible to determine the distribution wavelength as the target's range evolves, the interference pattern migrating along the vertical array, and the intensities measured changing from near-parabolic to near-linear to inverted near-parabolic shapes, and back. The interference wavelength can then be inferred as the target's range evolves (with an assumption of level flight).

The power distribution pattern can also be approximated by a second-order polynomial, in which the ratio of first (linear) and second (parabolic) terms are calculated, over an interval in target range, to provide an estimate of the elevation $\boldsymbol{\varepsilon}_{\mathbf{Tg}}$. If the array is appropriately close to the ground, a fourth-order polynomial fit can provide an elevation estimate from a single CPI. Curve fitting with low-order polynomials is more robust than that with sinusoids, given less than a half-period.

A procedure for integrating these approaches for low- and high-elevation targets is as follows, exemplified for the SRC configuration, with its 16-element vertical array extent.

For beams unaffected by SMP, resolution cells at each range may be formed by a weighted vector sum of array channels (or a Fourier transform) in the horizontal and vertical dimensions, including one or more beams below the horizon, followed by time- to frequency-domain transformation.

In the presence of SMP, a process stream is added in which azimuth beams are formed for each channel row and range; Fourier transforms in the time domain are applied to azimuth beams for coherent integration and Doppler separation, followed by the formation of vertical beams, amplitude comparisons and and an initial height estimation.

The scalar sum of powers at each element row at its azimuth maximum, at the appropriate range/Doppler cell, is then found. For a beam unaffected by SMP, its scalar power sum will be close to that of the vector sum at the target elevation, but greater than the vector sums for other elevations. However, for a target in a beam that is affected by destructive SMP, the power at the maximum vector sum will be degraded below the scalar sum, and the presence of SMP confirmed.

For elevations affected by SMP the vertical scalar power sums then support low-level cell discovery for elevations above about 1° for SRC. If the power correlation length is measured or estimated as above, the target elevation $\boldsymbol{\varepsilon}_{\mathbf{Tg}}$ can be derived:

$$\epsilon_{Tg} = \lambda_C / 2\,\lambda_{cv} \text{ radians (from Eq. 7.3)} \qquad (7.4)$$

Its slant range is known from the radar, and its height can be extrapolated from a flat reflecting surface near the radar:

For a target at H_{Tg} = 50 to 600 m and R_{Tg} = 5 000 m (for a SRC scenario), the power received by different vertically spaced elements or element rows varies as illustrated (rotated for visualisation into the horizontal dimension) in each graph of Figure 7.4B.

The target height as estimated by standard beamforming is illustrated by the point markers in Figure 7.4A, in which the beam heights are in the X-axis, and the relative beam powers obtained by explicit complex weighting or Fourier transformation are in the Y-axis, spaced at the actual target heights and excluding negative-elevation beam outputs.

Figure 7.4C illustrates the outputs for each of 32 part-sinusoid fits to the power distributions in Figure 7.4B. Above 4° this method converges with standard beamforming as described above. For elevations below 4° (350-m height at 5 km), the power distribution wavelength exceeds the height of the array and the scatter of heights is much less than with the beam-forming solution, in which multipath interference degrades accuracy.

This method, using complex part-sinusoid fits to the power distribution, appears to yield more stable estimates than that derived from weighted vector sums of the channel signals themselves.

Surface multipath physically affects both staring and scanning radars in a similar way because it occurs within the vertical beam patterns of both. Access to signals

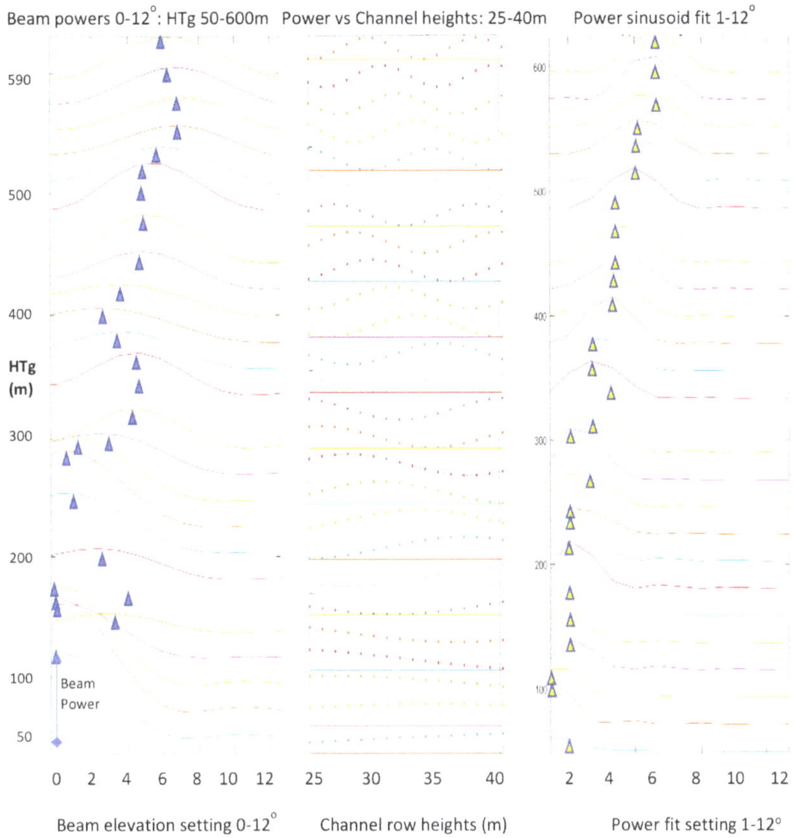

Figure 7.4 *A,B,C. Method of target height estimation from channel power variations (X axes: A: degrees, B: metres, C: degrees)*

generated as vector sums for each horizontal row of array elements provides either with a means of detecting and recovering from the occurrence of SMP, but may be more easily accommodated in a staring array, with independent access to element-level signals.

Nevertheless, at sufficiently low elevation and with a strong reflection, information about the target becomes too much degraded to support target capture or positional measurements, as for any microwave radar.

7.3.3 Azimuth multipath for HSR

The intentional wave propagation in radar consists of the direct, symmetric paths from transmitter to the target and back to the receiver; Path A in Figure 7.5. A reflection will also arise where a surface – either the ground or, in the azimuth case, a building – provides a secondary reflection (path B in the figure). For non-interfering azimuth multipath (NIMP) an apparent 'satellite' target is present in the direction of

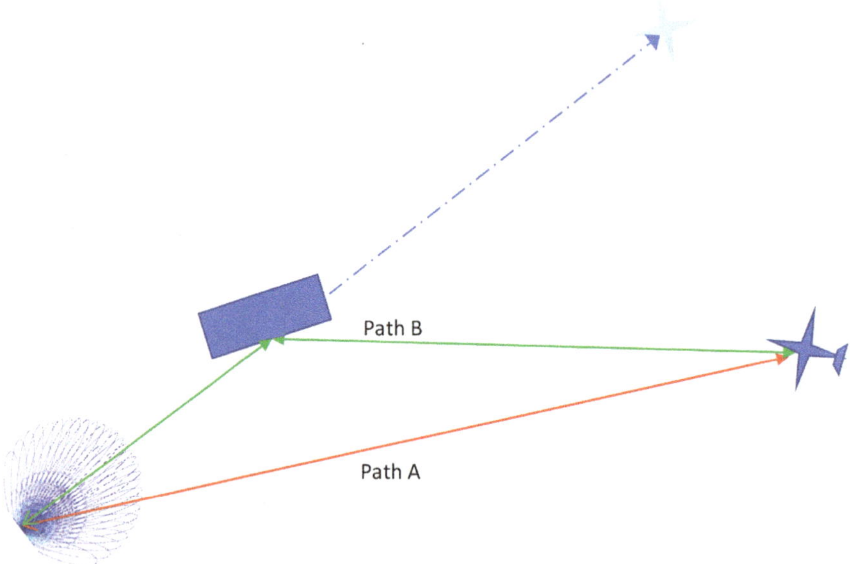

Figure 7.5 *Propagation paths in azimuth HSR multipath, with satellite targets, both in the target direction and the reflector direction*

the reflector, rather than towards the target, and at extended range due to the oblique geometry.

In the presence of azimuth multipath, four routes, AA, AB, BA and BB exist (Figure 7.6). Two additional satellites are formed, with four possible position solutions and Doppler depending on the aircraft's trajectory.

For staring radar routes AB and BA occur subject only to the secondary reflection coefficient, ρ_{sr}, and must be seen as structural aspects of the EUNIT solution.

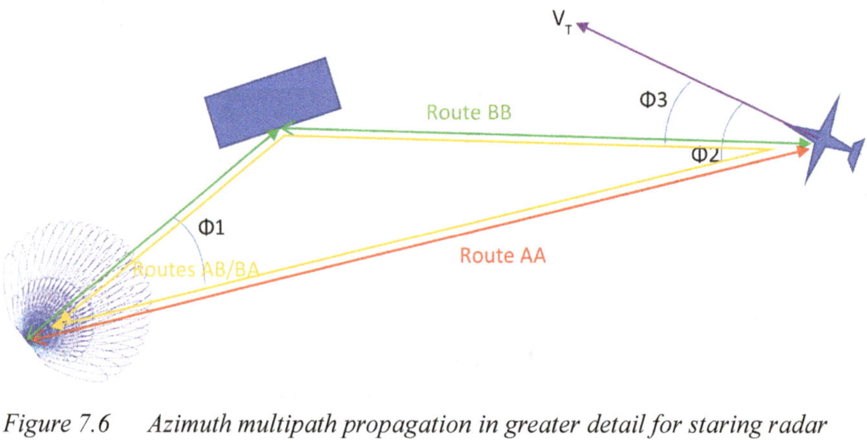

Figure 7.6 *Azimuth multipath propagation in greater detail for staring radar and target*

As such these two solutions occur synchronously and in a coherent relationship with AA, and with BB which is subject to twice the secondary reflection loss = ρ_{sr}^2. These propagation routes may be seen as yielding additional false target returns, but their coherence means that they also provide multi-look measures of the target.

In the scanning case, routes AB and BA again depend on the reflection coefficient (ρ_{sr}) at the secondary reflector but also incur a substantial loss related to sidelobes directed at the secondary reflector (AB) or at the target (BA). The multipath loss in each case is $\rho_{sr}.L_{sl}$. Case BB occurs via the main lobe in both directions but again with twice the secondary reflection loss = ρ_{sr}^2. The false targets are not synchronous or coherent with route AA. The subject of staring azimuth multipath is outlined below, and its recovery will be examined in Chapter 9.

For AMP, therefore, in many cases the dual paths are sufficiently different in either length or direction that the signals arriving at target or receiver do not interfere.

The effect of NIMP for HSR is that four separate targets may be expected; one (AA) at the correct and shortest range in the true target direction, the second as for BSR at a greater range but in the direction of the reflector, as at BB, but also AB and BA occur at identical mid-ranges and in each direction – towards the target and the reflector.

7.3.3.1 Azimuth multipath as a challenge to HSR surveillance

Figures 7.5 and 7.6 illustrate the alternate propagation paths and routes leading to azimuth multipath radar returns. In this case the reflector may be the wall of a large building, leading to 'satellite' targets via routes AB, BA and BB. The secondary reflection occurs in the building either on the transmit path, on the receive path or on both. Assuming the building is static the reflection will be coherent with the primary return, but each satellite return will be delayed and is typically resolved from it in apparent range. The term 'satellite' is used since these tend to occur at ranges and azimuths in a 'halo' beyond the primary target. These multiple returns per target must be recognised as a challenge to the HSR function. However, azimuth multipath returns are also a source of 'multilook' information and can be suppressed explicitly, as described in Chapter 9. In principle they can also be used to yield more precise and highly dimensional target information; however, their intermittent appearance makes this less than a robust source of information.

In Chapter 9 we shall address methods whereby correlated and coherent satellite targets, as provided under the EUNIT and the RT, can be resolved and reported or suppressed appropriately using extended target signal data; meanwhile, a summary of methods is given below.

In Figure 7.6 we introduce the velocity vector of the target, illustrating the relationships between the propagation paths and, with Table 7.1, their direction, range and Doppler effects.

Considering the four paths and their primary variables, we have in Table 7.1.

All objects in the CVoR can be subject to multiple propagation paths. For staring radar, additional asymmetric paths may exist and can give rise to target signals in the same two directions – that to the target and the reflector, and at different ranges and Doppler frequency. If these are not resolved and discriminated they threaten persistent false targets.

Multipath propagation might thereby threaten the basis of staring radar by undermining the unique pairings of target and radar returns that underlies all effective radar surveillance. However, the principle value of applying the EUNIT is that persistent observation with sufficient power and resolution can be expected to provide the information necessary for these different cases to be distinguished.

The Chain Home RDF (radar) system of the early 1940s was a staring radar, but it was highly effective in detecting and positioning its targets. Why was it not disabled by the incidence of multipath propagation?

Chain Home's absolute priority was to achieve early detection of bomber formations at long range. Later in the war, Chain Home itself (Ames Type 1) was found ineffective at low altitude, primarily due to the surface reflections over the North Sea and its low carrier frequency (about 50 MHz), and its successor (Ames Type 2 – or Chain Home Low) operated at a higher frequency (over 200 MHz) with a vertical array that allowed much lower propagation. Multipath must have been present, and will have led to intermittent observations, but it would have been less likely to suffer azimuth effects, with satellites arising from surface structures.

For HSR, as the legatee of Chain Home's adaptability, at frequencies typically used for air surveillance – L to X bands – we can determine the probable effectiveness of methods of NIMP suppression.

NIMP, like SMP, also involves not one scattering event at the target but two or more scattering events, at the target and at one or more scattering or reflecting objects. Except in extreme cases these differ from the strong reflection coefficient of shallow surface reflections. They incur a double space loss leading in most cases to significant contrast in amplitude between direct and NIMP radar returns. For Chain Home, the absence of very large vertical reflecting surfaces meant that the effect of azimuth multipath was less severe.

In the current context, where an association can be formed between direct and multipath targets, constructive approaches to multipath are available, as indicated above in the case of the array power distribution for surface multipath. Each multipath component is represented by a set of complex amplitudes, a Doppler frequency and an effective range and direction, and each constitutes a separate 'look' at the target during the CPI. Rather than constituting 'interference', more information

Table 7.1 Propagation distances and Doppler shifts

Path	Received direction	Range	Doppler	Resolution
AA	Target	Rt	2Vr (cos(Φ2))	R(min)
AB	Reflector	Rt + ΔR	Vr (cos(Φ2) + cos(Φ3))	Direction
BA	Target	Rt + ΔR	Vr (cos(Φ2) + cos(Φ3))	Doppler (precision)
BB	Reflector	Rt + 2 ΔR	2Vr (cos(Φ3))	Direction, Doppler, Range

may then be derived about the target, and also about the reflecting surface, whose position and orientation can be determined, and satellites identified and discounted.

Whether this effect is to be seen as a vulnerability or an opportunity depends on whether, to comply with the EUNIT, a HSR is configured to analyse a wide enough range of signals arriving from the CVoR. These include pairs of simultaneous but delayed satellite returns one in a different direction from the direct target azimuth, at equal ranges, and at equal Doppler shifts. Where the multipath components are resolved in range, Doppler or direction, their differences support discrimination and potentially finer measurements.

7.3.4 Exploitation of azimuth multipath

7.3.4.1 Interfering azimuth multipath

To avoid problems where azimuth multipaths are not resolved (i.e., where they interfere), the siting of the radar should take these effects into account. However, they need to be considered.

Interfering azimuth multipath (IAMP) leads to interference patterns across the receiving array. In a similar way to the method described above for surface multipath, but in which the form of the structure affects the pattern (as opposed to an effectively flat and level surface), IAMP offers additional detailed information about the CVoR.

Once this capability is available, HSRs may be expected to yield geometries for reflecting surfaces in their field of regard, as test targets or targets of opportunity provide multiple correlated measurements.

7.3.4.2 Non-interfering azimuth multipath

In the case of NIMP, the surveillance process may be designed to detect the presence of correlated target tracks, to record and compare them, to deduce and refine reflecting geometries. Evidence of the occurrence of multipath in either form (SMP or NIMP) can be found in the course of commissioning or early operations. However, the primary objective is to suppress false NIMP targets.

As we have seen in Section 7.3.3, NIMP provides three additional 'looks' at the target. These determine the presence of multipath scattering and cross-track or vector Doppler (obtaining the two cosine terms in Table 7.1). Figure 7.7 illustrates the measurement of position for azimuth scattering sources, using the target offsets of satellites, in range and direction. We shall address this further in Chapter 9.

Satellite targets due to multipath always occur at longer range than the 'host' target. They therefore cannot occur when the host is outside the CVoR. When they do occur, they occur in fours, including the host:

1. the host AA at range R'_h and bearing Az_h towards the target,
2. the two-way satellite BB, as for a scanning beam at range $R'_{g2} = R'_{h+} \Delta R'_g$ and bearing towards the reflector,
3. the one-way satellite BA scattered to the receiver by the target, at range $R'_{gt} = R'_h + \Delta R'_g/2$ and bearing towards the target and

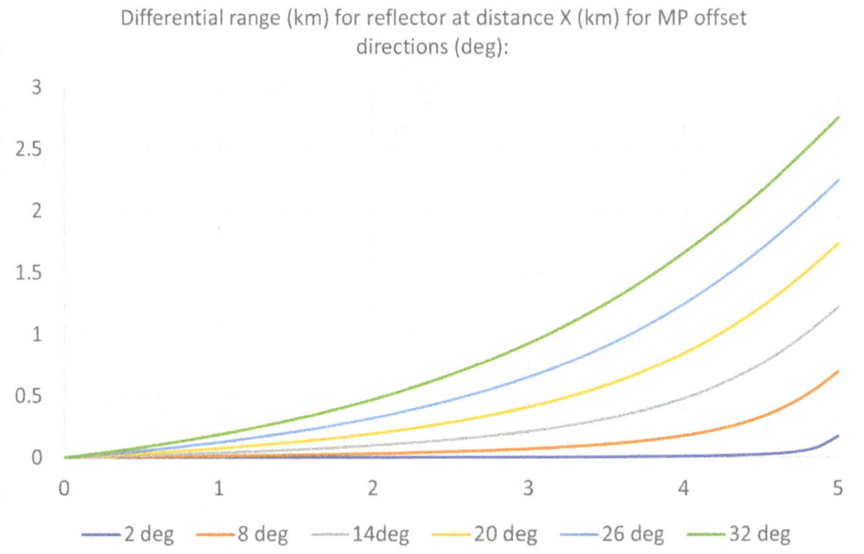

Figure 7.7 *Illustrating the differential range ΔR_g experienced from a reflector at various azimuth offsets from the primary target or the radar, for targets at different ranges*

4. the one-way satellite AB, scattered to the receiver by the reflector, at range $R'_{gr} = R'_{gr} = R'_{g2} - \Delta R'_g/2$, also equal to R'_{gt}, and bearing towards the reflector.

The differential range of the reflector either from the target or the radar can be determined from the target range, as illustrated in Figure 7.7, as a function of its azimuth offset.

Where a satellite target is captured these measurements are directly observable, providing evidence for satellites to be suppressed.

7.4 Target walk

In a basic form of staring radar, range cells are formed that exploit the signal bandwidth, beams are formed that exploit the receiver aperture in both azimuth and elevation, and Doppler resolution exploits the dwell time. Sensitivity is maintained by coherent integration over an extended dwell time, typically using a Fourier transform.

This capability is limited by targets moving with high radial speed and migrating between range cells during the integration interval. This is known as 'Range Walk', thereby limiting the effectiveness of integration and reducing sensitivity.

Similar effects can be expected in terms of transverse speeds and beamwidths at ranges where the cross-track resolution is comparable with the range resolution.

For Doppler measurements over long integration intervals the target's radial speed may also evolve, leading to the return migrating between Doppler bins, and its amplitude therefore being diluted by 'Doppler Walk'. This is directly related to the effects of target dynamics as introduced in Chapter 3, Section 3.11; Chapter 6, Section 6.5.3 and Section 7.2.2.

These effects might suggest irrecoverable disadvantages for HSR under these conditions. The truth depends on whether these disadvantages arise from limits on physical feasibility (if the EUNIT conditions cannot be met), or whether they can be countered by enhanced signal processing.

The limits on meeting EUNIT conditions are those of the radar range equation, plus continuity of coherent observations. Provided that the target motion remains within the range and speed specification from which the Nyquist-derived illumination and processing design are derived, the waveform will be sampled at a sufficient rate to acquire accurate phase migration information during the integration period.

It is also the case for HSR that although the integration window (the CPI) may contain substantial phase migration overall, we have seen in Section 7.2 that continuity of phase observation during the CPI allows Nyquist-consistent coherence over a wide range of dynamics. Appropriate processing can be applied, subject to limits on the higher orders of motion derivatives, to allow the effects of Target Walk to be measured and process gain to be recovered.

7.4.1 Doppler Walk recovery for ATC

Doppler Walk tests the susceptibility or sensitivity of HSR to radial acceleration of the target, and this term is used for the effect of target dynamics, as described in Chapters 3 and 6. Here we explore the extent of the effect and its recovery using the VH format also described earlier.

With a dwell time of 2 seconds, under constant radial acceleration of 3G, the change in radial speed at L band will be near 60 m/s. For the case of acceleration in a turn near tangential motion at 3G, the change in radial velocity will be the same, but centred near zero Doppler.

In both cases the target's Doppler return will be spread over several hundred Doppler bins, reducing the power per bin by up to 30 dB. Although a minimum (0 dBsm) target may be unlikely to pull this level of acceleration, there is a clear risk that sensitivity may be degraded unacceptably.

This apparent degradation is a key distinction between HSR and radars that dwell on any target for a much shorter period: the effect may appear similar in the modular spectrum to a random decoherence and in that case might result in a further challenge to the effectiveness of staring radar.

As referred to in Chapter 3 the complex amplitude sequence resulting from Newtonian target dynamics is quite distinct from that arising from random decoherence. Chapter 6 also introduced a simple form of non-coherent filtering to retain sensitivity. In Section 7.2.2 we indicated that, by Fourier transformation followed by analysis via a VH format, sensitivity can also be retained, and the information

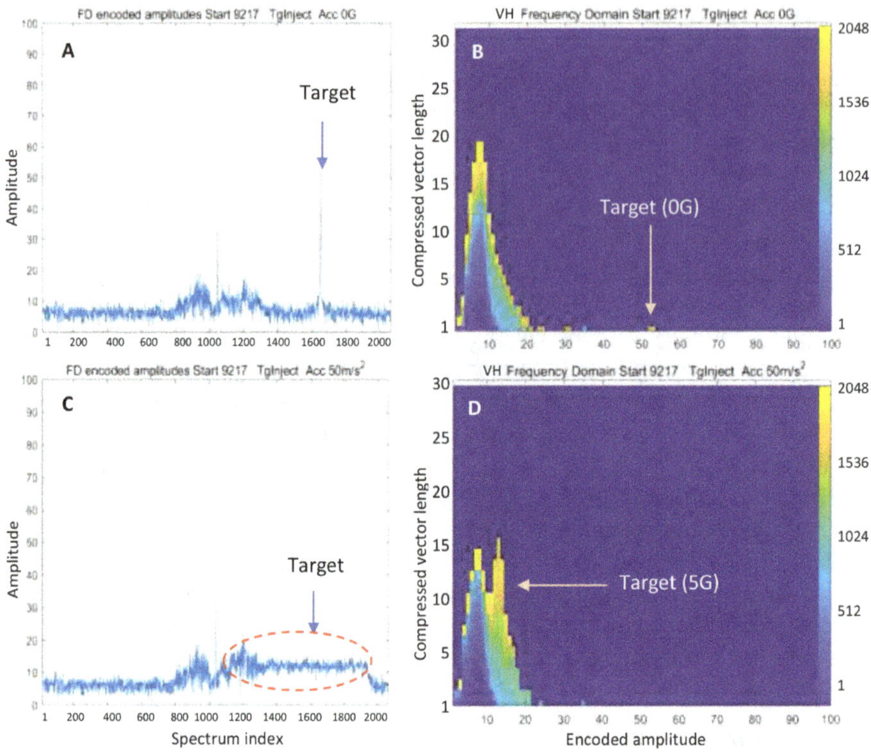

Figure 7.8 Encoded Doppler spectra and FD VH outputs for an injected target. Total VH points = 2 048.

content of the spectrum re-ordered in a versatile way, directly referencing rather than conditionally promoting amplitude features in the frequency domain.

To reiterate, in a VH data format, the X axis represents an encoded (compressed and integer-formatted) form of the signal amplitude, the Y axis is the number of occurrences and the colour code is either the time in a sample sequence or the frequency in a Fourier spectrum (2 048point here). Here, the data are in the frequency domain, and green is centred at zero Doppler, orange-yellow components are at increasing outbound speeds between 0 and 150 knots, while blue-indigo are incoming speeds over the same range, plus relevant ambiguities.

Forming and exploiting the VH for every cell is an intensive operation and has been discussed in more detail in Chapter 6. Its value will be in making the process of resolving different types of target, behaviour and interference more flexible than a system based on single, even adaptive, thresholds.

Figure 7.8 illustrates a modelled target on HSR cell noise and low frequency, suppressed clutter in two conditions: constant radial speed and outward acceleration at 5G (50 m/s^2).

At constant speed, in Figure 7.8A, the target is seen as a sharp peak, many times the encoded noise, which is centred at 6 on the encoded amplitude (X) scale.

When formatted as a VH (Figure 7.8B), the target in the zero-G case is coloured orange (outbound at 101 kt), and consists of two points at adjacent frequencies, clearly separated from the noise distribution.

When accelerating at 5G (Figure 7.8C), the return is spread over approximately 1 000 Doppler bins, reducing the apparent SNR to about 6–10 dB. Applying a typical CFAR threshold at noise plus 12–20 dB would result in a missed detection.

In the VH (Figure 7.8D), the population of points near amplitude 12, totalling about 1 000, is clearly identifiable. These points are linked by a contiguous address sequence and the successive and coherently phased Doppler components they address, and separable from the random-index noise residue. Rather than reverting to a threshold decision on this population to select points worthy of analysis, all points representing the target are here grouped according to arithmetic (not pre-defined logical) rules, describing signals representing the motions and dynamics of the target.

The amplitude axis (X) of the VH, as before, is used to divide the VH population into amplitude sectors on and across increasing-order address boundaries.

Simple means of recovering sensitivity are available. One is to measure the total spectrum power within a sequence of binary-spaced bandwidths (1, 2, 4, 8, etc. bins), and at binary-spaced centres, after subtracting the mean noise power from each spectrum sample, as described in Chapter 6.

Here, the estimates of speed and acceleration are not obtained simply by a smoothing filter acting on the modular spectrum, as described in Chapter 6, but by direct reference to the spectrum points contributing to the orange peak in Figure 7.8D. Multiplication of each squared VH abscissa (power) point by the ordinate (number of instances) yields the whole signal power within the appropriate range, at the cost of a manageable number of operations.

The combined probability of detection in either case is then no less than for the constant-speed peak. At the same time, measures of acceleration and mean radial speed will be available directly from the FD-indexed content of the VH ordinates. Using this data format, the process of target capture, and the measurements of speed and acceleration can be achieved without conditional, adaptive or iterative calculations.

More sophisticated uses of the VH, using the complex amplitude sequence and spectrum, will yield greater detail about the dynamics of the target, but this, and the growing state of knowledge about the range of exploitation of the EUNIT remains for later study.

7.4.2 Range Walk for ATC

Consider an SRC where range gates are spaced at 75 metres, target speeds are less than 100 m/s and the CPI is less than 500 ms. Targets should not migrate significantly

Range progression with time							
ᵛ Range gate							
7	**RG7(0-0.3)**						**RG7(1.8-2.1)**
6	**RG6(0-0.3)**	**RG6(0.3.-0.6)**				**RG6(1.5-1.8)**	**RG6(1.8-2.1)**
5	**RG5(0-0.3)**	**RG5(0.3.-0.6)**	**RG5(0.6.-0.9)**		**RG5(1.2-1.5)**	**RG5(1.5-1.8)**	**RG5(1.8-2.1)**
4	**RG4(0-0.3)**	**RG4(0.3.-0.6)**	**RG4(0.6.-0.9)**	**RG4(0.9-1.2)**	**RG4(1.2-1.5)**	**RG4(1.5-1.8)**	**RG4(1.8-2.1)**
3	**RG3(0-0.3)**	**RG3(0.3.-0.6)**	**RG3(0.6.-0.9)**		**RG3(1.2-1.5)**	**RG3(1.5-1.8)**	**RG3(1.8-2.1)**
2	**RG2(0-0.3)**	**RG2(0.3.-0.6)**				**RG2(1.5-1.8)**	**RG2(1.8-2.1)**
1	**RG1(0-0.3)**						**RG1(1.8-2.1)**
Time (sec) >	T0 = 0	0.3	0.6	0.9	1.2	1.5	1.8

Concatenation with speed (knots) and range for compound Range Gates (4, -600:+600kt)							
Speed (out) kt	700 > 500	500 > 300	300 > 100	100 > -100	-100 > -300	-300 > -500	-500>-700
ᵛ Range gate							
7	T7						T1:
6	T6:	T7				T1:	T2:
5	T5:	T5:T6:	T6:T7		T1:T2:	T2:T3:	T3:
4	T4:	T4:	T3:T4:T5	T1:2:3:4:5:6:7	T3:T4:T5	T4:	T4:
3	T3:	T2:T3:	T1:T2:		T6:T7	T5:T6:	T5:
2	T2:	T1:				T7	T6:
1	T1:						T7

Figure 7.9 A compound range gate centred at range RG4 concatenates time sequences as follows: (0–0.3 : 0.3–0.6 : 0.6–0.9 : 0.9–1.2 : 1.2–1.5 : 1.5–1.8 : 1.8–2.1) seconds, taken at successive ranges as illustrated

beyond range gate bounds, and coherent integration as described in Chapter 5 should maintain high levels of sensitivity.

For ATC the range resolution may be set at 150 m, with speeds up to 500 m/s, and significant Range Walk should be expected that will cause reduced sensitivity. For example, within a dwell time of 2 seconds and range resolution of 150 m, a target moving radially at 360 m/s (700 knots) will traverse nearly 5 range cells, reducing sensitivity by 7 dB.

A possibility to maintain sensitivity might be to extend the range resolution to 300 or 500 m. In that case degradation due to Range Walk is less probable, but it will directly impact resolution with respect to clutter and the accuracy of one of the most basic measurements made by the radar. We seek a method of maintaining sensitivity without degrading the radar's basic function, but accepting the cost of additional processing. This may be significant, and at a later stage, we shall count the cost of the several streams of HSR processing.

To maintain sensitivity in such a case will require a form of concatenation of range samples prior to coherent integration. To accommodate speeds between +/−700 kt, compound range gates indexed to range and speeds are necessary. These are formed by successive concatenation of shorter time intervals within the CPI. In the ATC case these would be 300-ms long, and the process can be described and visualised in Figure 7.9.

Since time-domain beam data are available within the **RAED** data structure, the formation of 'walking gates' is a simple, pre-definable operation. However, where this solution is necessary (such as for ATC), range gate calculations are multiplied by a significant factor as the different-speed versions are submitted in parallel to the Doppler process.

This process of range gate concatenations also has an additional value. Section 7.2.1 describes the probable appearance of Doppler ambiguities for an ATC configuration, and a method of resolution dependent on tracked changes in range, between CPIs. This process of cell concatenation in time allows different ambiguous speeds to be resolved directly, within a single CPI.

In summary, an ATC HSR is expected to capture targets through a process of cell discovery, as described in Chapter 6, but in which 'cells' include cells centred at a certain range gate but concatenated with inner and outer gates as above to accommodate the necessary range of radial speeds. The process described leads to a number of parallel coherent integrations by Fourier transform, and each of these may be analysed through a VH.

7.4.3 Range ambiguities

Regular pulse repetition intervals imply that ambiguities can occur in range measurements. For ATC a PRI of 1 ms determines a nominal unambiguous range of 150 km (80 nautical miles). A target at 151 km will generate an ambiguous-range return at an apparent range of 1 km. For this target to have an apparent cross-section of 0 dBsm at the ambiguous range of 1 km, it would need to have a true radar cross-section close to +90 dBsm, which is unlikely. In other words, the reduction of SNR with range means that targets beyond the CVoR boundary are unlikely to be detected.

However, SRC, designed to detect a target of −20 dBsm to a maximum unambiguous range of 5 km, will detect targets ambiguously that are still small. A target of RCS −8 dBsm but at a range of 10 km will yield the same return amplitude as the −20 dBsm target but will also appear, ambiguously, at 5 km. For this reason there will be a need both to slow and encode transmissions. Under appropriate conditions of PRI, range and sensitivity requirements, recovery of unambiguity in range can be achieved by appropriate phase or frequency-encoding of the transmit pulse, followed by reconciliation of the measured Doppler shift with the tracked range and range rate of a target. The accuracy required for successful reconciliation is a rapidly increasing function of frequency, and lower operating frequencies will reduce this constraint.

A configuration that may be more flexible in terms of range ambiguity will be discussed in Chapter 10, which arises in the case of a coherent network of HSRs. In these systems it may be possible to protect receivers from saturation during transmission, and in that case, transmissions can be received, monitored and controlled without disabling the receiving function. Where transmissions are transparent to the reception of target returns, several constraints disappear or become naturally manageable with respect to range and Doppler ambiguities. However, such a configuration depends on precise coordination of a system of surveillance sensors. Successful relevant tests have been performed, but this development appears to lie in the future.

7.4.4 Azimuth walk

Cross-track resolution, defined by azimuth beamwidth (radians) × range (m), is typically longer than range resolution, and therefore azimuth walk has a lower priority. For the ATC example, with range resolution of 150 m and azimuth resolution of 1.8°, cross-track resolution exceeds range resolution beyond about 5.5 km.

At ranges from 5.5 to 27.5 km (3 to 15 nautical miles) there can be some loss in the peak return due to azimuth walk, as well as due to geometric changes in radial speed, but at these reduced ranges target sensitivity is increased sufficiently under the range equation to compensate. Beyond 27.5 km (15 nautical miles) for ATC azimuth walk itself should not lead to desensitisation at speeds up to 500 m/s.

In the course of a CPI, changes in the direction to the target will lead to changes in the phase profile across the array at the target's Doppler frequency. These changes can be measured by the HSR, yielding a direct measure of the cross-track speed of the target, prior to track filtering, as follows.

The absolute phase Φ_{absTg} of a signal arriving is:

$$\Phi_{absTg}(n, m) = -sqrt((Rg_{Tg}.cos(Az_{Tg}) - N_{El}(n, m))^2 + (Rg_{Tg}.sin(Az_{Tg}) - E_{El}(n, m))^2 + (H_{Tg} - H_{El}(n, m))^2)) . 2\pi/\lambda, \tag{7.5}$$

where **n** is the index of elements within a row and **m** is the index within a column.

The azimuth direction of arrival of a target return at the array is given by:

$$Az_{Tg} = d/dm\ (\textstyle\sum(\text{Complex Weights}(Az, El, n, m)\ .*\ \Phi_{absTg}(n, m))) \tag{7.6}$$

Its rate of change with time is:

$$dAz_{Tg}/dt = d/dt(d/dm(\textstyle\sum(\text{Complex Weights}(Az, El, n, m)\ .*\ \Phi_{Tg}(n, m)))) \tag{7.7}$$

when arriving from a target at ground range Rg_{Tg}, measured from the array centroid, elevation El_{Tg} and North azimuth Az_{Tg}, at an array element at position **(n, m)** in the receiving array.

A method of deriving the rate of change of azimuth is to measure the trend of the change of Doppler phase with position across the array as a function of time. This is feasible but requires a process stream in which Doppler FFTs are applied to each column or subset of array elements, before azimuth beamforming.

7.4.5 Summary of adverse target conditions

The processes required to overcome performance challenges associated with target walk and multipath are relatively demanding in terms of processing power, but two considerations suggest that this should not prevent advances in target analysis; these are:

1. The presence of detailed target information carried by the target return signals but currently excluded by short and discontinuous CPIs and
2. Greatly increased and cheaper processing capacity.

The responses to target walk and multipath challenges, described above, are at an early stage of development. However, the forms of recovery that we have described are within the range of feasible designs and continue to indicate that HSR has capabilities not available without signal continuity. A successful result can support HSR sensors capable of a full surveillance function with respect to any Newtonian object moving within the CVoR and within the ATC specification, extracting all the available information under the EUNIT and reporting it to the extent provided by the computing resource.

A VH format, or its equivalent, is expected to remove obstacles to the efficient combination of highly parallel processing with conditionally directed processing associated with an adaptive single detection threshold.

Provided that these or similar processes can be accommodated within the affordable processing capacity, a number of advantages arise and can be maintained:

1. Appropriate treatment of Range Walk will allow sensitivity to be maintained, and for Doppler ambiguities to be resolved directly at any speed within the user-specified range.
2. Successful treatment of Range Walk will be significant for ATC, in high speed applications such as BMD surveillance, or for airborne HSR platforms.
3. Appropriate treatment of Doppler Walk will allow sensitivity to be maintained under radial acceleration. Measurement of the Doppler Walk rate will both identify the rate of acceleration and maintain the necessary gain of coherent integration.
4. Appropriate treatment of Azimuth Walk will allow direct measurements otherwise only available after track filtering has combined the results of several time-spaced observations.
5. Appropriate treatment of multipath offers the suppression of multipath and may support a developing map of clutter and oblique reflectors, including the ground surface and ground structures.
6. The presence of moving clutter such as wind turbines can be accommodated, and their effects on targets of interest suppressed.
7. Appropriate treatment of extended Doppler histories will offer multilook surveillance of targets, aiding in target discrimination, multilateral positioning, direct measurement of velocity and potentially imaging of the target.

7.5 Resilience under interference

Whatever its region or application, radar is used in support of activities that incur risks and require safety. Any form of radar that differs from what is established as the paradigm for air surveillance must justify its potential for resilience under a range of threats that may affect the radar receiver in terms of its probability of detection, its false alarm rate, or its failure to discriminate between targets of interest and those that are classed as clutter or confuser targets.

The threats we will consider in this section are:

1. saturation or distraction by external radio sources, including radar
2. noise degradation due to remote radio transmissions.

7.5.1 Channel saturation

Radar uses sensitive, linear receivers to observe small signals scattered by unknown targets from irradiation by a known transmitter, and all aspects of the radar's function depend on the receiver responding to signals scattered by targets of interest within its VoR.

All receivers that use electronic circuits to process signal data can only respond linearly up to a certain, usually sharply defined, limiting signal amplitude, and their function will be disabled during reception of signals above that limit.

A HSR deploys a large number of small receiving array elements, each of which feeds a separate receiver channel. This allows each receiver channel with its maximum amplitude level to maintain linear operation under a greater incident field strength than for a single-channel receiving beam. HSR receiver channels, as for those used in active array radars, will be susceptible to saturation by interferers of broadly similar power to those that would saturate a single-beam receiver via its sidelobes.

Calculation suggests that airborne megawatts of effective radiated power will be required to saturate an ATC HSR configuration. In the ATC configuration, the receiving elements are small dipoles with an effective area **Ae** of about 0.01 m^2. Typical receiver channels may saturate at an input level **Psat** of about 1 milliwatt, so the radiated power required to cause saturation can be calculated from the distance **Ds** to the source. The power-gain product required, **Pt.Gt,** is given by:

$$\mathbf{Pt} \,.\, \mathbf{Gt} > \mathbf{Psat} \,.\, 4\pi \,.\, \mathbf{Ds}^2 \,/\, \mathbf{Ae} \qquad (7.8)$$

At a distance of 10 km, saturation requires a power-gain product near 100 megawatts. Radiation via an antenna with 30 dB gain requires 100 kilowatts.

HSR incurs the special case of interference by other radars operating at a distance such that the HSR array may be saturated. For radars in service, this is avoided by using separate frequency assignments, so that interfering signals are rejected by front end radio-frequency filtering in the victim receiver.

For a HSR, the signal energy of an interfering BSR is incident only during the pulse duration and dwell time of its beam on the victim, and its saturating effect is limited to that period. Since the dwell time is a small fraction of the observing time (CPI), theory provides and experience has shown that the short and closely packed signals imposed during the dwell can be suppressed in the course of signal processing, without a significant effect on HSR performance. Some forms of radar (particularly staring FMCW) might prove a greater threat in this respect, despite operating at lower peak transmission power levels.

7.5.2 Noise degradation and suppression of radio interference

Chapter 6 has outlined the recognition of interference of different types with HSR.

For many present-day surveillance radars, frequency agility (even a choice of two frequencies) is a basic capability and provides a first line of defence against unexpected interference, with relatively little effect on operation. While it is possible for staring radar to operate at different frequencies, its focus on continuity of observation places a greater premium on frequency stability than is necessary with scanning radar, and this tends against any routine or rapid variability in frequency.

HSR's persistent coherence leads to other possibilities for interference suppression, despite that each HSR receiver is permanently open to reception of signals from any point within a large VoR sector, and it is possible for an external radio source to cause degradation of the noise level in every array channel, when the signal overlaps with the operating bandwidth. In situations where interference is not necessarily excluded by frequency assignments and device specifications, measures will be required to achieve resilience. Several standard approaches are available, and HSR offers additional capability in principle.

7.5.2.1 Adaptive null suppression

For HSR an interferer will create an apparently noisy signal in each receiver channel, degrading its sensitivity overall. However, HSR operates all its beams simultaneously. It is not subject to side-tracking via scanned sidelobes, and the direction of arrival of any non-saturating signal can be determined. An adaptive and agile null may be formed in that direction to suppress the direct interference, whatever its waveform, by adjusting the known complex weightings of array elements.

Adaptive nulling may be used to minimise the effect of a signal arriving in this known direction, but sidelobe residuals will exist in other beams. Accurate cancellation will be subject to drift in the complex gains of receiver channels and to phase noise effects; low phase noise will be essential, and ubiquitous interference suppression will be difficult.

7.5.2.2 Signal-specific suppression

A method of interference suppression is available for HSR that again arises from its persistent coherence.

The laws governing radio propagation do not provide for an interfering source to deflect the apparent direction of arrival of its signal at the intended victim receiver, and cannot, by exploiting scanned sidelobes, confuse the HSR's direction finding.

For HSR the interfering signal, arriving from the sole azimuth direction of its source, will be coherently correlated across all receiver channels. The staring aperture deploys receiving 'main beams' in all directions within the CVoR; the direction of the source and the signal itself can be isolated and measured accurately. This signal will be coherent with all element receptions from that direction, and is stored in the beam-wise RAED segment as a highly accurate complex amplitude sequence.

Knowing the direction accurately implies that for every range gate sample the actual signal sequence arriving at every channel from that source can be closely estimated and then suppressed optimally. The residual complex gain error for each is then found with a residual pointing error for the source.

For any pulse repetition interval the errors and offsets found are constant for each channel, for all range gates, and a high level of suppression is feasible. It will also suppress targets within its directional resolution in the same way as an adaptive null; it is expected to be an effective and deep countermeasure to the received interference, including its arising sidelobes; the method is process-intensive but probably not prohibitive.

7.5.2.3 CW interference

When a staring radar is subject to radio interference, the signal arrives in the form of a plane wave, which in the worst case occupies the whole operating frequency band. The planar wavefront falls on the staring array as does any incoming, distant-source signal. Its direction of arrival cannot be disguised, and assuming that its SNR is substantial; its directional error will be small.

Provided no saturation, the modelled receiver stores the in-band waveform incident in that direction as a function of delay and the pulse time sequence; $(\mathbf{S}_{int}(\alpha, \varepsilon, \tau, \mathbf{T}))$. Figure 7.10A shows amplitude modulated sinusoidal interference at the beam level, modelled in a beam pointing directly at the source. Figure 7.10B illustrates the signal at a single receiver channel (in this case 1 of 1 536 within the ATC receiving array) to generate that beam result, and Figure 7.10C shows the result after channel-wise, complex adaptive cancellation. Signals arriving in other directions are present in $\mathbf{S}_{int}(\alpha, \varepsilon, \tau, \mathbf{T})$, but only via sidelobes, and will not be cancelled.

Following cancellation, the channel is again dominated by modelled thermal noise. Here as usual in-phase and quadrature components are red and yellow, and the amplitude is plotted in blue. This is a test with modelled channel errors but uses a method that is consistent with a real array. Figure 7.11 illustrates a more complex interfering signal and its modelled suppression.

7.5.2.4 Radar interference

A radar sensor can suffer interference from other radar sensors within its frequency band, as has been introduced in Chapter 6 with respect to interference identification.

Such interference can occur from scanning or staring radars. Scanning radars may cause saturation of a staring array, but for short periods within any HSR CPI, in which case the effect may be accommodated.

HSR, deploying receiving beams in all directions, should not suffer from sidelobe distraction. It should receive and locate high-amplitude radio, radiodetermination signals or countermeasures, classify them appropriately (Chapter 6) and use similar methods as in Section 7.5.2.3 to suppress them by signal-specific, channel-wise subtraction.

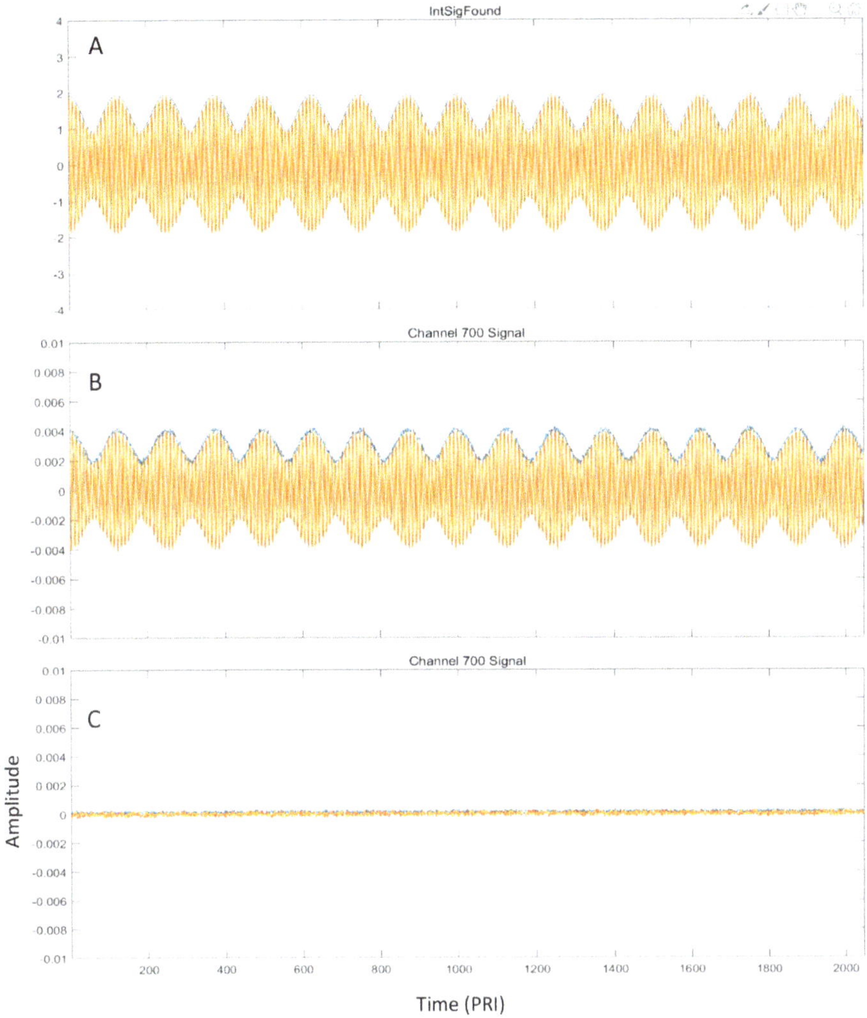

Figure 7.10 A, B and C illustrating adaptive nulling of an amplitude-modulated signal

7.5.3 Suppression in the presence of multipath

The previous discussion refers to the case where single plane waves are incident from target to the receiving array. Without multipath, these interfering signals may be accurately nulled by a HSR receiver, with residual errors dominated by phase noise, and at a level that should leave performance close to specification.

This is realistic for interfering sources in the absence of multipath, but it cannot cover all cases of interference, in which multipath propagation plays a significant part.

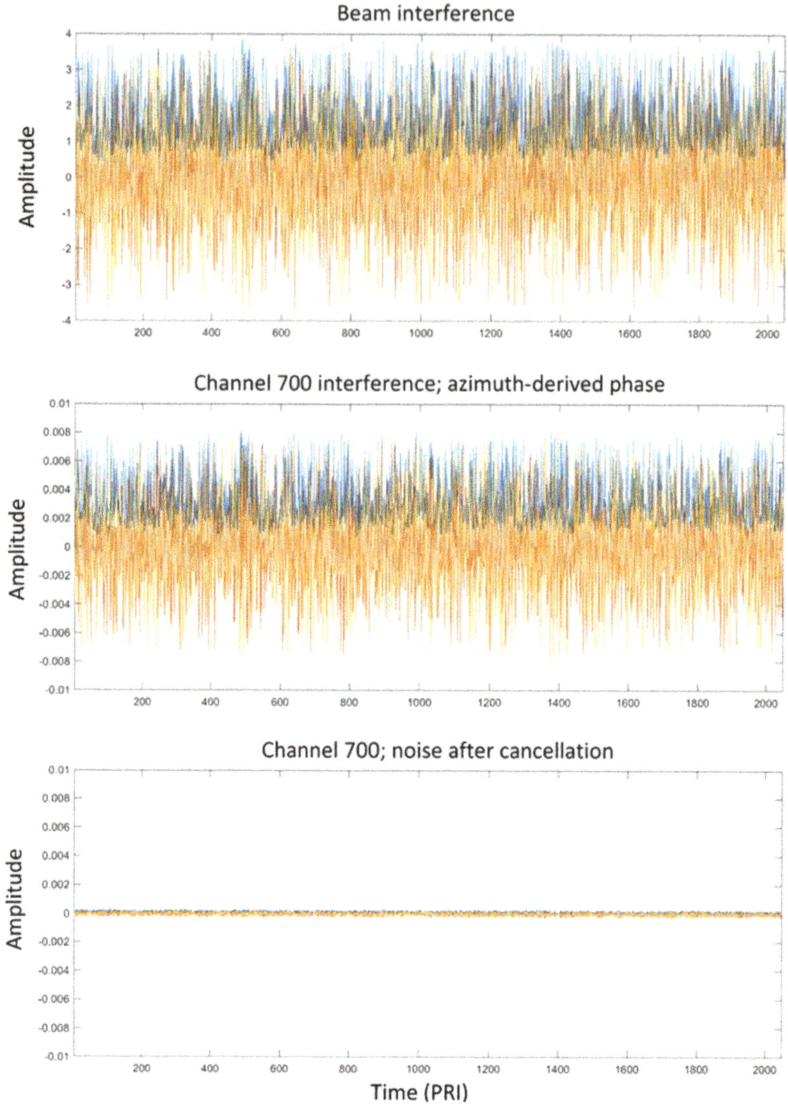

Figure 7.11 *A, B and C illustrate the nulling of a random phase-modulated broadcast signal, its reception at an array element and the result after signal-specific subtraction*

For a low-elevation source, propagation to the HSR receiver includes a direct component plus a surface multipath component. To assess the effects of interference from low-altitude or surface sources we must include those of surface reflections (multipath D in Figure 3.4/Figure 7.2).

The arriving wavefront consists of at least two near-planar waves, one direct from the source; the other a grazing reflection from the earth surface with a small delay after the direct wave that is a function of target elevation and of height on the array. This adds complexity to counter-countermeasures.

The time series received by the HSR from an interfering source, and to be used in its suppression, is the complex time-domain waveform found in the sequence of range gates, after beamforming in the azimuth and elevation directions, for every pulse interval, and including the effects of both direct and multipath propagation. The waveform to be used is that received in the peak azimuth/elevation beam. Its direction of arrival in azimuth is estimated by conventional beamforming, but in elevation the method indicated in Section 7.3.2 may be preferable.

This direction can be used to calculate phase offsets expected for vertically spaced array elements. However, the primary effect of surface multipath will be to add an error term in the complex amplitude of the same waveform arriving at each element. The gain-corrected version of the signal received from the direction of interference will still be an amplitude- and phase-modulated version of the peak beam signal. It can still be used in achieving minimum channel residues if treated element-by-element, and distinguishes between signal-specific suppression and adaptive nulling.

This form of suppression needs to occur within the time frame of one pulse interval, and after beamforming – i.e., within 1 ms, and is expected to add several billion arithmetic operations for each processor handling CWVS beamforming.

7.6 Processing burden

Extended computer processing results in increased cost. At the same time, appropriately extended processing, exploiting all the arriving signal information, will necessarily lead to improved information delivery, which is the radar's fundamental purpose.

The process functions to be included in this estimate of the burden for both SRC and ATC are:

1. channel-wise time domain assembly
2. channel-wise Fourier transform
3. time-domain beamforming
4. beam-wise Fourier transform
5. RAED feed, access and management
6. VH formation and cell discovery.

The process burdens associated with each function are estimated as follows in Table 7.2.

SRC uses a 8 × 16-channel receiving array at S band. Each channel is sampled at 3 MHz, yielding 100 range gates spaced at 50 m, up to 5 km. ATC uses a 12-facet array of 8 × 16 receiving elements. Each channel is sampled at 1 MHz, yielding 900 range gates spaced at 150 metres, up to 135 km.

Table 7.2 Base processing rates and memory capacities for SRC and ATC

Processing and memory for SRC and ATC				
	SRC	**SRC**	**ATC**	**ATC**
RAED segments	**GFlops**	**MBytes**	**GFlops**	**GBytes**
Channel TD	0.17	210	2.765	23
Channel FD	2.00	210	30	23
Beam TD	87.00	839	7373	8
Beam FD (FFT of TD beams)	8.00	839	11	8
	GFlops	**GBytes**	**GFlops**	**GBytes**
Totals	97.00	2	7417	61
VH formation	**GFlops**	**GBytes**	**GFlops**	**GBytes**
VHs TD	3.00	6.71	46	377
VHs TD	3.00	6.71	46	377
VH total	6.00	13	92	755

These rates are based on raw calculations (complex multiplies required for CWVS, etc.), and must be multiplied by a significant factor (probably more than 2 and less than 10) to estimate the effective total burden with Range Walk, interference suppression and target capture. ATC may require 50 trillion floating point operations overall, at a cost in the region of one quarter of the array components. Shared between 12 GPUs, this represents 4 TeraFlops and memory capacity of 70 GB per unit.

For either SRC or ATC, the demands arising from increasingly detailed target analysis are different from those imposed by, for example, a need to improve range or resolution. Those needs must result in a re-scaling of the radar, potentially affecting all its components, whether transmitter power, antenna size, operating frequency, number of array elements or operating bandwidth, all of which are increasingly expensive with time. Provided that the CVoR is properly observed, greatly improved target analysis can be achieved by appropriate processing, which continues to get cheaper with time.

The processing required for beamforming, Doppler integration and VH-oriented downstream data ordering and analysis is compatible with highly parallel processing engines such as graphics processors. This technology has been advancing rapidly, and graphics processor cards are now available quoting many trillion floating point operations (TeraFlops, 'TF') for less than £1 000.

The requirement for SRC is easily met by a single computer with a high-speed parallel-processing engine such a s GPU. ATC is unlikely to be achieved by a single machine. A set of similar machines would be networked with the receiving array, keeping the burden for each GPU, which needs to consist of largely non-conditional processes, to a few TF.

Processing burdens for beamforming and Doppler integration in a HSR for this application and at a number of different operating frequencies are shown as examples in Table 7.3. To maintain sensitivity the effective area of the receiving array is

Table 7.3 Complex-weighted vector sum beam-forming process burdens vs operating frequency at the scale of ATC

	L band	S band	C band	X band
Frequency (MHz)	1 350	2 700	5 400	9 500
Array rows	32	64	128	256
Array columns	96	192	384	768
Azimuth beams	200	400	800	1 600
Elevation beams	16	32	64	128
Range gates	750	750	750	750
CPI (seconds)	2	2	2	2
Min. ops/CPI (TF)	15	240	1 920	3 840

maintained constant in the table. As the frequency increases, angular resolution will improve for the same array scale.

There are different ways of balancing the effectiveness and the cost of such a radar and choosing its design parameters. Here the choice of operating at lower or higher frequencies may indeed be driven by the large processing burden. If the primary requirement is angular resolution rather than sensitivity, performance may be achieved with decreasing sizes of array at higher frequency; however, to maintain sensitivity as well the power of transmission must be increased.

Processing aspects other than beamforming, and in the presence of adverse target conditions or interference, are expected to multiply the total burden. The scale of increase will depend on the target specifications but will be proportional to the square of the operating frequency.

7.7 The balance of HSR vulnerability and resilience

This chapter has shown that potential vulnerabilities in the development and cost-effectiveness of staring holographic radar can be overcome when continuous periods of observation meet the constraints of the EUNIT, and solutions can be generated by sufficient processing resources. Beyond overcoming the pitfalls, these conditions suggest ways in which radar surveillance using HSR may transcend the conceived purposes and capabilities of the beam-scanning form of radar.

This indicates a direction of travel for HSR technology such that the normal surveillance process of detecting, tracking and reporting targets of interest to the user may be updated. A process of cell discovery, target capture, information ordering and analysis may be introduced, with the aim of assessing all aspects of the VoR with regard to the surveillance requirements. Issues of target dynamics and surface and target complexity and classification may, as predicted from the electromagnetic uniqueness theorem, be accessible to an HSR, enabled by methods such as vector histograms.

Table 7.4 Outline assessment of field deployment of radar developments (2014)

Modes	Decade	JSTARS	Global Hawk	LSRS	F-22
Stretch SAR	1950s	Y	Y	Y	Y
Vehicle GMTI	1970s	Y	Y	Y	Y
Maritime MTI	1980s	Y	Y	Y	Y
Inverse SAR	1980s	Y	Y	Y	Y
Bi/Multistatics	1990s	X	X	X	X
Real time ECCM	1990s	X	X	X	X
MIMO for GMTI	1990s	X	X	X	X
Sense and Adapt	1990s	X	X	X	X
Feature-aided track	1990s	X	X	X	X
Target-matched WF	1990s	X	X	X	X
Dismount GMTI	2000s	X	X	X	X

Investment in improvements in radar technology has increased rapidly in recent years. However, in some cases the acceleration in performance of fielded systems is not seen as matching these investments, which have tended to focus on radar electronics yielding scanning and signal processing agility. The aim has been to maximise attention on targets of interest while dividing the limited time between multiple functions.

This book addresses the possibility that a logjam has arisen in radar development not because the radar is insufficiently agile but because the scanning mode of operation, enforcing intermittent target observation, itself excludes most of the information that is to be found encoded in continually scattered target returns.

Table 7.4 is based on an illustration by the US Defence Advanced Research Projects Administration at a radar conference in 2014 of the reduced rate of implementation of 'Advanced RF Modes' in radar. The present situation is an advance on this, but its message is significant.

A notable aspect of this is that prior to the 1990s the focus was on intensive decoding of incoming complex signals in SAR imaging, ISAR and pulse-Doppler radar. Since then the focus has shifted more to the structure, agile electronics and signal processing methods within the radar, rather than to enhancing the decoding of input information. For HSR, performance comes with continuity of input information.

We shall see in Chapter 10 that aspects of HSR architecture and performance may lend themselves to operation of HSR not as single, independent surveillance sensors, but as nodes in a network of co-operative sensors, distributed over an extended field of surveillance. We shall show that under certain conditions such networks can avoid the potential for mutual interference that arises between neighbouring primary radars, and, in a similar way to the potential for exploitation of Doppler Walk and multipath, performance may be enhanced over that of singular sensors.

References

[1] Scheer J.A. (ed.). *Coherent Radar Performance Estimation*. Artech House Publishers; 1993.

Chapter 8
Coherent target histories

Chapter 3 introduced the physical basis for the expectation that EM signals scattered by objects within a Volume of Regard, irradiated by a known, persistent EM field, encode and communicate information about those objects beyond their presence and position.

Chapter 6 showed how the presence or absence of that information can be decided for each resolvable spatial cell in the volume and gave evidence that a significantly extended CPI goes beyond the recovery of sensitivity sacrificed by spreading the radiation throughout the volume, to yield finer details, with greater fidelity when finding, classifying and reporting a radar target.

Chapter 7 shows where challenges occur, whether and how those challenges can be met when the contents of the VoR change and combine under the constraints of Newtonian physics and dynamics, and examples where challenges may morph into opportunities when these changes and combinations are recognised through the analysis of the signal contents.

The evidence indicates that continual, regular observations, made available by staring radar, offer benefits in signal quality and surveillance capability. The next step is to explore what is to be gained by further continuing coherent observations. This chapter focuses on extended observation of targets known to be of interest, and their potential for deeper analysis, including target dynamics and suppression of repetitive clutter. In Chapter 9, we shall consider extended observation for analysis and suppression of sources of clutter and satellite returns by means of multipath propagation. In Chapter 10 we shall explore extending the spatial scale of coherent surveillance.

8.1 Cell status and classes of information

Chapter 6 outlined the analysis of signals acquired during the formation of resolution cells defined in range, azimuth and elevation, and with both time- and frequency-domain signal datasets. Vector histograms were used as non-conditional means of directing information-rich radar signals, providing a means of cell classification and the discovery of target-bearing cells.

Single-CPI datasets allow each cell to be classified as members of eight classes, repeated here for reference:

1. Noise plus static clutter within specification (NSC)
2. Noise out of specification (NXS)
3. Noise plus static clutter-related degradation (NCD)
4. Noise plus repetitive clutter degradation (NRD)
5. NSC plus Newtonian targets (CST) and with multipath (CMT)
6. NCD plus Newtonian targets (CDT)
7. NRD plus Newtonian, dynamic targets (RDT)

The reference here to 'Newtonian targets' places the focus on signal information consistent with the dynamics of targets with persistent mass and finite aerodynamic agility and motive power.

This chapter focuses on cells qualifying as CST, CDT or RDT, and on the effects of continuing coherent processing of targets. NXS cells contribute to system performance monitoring but do not contribute to target capture and reporting and have been addressed in some detail in Chapter 6. Chapter 9 will focus on CMT cells.

8.1.1 Concatenation of CPIs

A HSR illuminates its whole CVoR continually. This allows for an extended CPI to be assembled and processed to achieve the required sensitivity and to accommodate target dynamics, as described in Chapters 6 and 7. It also allows information acquired during one CPI to be reinforced or adjusted by extension. Subject to the capability for extracting dynamic trajectory data and extending Range Walk and Doppler Walk processes, overall performance may then be improved with respect to fading and multipath effects.

These extended processes may require conditional selection, and none is likely to be applied to all resolution cells in parallel. They may be applied for cells already discovered, during the parallel cell discovery process, to contain signals indicative of targets of interest, or for cells known to contain either items of fixed and repetitive clutter or qualifying targets of interest.

CPI concatenation assumes that signal data are available for all cells in an unwindowed form such that a current cell can be extended or shifted over one or more past CPIs, and its content are evaluated. Further concatenation is then possible along with discovered target data and trajectory focusing parameters.

A key consideration for concatenation of HSR data is how the result is to be accommodated in and accessed from memory. The volume required for all resolution cells has been estimated at 30 GBytes for SRC for a half-second CPI, and 1.7 TBytes for ATC for a 2-second CPI. This would determine the scale of the **RAED** data structure considered below and in other chapters.

To multiply this whole volume for each extension of the CPI, for all cells, appears demanding and needs a limit. Functions that benefit from CPI concatenation probably need to be restricted to cells known to include persistent target clusters (as shown in Sections 8.2.1 or 8.4.1), isolated, intense clutter features or targets qualifying for precision analysis, with memory apportioned appropriately.

8.1.2 Concatenated cell processing

A staring radar generates Doppler spectra with thousands of bin indices and an extended history whose retention of phase information makes it adaptable to multiple forms of analysis: for micro-motion, imaging, dynamic analysis, classification and potentially identification. Aerial target trajectories are guided by their pilot or autopilot or, for unmanned air systems (UASs), interact with their controller. Their dynamics are constrained by Newtonian physics, their airframes and appropriate propulsion and control, but allowing for variability due to air motion.

Chapter 6 describes how cell signal analysis leads to the identification of cells that appear to contain targets of interest (classified CST, CDT or RDT). For CST cells, adequate performance in target capture and reporting is expected. For CDT, degradation arises from large static clutter combined with phase or gain noise, as also described in Chapter 6. The benefits of cell concatenation are expected to arise primarily for targets and clutter in the RDT class. Two examples are where (1) traffic concentrations, which essentially consist of persistent, mobile targets that as a whole are not of interest, and (2) certain resolution cells are affected by fixed but repetitive, moving clutter such as wind turbines.

Here a form of target and clutter tracking is introduced in which concatenated signals are resolved, as continuing complex amplitude sequences, which are consistent with a scattering object likely to obey (1) Newtonian dynamics, (2) extended Nyquist sampling criteria and (3) to be within or to overlap with the target specification. This is in contrast with the use of a tracker to form associations and interpolate between temporally separated, non-coherent positional samples.

Over an extended period, resolution cell contents provide the basis for an assessment of the CVoR environment as a whole. A continuing summary of cell analysis and discovery will make it possible to identify volumes that are subject to wind turbulence (including approach and departure sectors), to bird populations, to effects of buildings and to surface transport, and at closer ranges to presence and movements of unmanned air vehicles (UAVs). This assessment can be maintained as a continuing function of the staring surveillance radar, and some aspects will benefit from extended CPIs.

Once a first analysis has been carried out, leading to the discovery of resolution cells that conform with models of a target of interest, including dynamics, greater sensitivity and higher resolution can be achieved by further extension of the data, including direct analysis of aircraft manoeuvres. A short example of CPI concatenation is illustrated in Figure 8.1, in which an aircraft is modelled during an evasive manoeuvre. As a first illustration of CPI concatenation, Figures 8.1–8.4 refer to a simple aerial 'jink', lasting several seconds, in which the target response includes such a manoeuvre.

8.1.3 Aerial manoeuvring

In the example, the aircraft is modelled flying with a constant positive (receding) radial speed **Vrad** for 2 seconds, then radially decelerates (by a roll manoeuvre) for 2 seconds towards the radar and, then returns by a reverse roll to its previous

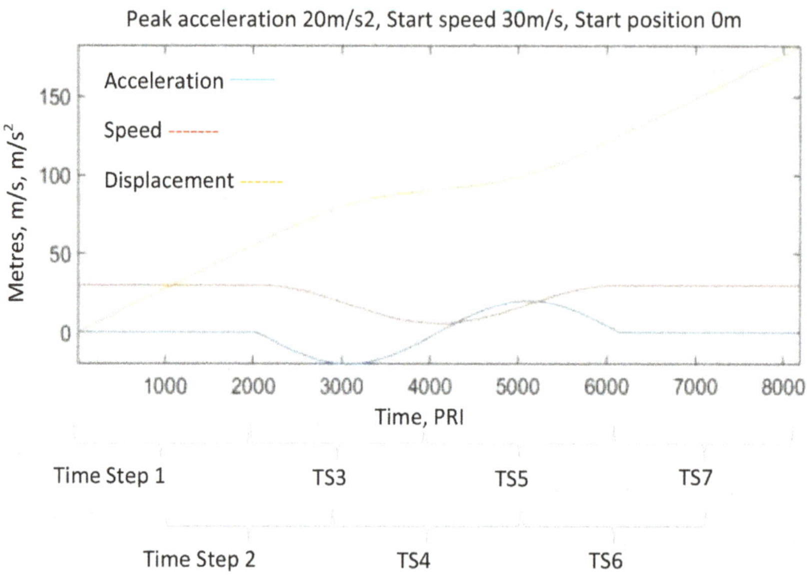

Figure 8.1 Aircraft manoeuvre: acceleration, velocity and position vs time

direction and speed, but displaced. This sequence occupies four CPIs as defined for the ATC configuration. This example is based on the ATC concept since its PRI is significantly longer than for SRC, making it more susceptible in principle to dynamics. Here four 2 048-point sequences are concatenated to form an 8-second sequence, and spectra are derived for 7 successive time steps, each lasting 2 seconds and overlapping by 50%.

In Figure 8.1, the abscissa is the time index in terms of the number of concatenated pulse intervals. The ordinate is shown in metres (displacement: orange), m/s (radial velocity: red) and m/s^2 (acceleration: blue). The aircraft accelerates at a peak of +/−2G, across its track.

Signals associated with this manoeuvre are illustrated in Figure 8.2A-G, the modular Doppler spectrum, Figure 8.3A-G, the VH formatting, and Figure 8.4A-G, estimates of radial speed and acceleration derived from modelled signal values for the 7-step, 8-second period. The target model is set at a nominal level close to a minimum ATC target. The total signal power margin over the cell noise power is within +/−2% for each case of acceleration through the sequence; it does not vary significantly with acceleration. With respect to cell discovery, although the peak amplitudes are reduced at Doppler indices within the acceleration profile, Newtonian motion assures that they are contiguous, and that the total power exceeds the total noise by the same margin. The accelerating Doppler does not itself yield random phases or multiplicative noise.

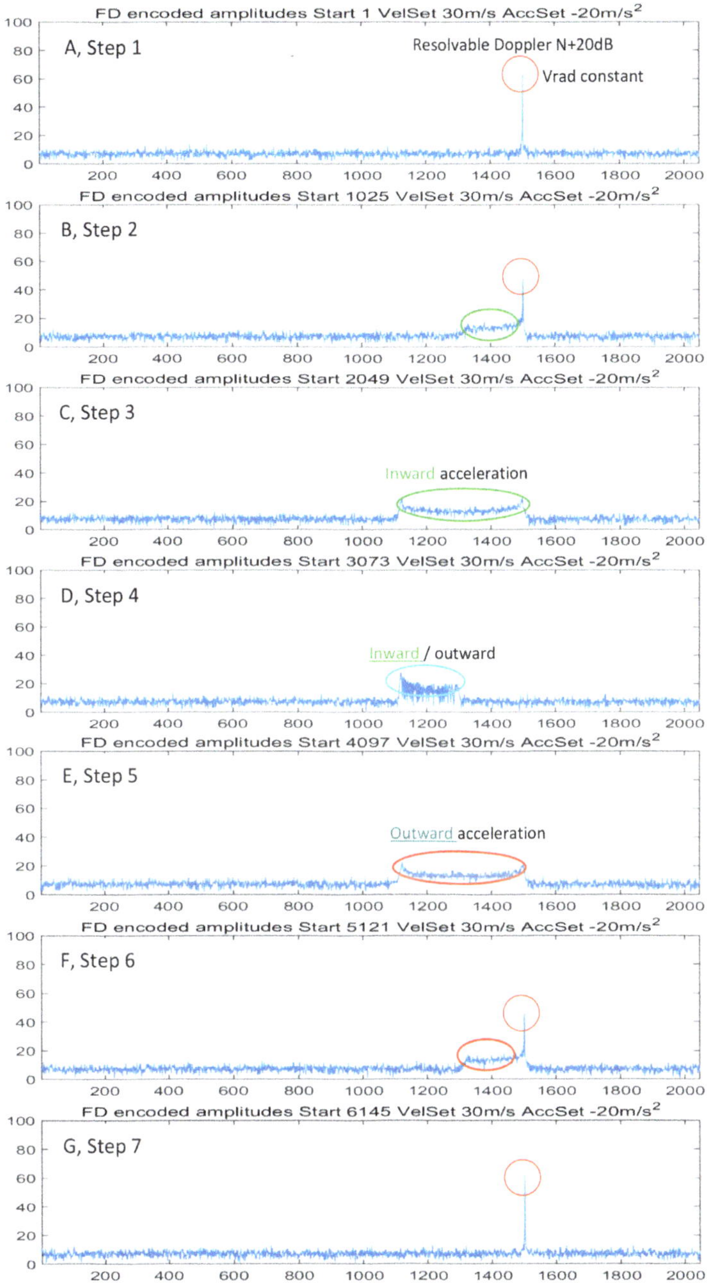

Figure 8.2 *A–G. Modelled, modular Doppler spectra for a target receding at 30 m/s, then undergoing an inward deviation of about 50 m over a 4-second period. Signal energy is conserved, the minimum target remains detectable and the manoeuvre can be measured.*

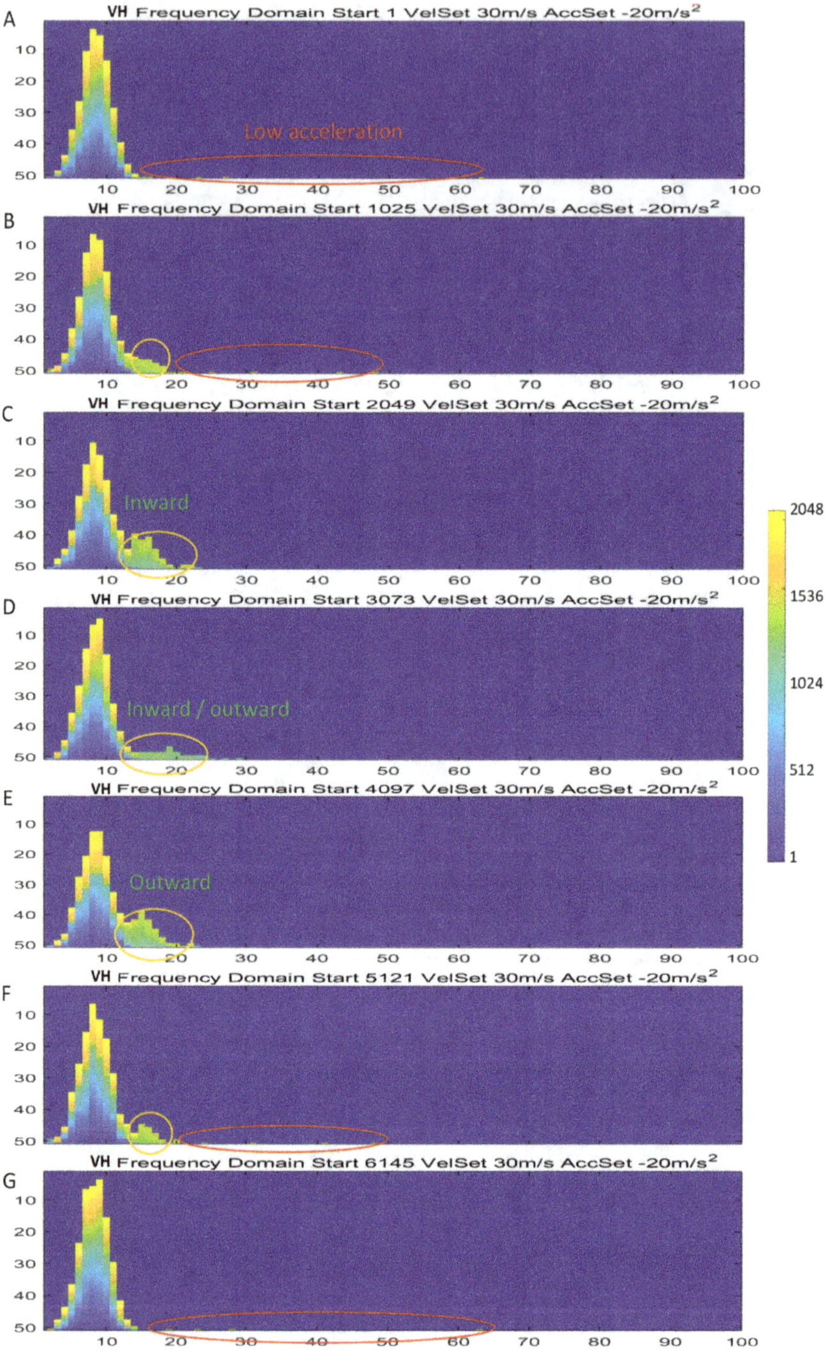

Figure 8.3 A–G. VH formats for distribution analysis, spectrum reference and dynamics estimation

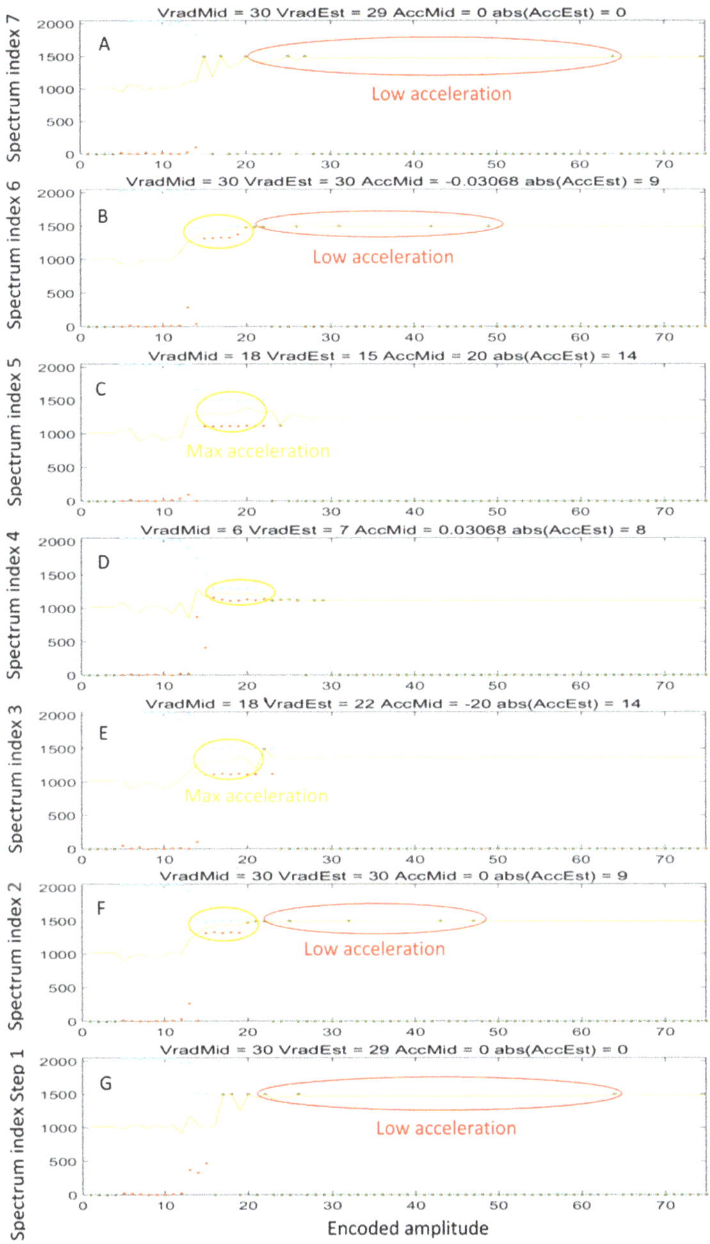

Figure 8.4 *A–G. Dynamics analysis for 8-second concatenated signal history, derived from the VHs in Figure 8.3. Signals at abscissas 1–14 are noise-dominated, with a mean Doppler abscissa near 1 024. Those from 14 to 22 illustrate measurable acceleration over each span; higher abscissas represent the single Doppler peak.*

The estimates of velocity and acceleration are derived from signal levels as functions of frequency. **Vrad** is derived from the mean frequency of the excess signal, and the acceleration is derived from the width of the excess signal spectrum. Based on these estimates, a precise interpolation of **Vrad** and **Arad** can also be found with a small range of tests.

For time Span 1, the acceleration is zero and receding speed is constant at 30 m/s. Acceleration begins during Span 2; a sinusoidal excursion inward after 2 seconds, reaching a peak of -20 m/s^2 after 3 seconds and reversing after 4 seconds, with a peak outward at 20 m/s^2 after 5 seconds, and returning to constant speed after 6 seconds. This manoeuvre results in an evasive displacement close to 50 m.

Random, short-duration disturbances emulating air turbulence are included in the model. They show that disturbances of small fractions of a wavelength will be distinguishable from the effects of additive system noise and would have characteristics similar to those of phase noise but do not significantly affect the estimates.

The plots in Figure 8.4 are first and experimental visualisations of the cell information content with respect to acceleration of a single target, derived from the VH. The abscissa is encoded amplitude as in the vector histogram. The ordinate is the Doppler frequency. Cyan points are plotted at the maximum frequency represented at a given amplitude, and red points at the minimum. The orange line indicates the estimate of a mean Doppler shift at each amplitude, and the acceleration is indicated by the spread of Doppler values at amplitudes above the noise.

The encoded noise spectrum begins at amplitude 4. For amplitudes within the noise spectrum (4–14), maxima are close to index 2 048 and minima near 1, illustrating that noise is present and can dominate at any apparent Doppler frequency.

At amplitudes above the noise spectrum (14–75) in Step 1, there is a single, narrow peak at 65 near Doppler index 1 500.

During Step 2, the turn begins, the 1 500 peak index is reduced from 65 to 50, and the spectrum spreads downwards over 200 indices, at reducing amplitudes as acceleration increases.

Step 3 includes continuing inward acceleration, reaching and passing the peak; however, its signal energy is spread over almost 400 indices between amplitude indices 14 and 22 and maintains full detectability against the mean noise over that spread.

Step 4 sees the reversal of the accelerating Doppler signature, with the maximum near Doppler index 1 100.

Steps 5, 6 and 7 illustrate the reversal of the manoeuvre.

Figure 8.4A–G shows, in the upper title, the speed value set at the mid-point of the span and the estimates derived from the figure data for speed and acceleration.

This illustrates work in progress on methods to extract a full dynamic measure of target trajectories. It shows that with an extended CPI, the process retains overall sensitivity, and the concatenated signal record provides the ability to estimate target dynamics.

A wider range of manoeuvres is considered and illustrated in Appendix 1.

The ability to measure small-scale deviations from a smooth trajectory (at the scale of centimetres) will have consequences where the requirements for surveillance go beyond traffic control and enter the field of aircraft performance and aircrew monitoring. More significantly, a manoeuvre such as that depicted will involve steep bank manoeuvres, which themselves will provide an entry, for a staring radar, into target imaging via an ISAR process.

This example aims to illustrate how extended target observation, far from being a risk related to target decoherence, can yield additional performance and functionality. If an aircraft accelerates or deviates, the user needs to follow it persistently rather than be blind to it for several seconds.

The next step is to consider how the HSR interacts with the surveillance environment it encounters within the CVoR.

8.2 The VoR environment (NSC)

The VoR covers many resolution cells. For SRC and ATC, that number would be 12 800, and 3 552 000, respectively. The primary objective is to find the status of each cell and derive the conditions under which it is operating, as discussed in Chapter 6.

Moving objects that do not constitute RToIs but also do not introduce random noise may nevertheless scatter returns from the radar transmission and may therefore be relevant either to the airspace or to the operation of the radar. Examples include:

1. Road and rail traffic and transport infrastructure
2. Sea clutter
3. Weather targets and turbulence
4. Birds.

The effects of high-intensity scattering from clutter objects in the sense of multiplicative noise (phase noise or gain noise) have been discussed in Chapter 6 and a means of recovery is introduced.

'Objects' that affect cells well resolved from the surface include turbulent weather features and birds.

Features that result in Doppler signals that overlap with the radial speeds of aircraft are surface vehicles, sea clutter and wind turbines.

8.2.1 Road and rail traffic

The SRC configuration, with a need to observe objects near the ground surface, with small scattering cross-sections, low speeds and potentially high agility, needs to discriminate reliably between RToIs and TNoIs. Here an airport environment presents a good example for the necessary performance aspects, and we consider the case of a roundabout or traffic circle in an airport neighbourhood.

These traffic flows are dynamic in every case; some vehicles stop, but all accelerate and turn in one of a number of ways, with the objective of taking one of the route choices quickly and accelerating into it, subject to the rule of the road and a typical speed limit of 30 miles per hour.

Road vehicles are large compared with the half wavelength at the frequencies of either ATC or SRC, and their scattering cross-sections may vary from about 1 (for a motorcycle) to 100 m^2 (0 to +20 dBsm), with deep nulls and strong peaks. A total variability of 40 dB in RCS is very probable.

For a HSR near the airport, a low-elevation spatial resolution cell will contain the junction, and many targets in this range will occur, with several at any one time. A key variable will be the range of radial speeds and accelerations as seen from the radar. Although the traffic flow is not accurately controlled, the motion of each vehicle is constrained by weight, power and safety regulations.

We have shown that radar signals scattered by moving, inertial targets carry coded information about their motion, and we seek to establish whether and how a staring radar such as SRC might succeed in resolving traffic targets from UAVs.

A traffic scenario with a car and motorcycle at a road junction is illustrated in Figure 8.5, and a set of positions, speeds and accelerations are shown in Figure 8.6. The car starts at 25 m from the roundabout receding towards it at 15 m/s. It starts to slow at 4 seconds, peaking at −7 m/s^2 (0.7G). The motorcycle starts at 90 m and 20 m/s, also slowing at up to 7 m/s^2 from 6 seconds. Both stop and accelerate away. Here, we illustrate a process of acceleration matching for the car by successive phase modulation profiles. The Doppler spectrum is shown in Figure 8.7, with acceleration compensation in Figure 8.8 and showing the car's acceleration, after matching, in Figure 8.9.

The process of matching acceleration using phase modulations of the complex waveform used above is accurate but computer-intensive, especially when variable acceleration must be accommodated. The acceleration-spread spectrum for the car is shown in Figure 8.10. Here the car component is perceptibly broadened, but VH formatting allows sensitivity to be maintained and speed and acceleration to be estimated (Figure 8.11).

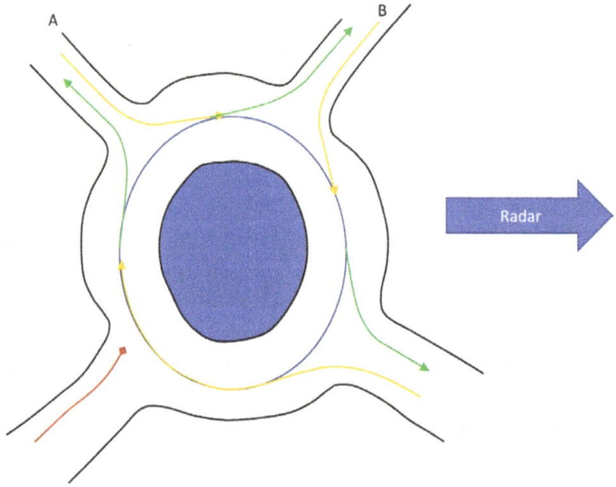

Figure 8.5 Positions and flows of traffic at a road junction

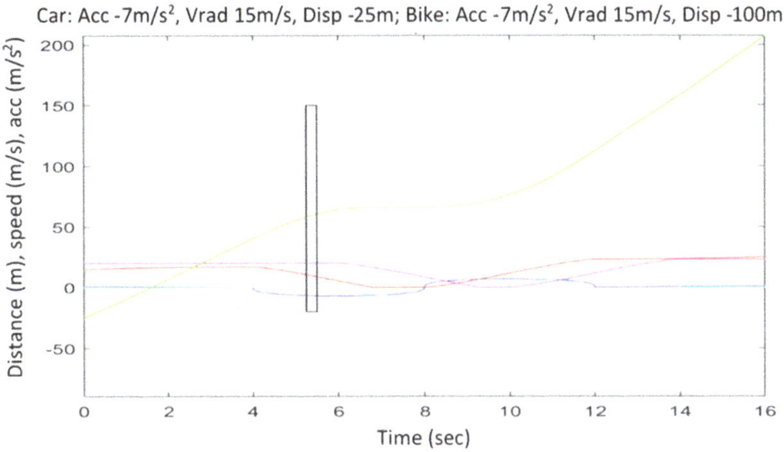

Figure 8.6 *Modelling acceleration, speed and distance for two vehicles at the roundabout. The boxed outline identifies the time span referred to in the discussion. Acceleration: Blue – car; Cyan – bike. Speed: red – car; magenta – bike. Distance: green – car; Lt blue – bike.*

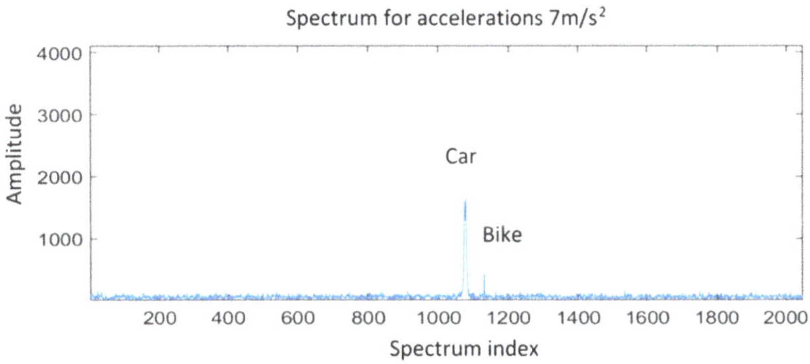

Figure 8.7 *SRC Doppler spectrum over a CPI after 5 seconds is illustrated above. The car has begun to decelerate (red trace in Figure 8.6) but the motorcycle, with a smaller cross-section, has not.*

Figure 8.12 uses the same depiction of Doppler values and spreads as in Figure 8.4A–G, illustrating the estimation of speed and acceleration, in this case of the major target, in a way that is affordable for use in every resolution cell. This format is adaptable for multiple Doppler-resolved targets.

For a resolution cell that is expected to generate excess target activity, such as the road junction in this SRC case, the motion estimates from the VH can then be used to narrow the search volume for complex waveform compensation as in Figures 8.7–8.9, also allowing for variable acceleration.

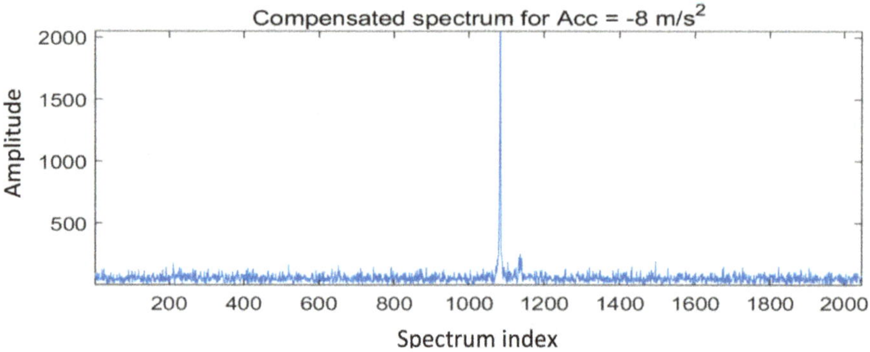

Figure 8.8 *After a variable-acceleration phase compensation at −8 m/s², the car is enhanced and the motorcycle diminished in the Doppler spectrum*

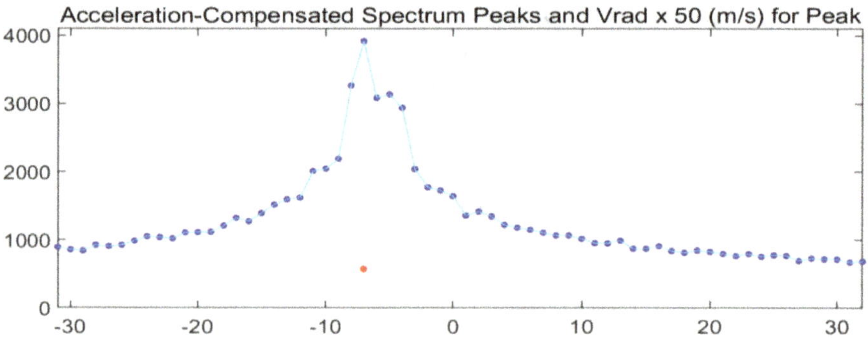

Figure 8.9 *The red dot denotes* ***Vrad*** *~12 m/s at −8 m/s² target acceleration match*

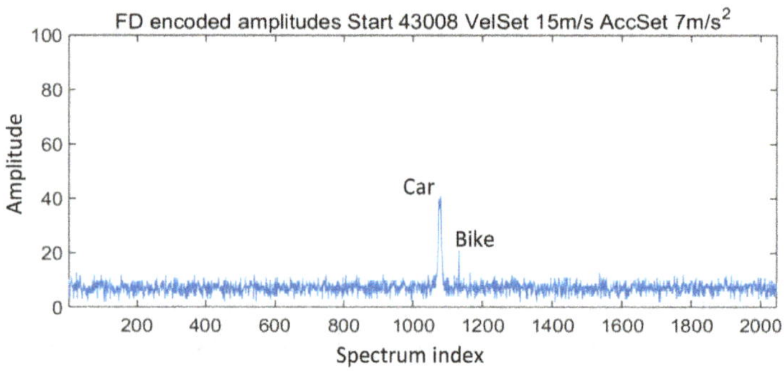

Figure 8.10 *The amplitude-compressed vehicle Doppler spectrum*

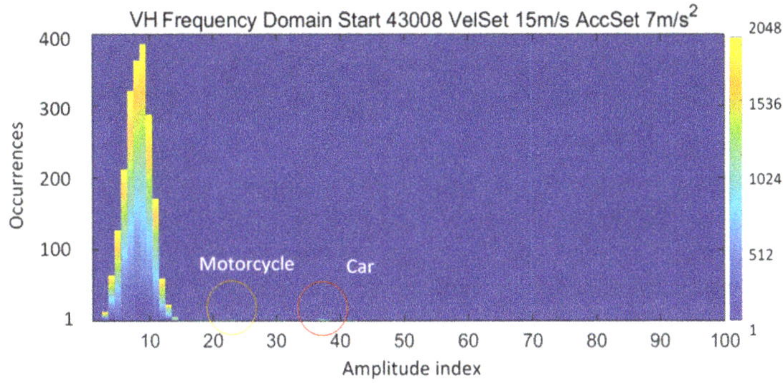

Figure 8.11 VH for amplitude-compressed vehicles

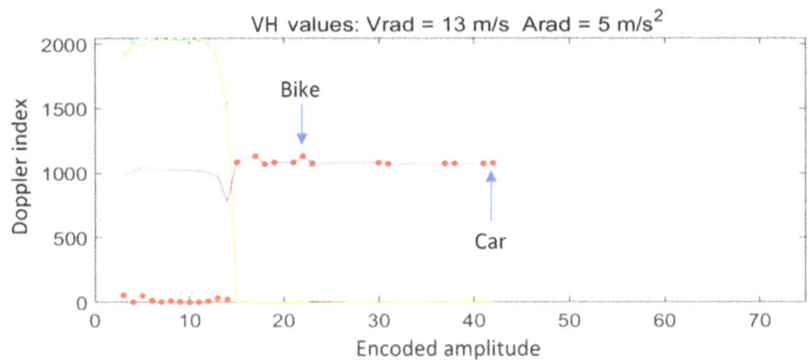

Figure 8.12 Estimation of speed and acceleration for the car target

Continuity of observation between CPIs in this extended sequence allows increasing intensity of surveillance in locations of known interest and characterisation of particular clutter features.

8.2.2 Sea clutter

The ocean surface is rough and mobile and is well known to present scattering targets that can be reported by radar. The Huygens model used here to simulate scattering targets is not well-conditioned for the huge extent of scattering from the sea and needs research to bound the form of its radar returns; however, these returns are known to originate at the surface to cover wide areas, to occupy a relatively small Doppler region (when measured from a static HSR) and to follow a K-distribution within relatively small bounds.

These characteristics make sea clutter relatively amenable to suppression by a staring radar or else to exploitation by generating imagery or quantitative surface analysis as part of marine surveillance.

The issue of sea clutter will not be addressed here in further detail, but experience with HSR staring over the sea has shown that its vertical accuracy and

Doppler resolution can be effective in suppressing sea clutter under a wide range of circumstances.

8.2.3 Weather features

Large, smoothly moving volumes of air have very small 'radar cross-section densities' (radar cross-section per unit volume) at any radio wavelength scale at which air motion is smooth. Turbulent air has a cross-section density that varies with wavelength, with a peak where a half-wavelength converges with the diameter of features present in the cell. The features themselves are highly variable but at some scales affect the use of airspace significantly.

Water droplets also scatter EM waves. Airborne droplets have an insignificant scattering cross-section for half-wavelengths far greater than the droplet diameter. Large droplets fall in the atmosphere in the form of horizontal oblate discs and scatter at a maximum when their diameter equals the RF half-wavelength in water (one-ninth the figure in air), their cross-section reducing for wavelengths greater than the peak, inversely as the fourth power of wavelength. Rainfall consists of random and randomly fluctuating volumes of droplet populations; its cross-section density varies according to that population and frequency, while its loss factor in propagation varies with its range extent, and with the cross-section density plus absorption losses in the fluid.

Table 8.1 illustrates the effects of weather on radar, in terms of attenuation and backscatter, and the reason why long-range surveillance radars operate at lower frequencies.

Radars applied to weather monitoring tend to be at a higher frequency and with dual polarisation to combine sufficiently long range with sensitivity to smaller droplets, and discriminating a wider range of weather conditions. HSR can be designed for weather surveillance, using polarimetric methods.

Signals scattered by atmospheric effects such as translating or blustery winds and turbulence, associated clouds and rainfall have a range of Doppler characteristics. With a short dwell time, these tend only to differentiate between moving and stationary targets, but an extended CPI allows the actual distribution of speeds to be evaluated.

For SRC or ATC, these features are usually part of the TNoI population, discriminated by randomly spread Doppler shifts, but the range of application will be greater where an extended CPI is possible. High-performance weather radars such as NexRad or the research radar at Chilbolton, UK, use a narrow beam approach with a long scan period to resolve Doppler profiles, but future weather mapping would benefit from a staring configuration.

8.2.4 Birds

Birds and flocks of birds do not typically have radar cross-sections that compete with manned aircraft. They can scatter to an extent comparable with UASs, and means are necessary to discriminate between these targets. Birds can also qualify as targets of interest to specialist observers.

Table 8.1 Attenuation and backscatter versus frequency, rainfall rate and range

				Attenuation (dB)								Backscatter cross-section per resolution element vs range and rainfall rate							
S band					RR (mm/hr)							RCS/m^3	6.31E-10	1.00E-08	1.58E-07	2.51E-06	3.98E-05	6.31E-04	1.00E-02
11	2.73E+09	0.01		R (km)	1	2	5	10	20	50	100	km	Cross-section per res element						
RCS scaling vs S band		0.02	3.43E-04	1	2.07E-02	2.13E-02	2.33E-02	2.64E-02	3.25E-02	5.05E-02	7.97E-02	1	1.46E-04	2.31E-03	3.66E-02	5.80E-01	9.19E+00	1.46E+02	2.31E+03
1	0	0.04		2	4.14E-02	4.27E-02	4.65E-02	5.28E-02	6.51E-02	1.01E-01	1.59E-01	2	5.83E-04	9.23E-03	1.46E-01	2.32E+00	3.68E+01	5.83E+02	9.23E+03
Az beam		0.1		5	1.03E-01	1.07E-01	1.16E-01	1.32E-01	1.63E-01	2.53E-01	3.99E-01	5	3.64E-03	5.77E-02	9.15E-01	1.45E+01	2.30E+02	3.64E+03	5.77E+04
1		0.2		10	2.07E-01	2.13E-01	2.33E-01	2.64E-01	3.25E-01	5.05E-01	7.97E-01	10	1.46E-02	2.31E-01	3.66E+00	5.80E+01	9.91E+02	1.46E+04	2.31E+05
El beam		0.4		20	4.14E-01	4.27E-01	4.65E-01	5.28E-01	6.51E-01	1.01E+00	1.59E+00	20	5.83E-02	9.23E-01	1.46E+01	2.32E+02	3.68E+03	5.83E+04	9.23E+05
5		1		50	1.03E+00	1.07E+00	1.16E+00	1.32E+00	1.63E+00	2.53E+00	3.99E+00	50	3.64E-01	5.77E+00	9.15E+01	1.45E+03	2.30E+04	3.64E+05	5.77E+06
		2		100	2.07E+00	2.13E+00	2.33E+00	2.64E+00	3.25E+00	5.05E+00	7.97E+00	100	1.46E+00	2.31E+01	3.66E+02	5.80E+03	9.19E+04	1.46E+06	2.31E+07
		4		200	4.14E+00	4.27E+00	4.65E+00	5.28E+00	6.51E+00	1.01E+01	1.59E+01	200	5.83E+00	9.23E+01	1.46E+03	2.32E+04	3.68E+05	5.83E+06	9.23E+07
		10		500	1.03E+01	1.07E+01	1.16E+01	1.32E+01	1.63E+01	2.53E+01	3.99E+01	500	3.64E+01	5.77E+02	9.15E+03	1.45E+05	2.30E+06	3.64E+07	5.77E+08
				Attenuation (dB)								Backscatter cross-section per resolution element vs range and rainfall rate							
L band					RR (mm/hr)							RCS/m^3	3.03E-11	4.80E-10	7.61E-09	1.21E-07	1.91E-06	3.03E-05	4.80E-04
23.5	1.28E+09	0.005		R (km)	1	2	5	10	20	50	100	km	Cross-section per res element						
RCS scaling vs S band		0.01	6.13E-05	1	1.01E-02	1.02E-02	1.06E-02	1.12E-02	1.25E-02	1.61E-02	2.23E-02	1	1.05E-05	1.66E-04	2.63E-03	4.18E-02	6.62E-01	1.05E+01	1.66E+02
0.04801	−13.187	0.02		2	2.02E-02	2.05E-02	2.12E-02	2.25E-02	2.49E-02	3.23E-02	4.45E-02	2	4.20E-05	6.65E-04	1.05E-02	1.67E-01	2.65E+00	4.20E+01	6.65E+02
Az beam		0.05		5	5.06E-02	5.12E-02	5.31E-02	5.61E-02	6.23E-02	8.06E-02	1.11E-01	5	2.62E-04	4.16E-03	6.59E-02	1.04E+00	1.65E+01	2.62E+02	4.16E+03
1.5		0.1		10	1.01E-01	1.02E-01	1.06E-01	1.12E-02	1.25E-01	1.61E-01	2.23E-01	10	1.05E-03	1.66E-02	2.63E-01	4.18E+00	6.62E+01	1.05E+03	1.66E+04
El beam		0.2		20	2.02E-01	2.05E-01	2.12E-01	2.25E-01	2.49E-01	3.23E-01	4.45E-01	20	4.20E-03	6.65E-02	1.05E+00	1.67E+01	2.65E+02	4.20E+03	6.65E+04
5		0.5		50	5.06E-01	5.12E-01	5.31E-01	5.61E-01	6.23E-01	8.06E-01	1.11E+00	50	2.62E-02	4.16E-01	6.59E+00	1.04E+02	1.65E+03	2.62E+04	4.15E+05
		1		100	1.01E+00	1.02E+00	1.06E+00	1.12E+00	1.25E+00	1.61E+00	2.23E+00	100	1.05E-01	1.66E+00	2.63E+01	4.18E+02	6.62E+03	1.05E+05	1.66E+06
9		2		200	2.02E+00	2.05E+00	2.12E+00	2.25E+00	2.49E+00	3.23E+00	4.45E+00	200	4.20E-01	6.65E+00	1.05E+02	1.67E+03	2.65E+04	4.20E+05	6.65E+06
		5		500	5.06E+00	5.12E+00	5.31E+00	5.61E+00	6.23E+00	8.06E+00	1.11E+00	500	2.62E+00	4.16E+01	6.59E+02	1.04E+04	1.65E+05	2.62E+06	4.16E+07

Discrimination between UAS and birds has been referred to in Chapter 6. Counter-drone surveillance, unmanned air traffic management and avian research are key applications for the SRC configuration, with RToIs near the surface and at ranges up to several kilometres and with an essential requirement for discrimination between the different types of targets.

Provided that staring radar responses of birds meet the criteria we have described under the EM Uniqueness Theorem, then the movements of birds can be measured. We do not currently know whether those measurements will lead to identifying species, assessments of behaviour, feeding or predation, population changes or how to predict incursions of birds into flight paths, but the existence of such data presents a starting point for several applications.

Figure 8.13 adds to Figure 6.24 exemplifying a visual inspection of the sequence of signals and their possible interpretation.

The question arises – what is the information that may identify this cell as class NSC, but containing birds as opposed to RToIs such as UAVs or manned aircraft?

The graphs constitute a sequence of successive and non-overlapping CPIs in the frequency domain at intervals near 500 milliseconds, in a single range cell, with range resolution about 150 m. The zero-Doppler bin is at index 1 024, and the spectrum is zeroed from 1 020 to 1 028.

These signals present a high signal-to-noise ratio, with a number of small constituents in addition to 21 larger signal features that vary significantly in spectral frequency and spread between bins 920 and 1 070. The signals model the Doppler shifts (radial speed) and modulations (wing motions and accelerations) arising from motions of specific objects. In that case, these bins represent objects with varying radial speeds, plus modulations arising from their motions, accelerations or changes in shape.

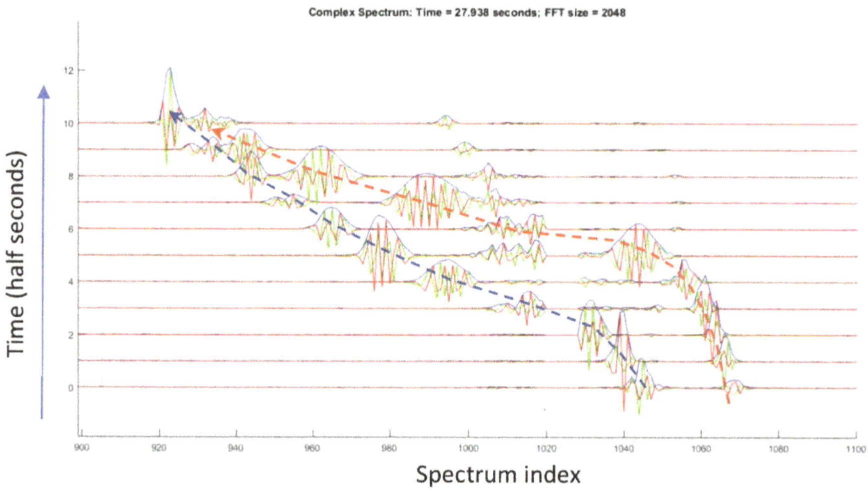

Figure 8.13 Radar responses likely to be scattered by birds. 'Tracks' are visual sketches.

The model signal features themselves change considerably in amplitude and frequency spread, as the CPIs pass.

They do not show associated modulation spurs as seen in the Doppler signatures of both rotary and fixed-wing drones (Figures 6.24 and 6.25).

The radial speeds vary from 11 m/s (receding) to 25 m/s (approaching) over a period of 10 seconds, and if we identify apparently consecutive returns from objects in two distinct trajectories (A and B), one changes quite steadily from +5 m/s to −25 m/s over 5 seconds, while the other accelerates from a higher receding speed to approach at 25 m/s over only 2.5 seconds.

These signals represent one or more scattering targets, capable of acceleration up to 12 m/s^2, or 1.2G.

These speeds and accelerations, combined with the variability of Doppler signatures and with the lack of associated modulation spurs, model substantial birds, one of which is agile.

Their final common radial speed models a successful convergence in flight.

Certain aspects of measured trajectories can be seen to provide a basis for classification, especially where the CPI can be extended to yield dynamics as described above, in addition to the ability to discover modulation spurs in a drone's backscatter signature, providing strong discrimination.

For air traffic surveillance, at ranges that are neither too long to observe these features nor close enough to be obscured by transmitter saturation, nor dependent on behavioural conjectures, analysis of modulation spurs is the most robust source of classification. The radar cross-sections of drones vary widely, but many are as low as −20 dBsm. Their rotors tend to have cross-sections a further 20 dB smaller at −40 dBsm and are unlikely to be observable to the maximum range at which the airframes are seen. For that reason, the ability to extend the duration of the CPI has additional significance.

Many air surveillance radars such as ATC are specified to detect targets near 0 dBsm at ranges up to 110 km. In the absence of transmitter saturation, this would suggest that UAS rotors should be detectable up to about 11 km, which is an effective range for UASs and infrastructure such as an airport. This may be countered by the fact that for solid state radars, long range is achieved by pulse compression that leads to receiver desensitisation up to several miles.

Such radars achieve shorter minimum ranges by transmitting a shorter secondary pulse with a blind range as SRC. With the same peak power as the primary transmission, this may maintain range of −40 dBsm targets to about 4.5 km, a useful range for UASs.

This degree of sensitivity is sufficient that process adaptations feasible within a staring radar may make such applications attractive and affordable within a surveillance radar.

8.2.5 Interrelation of cells

Signals representing activity in any known collection of cells are likely to have mutual significance, and potentially to support characterising the surveillance environment in terms of clutter, phase noise, interference, etc.

Neighbouring range cells within the same beam structure operate precisely coherently. Neighbouring azimuth and elevation cells will also be closely coherent but may differ to some extent as a result of sidelobes and multipath propagation. In the same way that neighbouring range cells can be coordinated to take account of Range Walk, neighbouring cells in azimuth and elevation can be used in a linear process to accommodate angular translation.

For geographically fixed clutter, cells geometrically close to the actual clutter location may be ‘clarified’ by reference to the ‘home’ clutter cell (see Section 8.4). Particular examples where this may be relevant are airport approaches, critical and military infrastructure and extended areas of wind power generation.

For populations of targets related to airports, aircraft approach, landing and departure, patterns and behaviour can be monitored using an ATC configuration. Bird populations or activities by UAVs, whether for unmanned air transport or defensive counter-UAV operations, can be followed as a volume-monitoring activity, given appropriate radar characteristics (as SRC, for example). Migration of flocks or formations can be correlated from one cell to the next, and behaviours assessed on a continuing basis if required.

In Sections 8.3 and 8.4, different targets are monitored over extended periods to provide the basis for analysis of behaviour and suppression where clutter is known to be a direct threat.

8.3 Target analysis, history and recovery (CST)

With successive, concatenated CPIs, multicellular target analysis offers an approach where targets can be followed through any succession of resolution cells, with increasing detail and precision as time progresses. Newtonian mechanics and persistent observation ensure that targets do not in fact hop across resolution cells.

8.3.1 Target detail

The distinction between the treatment of targets and clutter is that RToIs are recorded with maximum detail for current and subsequent reporting, whereas signal components qualifying as clutter are recorded sufficiently to support their suppression over time. In practice, and given a need to monitor the status of clutter sources, these may not differ greatly in their complexity and demand for computing power.

For example, we are interested in and might report on the length and rate of rotation of helicopter rotors (Figure 8.14). Wind turbine rotors may only need retention to support suppression of their next flash, but their continuing status must be monitored. Should the radar be used as an instrument for monitoring the condition of a wind farm, the turbine returns would then migrate from the clutter record into the RToI archive.

A typical small helicopter might have rotor blades about 8 m in length, rotating at 200 rpm, and generate a succession of intense reflections as they pass through the normal to the radar direction. This is modelled and illustrated in Figures 8.14 and 8.15.

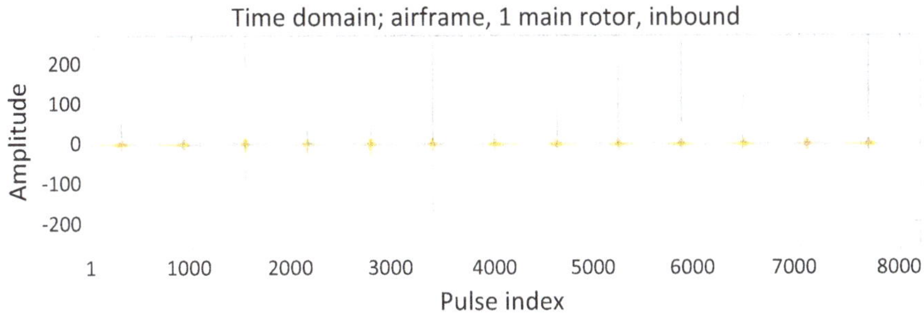

Figure 8.14 *Time domain model sequence for straight 8 m helicopter rotor blades*

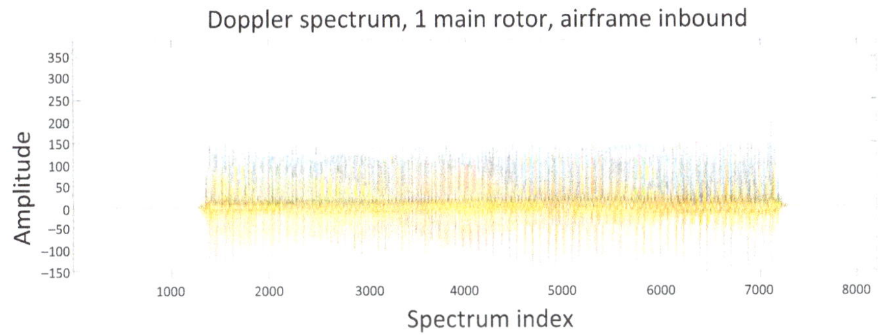

Figure 8.15 *Doppler spectrum and I/Q for a straight 8 m 2-blade helicopter rotor*

A helicopter is certainly a RToI. Its number of rotors, their length and tip speed are significant items of surveillance information. Here for a two-bladed rotor, the second harmonic of the rotor rate determines the intervals between blade 'flashes', and the outer limits of the rotor spectrum are determined by the rotor rate and blade length.

8.3.2 Target fading

Target fading results from the fact that the radar cross-sections of all aircraft vary strongly with the target-to-radar relative orientation, as different scattering centres in the target are seen from different directions and interfere with each other constructively or destructively as their relative ranges evolve in phase.

Huygens scattering models were introduced in Chapter 3, Section 3.1. A model can be constructed for any shape of target and will yield a reasonable approximation to the variation of cross-section with look direction. Chapter 10 will also refer to target fading for the case of multistatic radar, in which the directions of irradiation and scattering are different.

8.3.3 Target imaging

Extended observation of an aircraft through a tangential pass or a turning manoeuvre will provide a range of viewing angles that effectively enable Inverse Synthetic Aperture imaging. Turning manoeuvres will also require banking, in which case, some level of vertical aperture synthesis will be possible, although complex.

Images derived from either of these manoeuvres cannot be expected to be picture-quality but may lead to estimates of the target's scale in each dimension. This is a large subject for future research.

8.3.4 Extended target behaviour

Holographic staring radar enables detailed and continuous observation. In Figure 8.16, a modelled sequence of Doppler spectra for successive time spans represents an approaching, single-rotor, fixed-wing UAV.

This is a 2 048-point spectrum, and the low-Doppler bins near 1 024 are attenuated. Time runs upward over a period near 5 seconds.

Again, blue traces represent modular amplitudes; red and orange graphs illustrate the in-phase and quadrature components of the signal at the staring array. The larger-amplitude signals running from indices 645 to 690 are airframe returns. Associated with them and equally spaced on either side of them, the process finds a series of spectrum spurs that represent harmonics of the rotor blade rate. Their positions meet the symmetry test for amplitude modulation. The drone airframe is slowing in its approach. At the same time, the rotor spur separations are narrowing and indicate that this fixed-wing UAV is preparing to descend. A rotary-wing drone with this change in rotor rate (with fixed pitch) could have difficulty staying in flight.

Sequences of this kind offer target identity and behavioural evidence that, provided continuity, naturally grows in significance with time. A single CPI extending

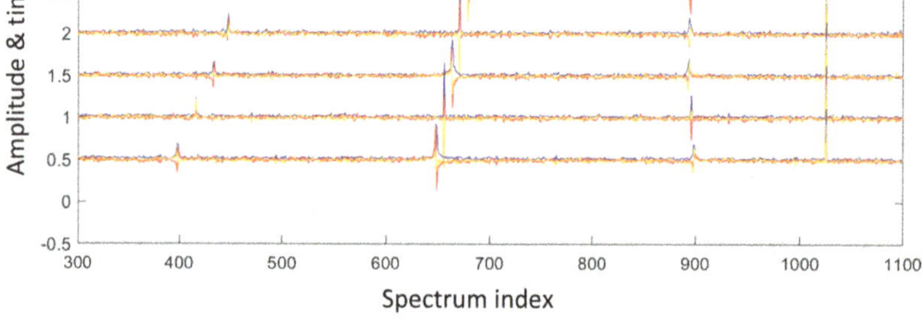

Figure 8.16 Drone signatures and behaviour over time

over this period might appear confusing; nevertheless, it will contain all the information necessary to measure the dynamic history.

8.3.5 Target recovery

CPI extension can also offer maintenance of detection under target fading or multipath, or resolution of multipath returns. Under the condition where a target return decreases to the extent that it escapes capture, a further-extended CPI can increase sensitivity, can separate close Doppler components and can broaden the range of compensation cases.

Each HSR resolution cell is expected to generate some equivalent to the vector histogram, in which the distribution of signals in time and frequency is linked to components within the distribution and their evolution. A succession of VHs then provides access to a direct sequence for each component in time or frequency, in amplitude and phase.

During a fading sequence due to either target fading or multipath, using VH format, a component does not simply disappear below a single threshold. The VH has the equivalent of many thresholds, and while a single threshold might be set at 15–20 dB above the RMS noise, the VH maintains signal contact down to the noise level itself. In the case of Figure 8.10, the modelled 'car' (unencoded it is 23 dB above noise) starts at level 40 compared with RMS noise near 10 and would have to pass through 20 levels before becoming unmeasurable. That would require an extended –20 dB fade and justifies extended complex compensation. Target recovery, or contact maintenance, becomes a reality, given appropriate computing resources.

8.4 Repetitive clutter analysis: wind turbines (NRD, RDT)

Wind turbine generators are being installed across the world as a primary source of renewable electric energy. They range from single installation to extensive wind farms, both on and off shore.

Wind farms may contain many turbines, recent versions of which may be over 300 m tall. Although fixed in position, they include large components that move at speeds up to 100 m/s. Turbine towers and blades have radar cross-sections that may exceed 10 000 m^2. They can therefore appear to surveillance radar as targets of interest. They are usually 3-bladed, and to generate electricity, they rotate at intervals between 3 and 10 seconds, creating highly but not exactly repetitive radar responses. Cells affected by wind turbines are classed as NRD, and the key objective is to separate RDT cells, where targets are present.

The particular difficulty suffered by beam-scanning radars is that their operation imposes a regular 4-second to 12-second time interval between successive observations of any target, asynchronously with any wind turbine. Turbines in a wind farm are not themselves synchronised, and their motions vary with wind speed and direction, and with energy demand. For a BSR, turbine responses are therefore unpredictable. However, turbine responses are intense, and those from separate turbines can occur in sequences, which may appear as a track across a wind farm. Mitigation of these events is a major and continuing focus of radar development, largely on the basis of reducing sensitivity at the site of a turbine while aiming to retain P(d) for aircraft.

Under the EUNIT, provided that turbines move under Newtonian constraints and are observed persistently with appropriate resolution, it appears possible to separate aircraft responses from turbines by linear matched filtering. For a staring radar, the turbine generates a continuing sequence of blade flashes and very slow-Doppler tower responses, which are amenable to modelling, analysis and such linear processing.

The approach available with HSR is that, rather than attempting to minimise the effect of wind turbines on radar returns, a HSR acquires the maximum possible information about both the targets and the turbines in order to separate them by resolving each precise set of information and its evolution.

We assume a HSR located on the ground in a fixed, known position with a wind farm within its CVoR and that its memory provision is sufficient to retain complex signal data for all affected resolution cells over a period of many seconds.

Extensive experience now exists with mitigation of the effects of wind turbines using HSR. It records good examples of how extended time sequences of persistently sampled signals can allow aircraft and turbine signatures, with their particular spectral and temporal characteristics, to be resolved and differently reported or suppressed, and interrelated between cells.

To explore this potential for HSR to resolve aircraft from moving turbine clutter a Huygens model provides controlled illustrations of wind turbine responses. We can describe processes for their suppression, based on the ATC configuration, following the introduction in Chapter 6, Section 6.4.5.

8.4.1 Wind Turbine Generators

We base the model on a shape assumed for a turbine blade and then use that to construct a 3-bladed simulation of a turbine, whose radar response is illustrated first in Figure 8.17. A more detailed model is shown in Figure 8.18 in the time domain, and in Figure 8.19 an aircraft pass is included in the time domain, at low amplitude, yielding the Doppler spectrum shown in Figure 8.20. The Doppler spectrum of Model 2 in isolation is shown in Figure 8.21.

Wind turbine blades vary in width and curvature, and this model uses a near-straight leading-edged shape. It is also provided that the internal blade structure will be more substantial near the blade root, and that that there are some components (lights, caps, etc.) associated with the tips themselves.

This simple model suggests a sharply peaked scattering waveform shown in Figure 8.17, with three approaching-edge flashes and three receding. Each flash has a double peak, which can be associated with the leading and trailing edges for each blade.

Doppler spectra derived from the blade flashes are shown in sequence in Figure 8.21. These contain notably sinusoidal features, which also might oversimplify the turbine response.

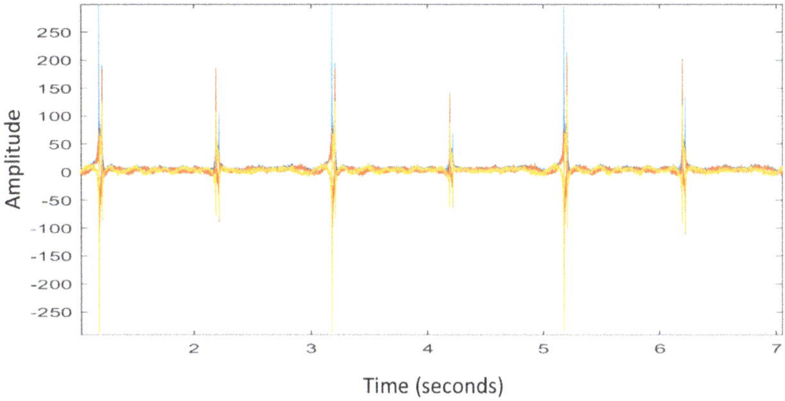

Figure 8.17 *TD sequence for 3-bladed turbine model 1. Amplitude (blue) and in-phase and quadrature (red and orange), covering 6 seconds at a PRF of 1 024 Hz.*

To assess expected suppression capabilities, a range of model parameters has been used for comparisons, and illustrated in Figures 8.17 and 8.18 (in isolation) and Figure 8.19 (with aircraft pass) in the time domain.

These model responses yield Doppler spectra by Fourier transformation, as illustrated in Figure 8.20 (with target Doppler spectra) and Figure 8.21 (without). Figures 8.22 and 8.23 show single Doppler responses in greater detail. The VH form is in Figure 8.24.

A model of this kind provides the opportunity to introduce targets of varying sizes, trajectories, scattering properties, etc., before testing different processing approaches.

The response as the target passes this range gate (Figure 8.20), after Fourier transformation, exceeds the turbine Doppler amplitude, but in many cases, this will not be so. A threshold set on the basis of a variable turbine Doppler signature will

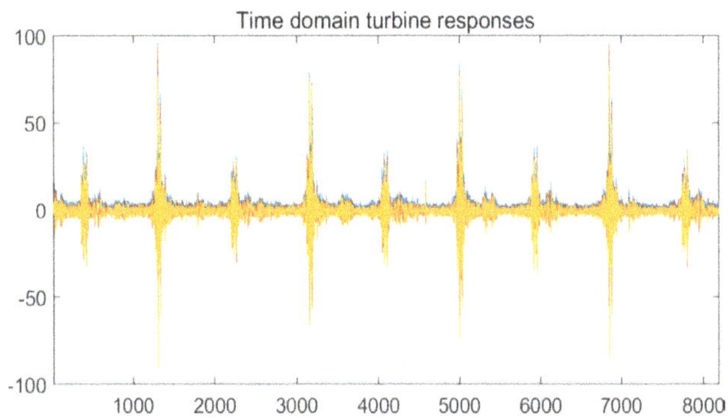

Figure 8.18 *TD sequence turbine model 3, pointing obliquely, rotating in 5 600 pulse intervals*

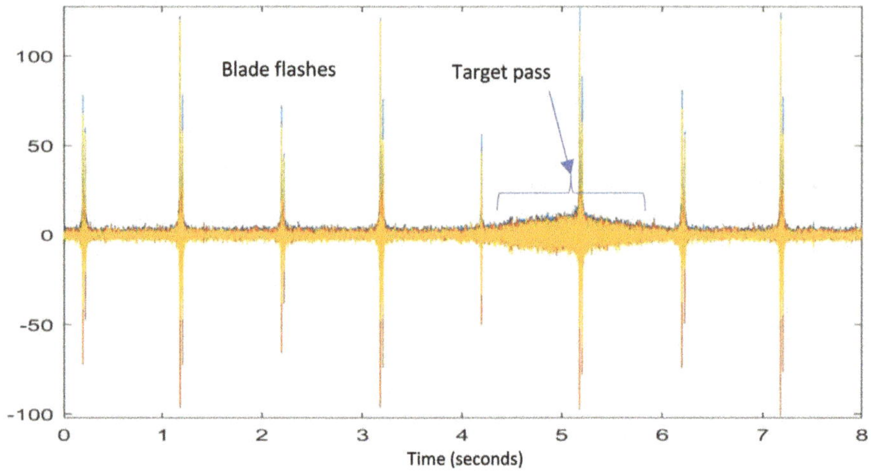

Figure 8.19 Time domain IQ sequence: model 2 turbine and aircraft passing through a 150 m resolution range gate

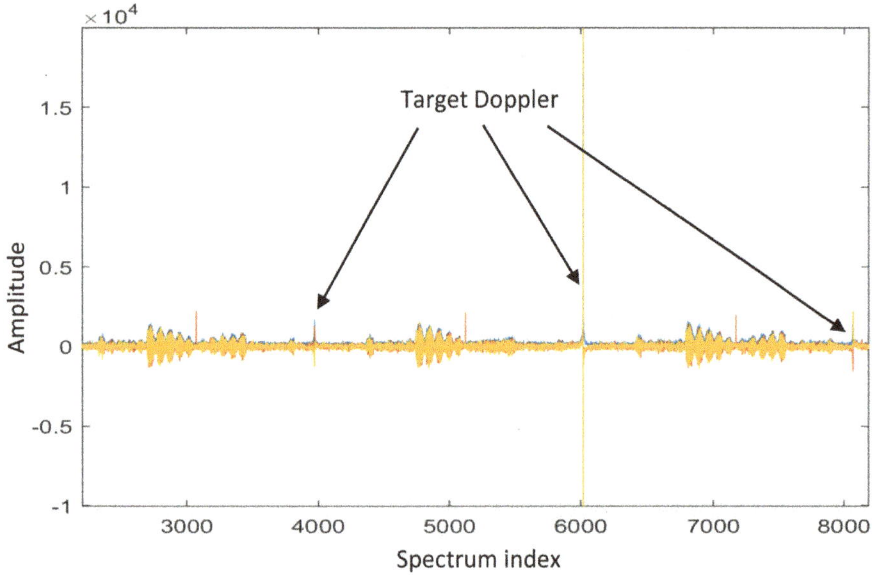

Figure 8.20 Incorporating the Doppler response to a target receding at constant speed in three successive time intervals, at increasing range: entering, occupying and leaving a range gate

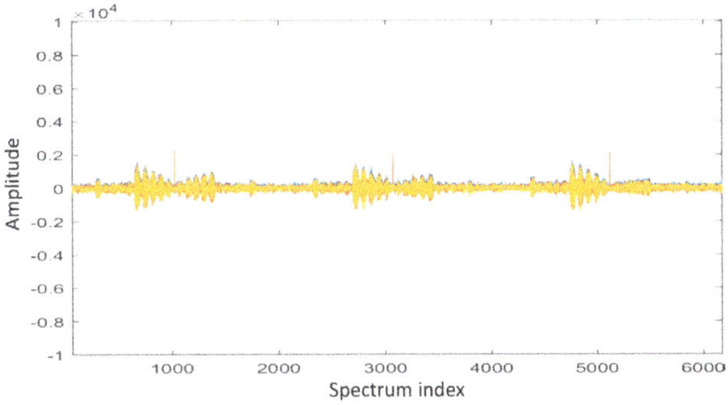

Figure 8.21 *Sequence of Doppler spectra: three successive model 1 blade passes in concatenated frequency spans from −500 to +500 Hz. Blades 1,2,3 leading and trailing edges have small differences.*

result in unpredictable sensitivity and target capture performance. The response of a target aircraft clearly differs in form from that of turbines. Sections 8.4.2–8.4.4 use these models to explore the capacity of HSR to resolve or suppress different return signals linearly, exploiting the information acquired under the EUNIT.

In Figure 8.22, the approaching return is from blade 1 and the receding edge from blade 3. Zero-Doppler clutter is small after suppression. The approaching and receding blade responses are separated between the left and right sides of the

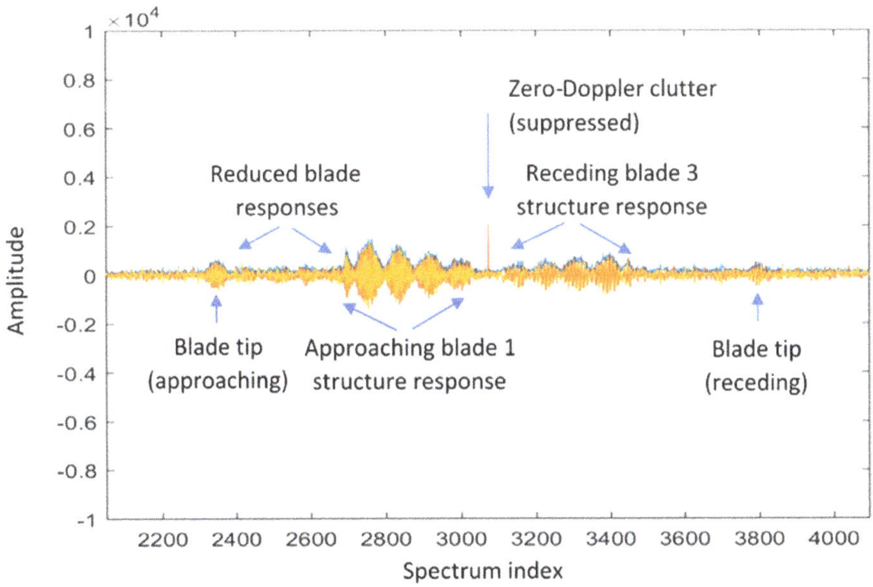

Figure 8.22 *Fourier transform responses from model 2 approaching and receding blades*

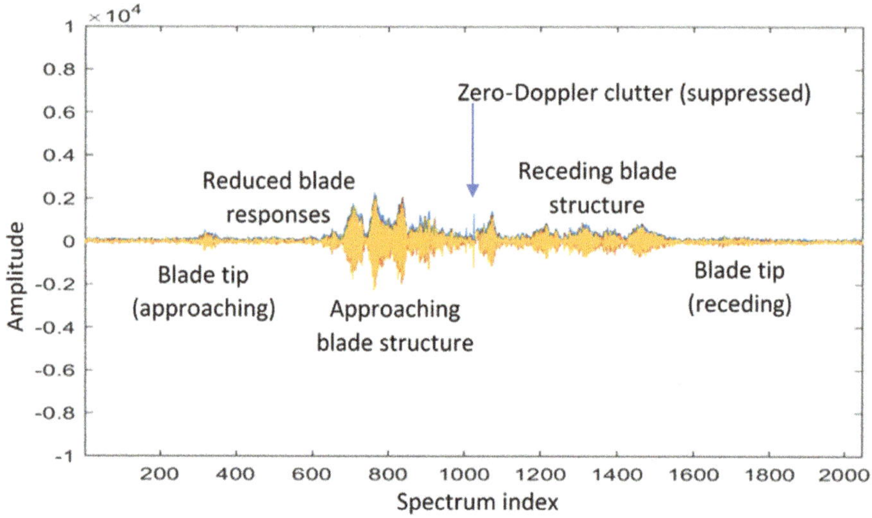

Figure 8.23 *Fourier transform for model 3. Zero-Doppler clutter is attenuated. Approaching and receding Doppler components represent a more highly structured blade.*

spectrum; suppressed tower clutter is near index 3 072; an inner region of increasing amplitudes with increasing approaching and receding Doppler shifts between 2 700 and 3 450; reduced returns from the outer blade section are between 2 400–2 700 and 3 450–3 750, and responses from each blade tip lie near indices 2 350 and 3 800.

Amplitudes again are greater on the approaching side in Figure 8.23. This model is a starting point for signature exploration.

By concatenating successive CPIs in the ATC case, a modelled history of turbine flashes can be assembled as complex values in the time domain, extending to many seconds, as illustrated in Figure 8.17. This can be extended as necessary to

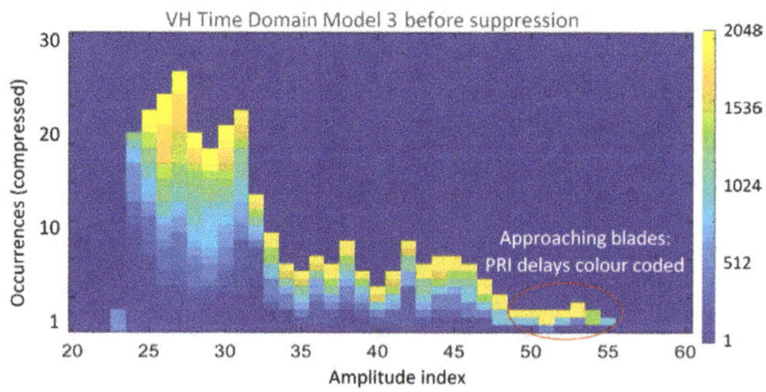

Figure 8.24 *Time domain VH format for model 3 turbine flash timing selection. Pulse # colour coded.*

find turbine sequences, down to blade tip speeds below typical aircraft speeds – say, 5 m/s.

8.4.2 Time domain suppression

Information derived from these signals offers the opportunity to fully characterise those related to turbines as opposed to those related to aircraft targets, to support analytical methods of discrimination and then separation of the two forms.

A VH format can be used simply to measure an extended time window where the higher-amplitude components will contain the major turbine flashes, with their time indices directly yielding the flash sequence for an estimate of the timing of the next flash event, and is illustrated in Figure 8.24.

It offers an efficient basis for locating and attenuating short-duration events in the time domain, without significant loss for a target with a longer dwell time in the resolution cell, as was introduced in Section 6.4.5.

By contrast with short flashes shown in Figures 8.17–8.19, an example aircraft passage takes a number of seconds; in this case 2 000 pulse sample intervals illustrated in Figure 8.19.

The modelled timing, shape and level of stability of turbine returns demonstrate that signal histories over seconds or longer can be expected to show correlations that differ from any expectation for aircraft, helicopters or drones.

The application of a suppression filter in the time domain can achieve a reduction of at least 10 dB in the level of wind turbine interference, illustrated in Figure 8.25.

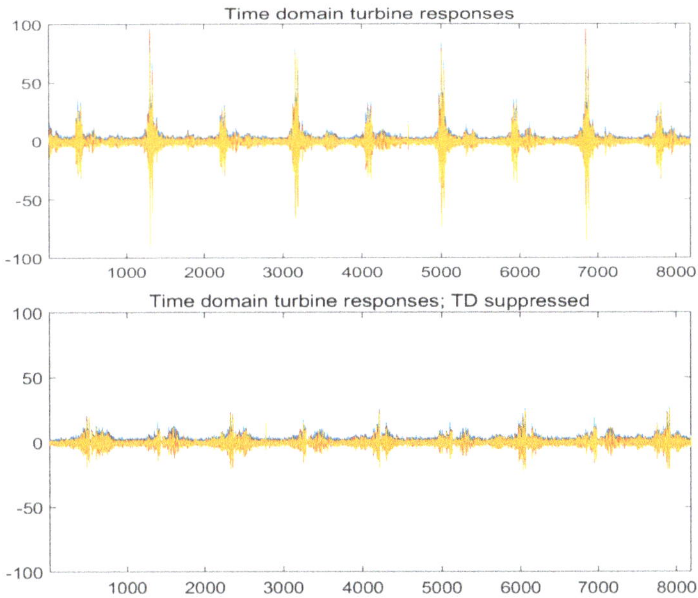

Figure 8.25 Time domain suppression can provide over 10 dB peak suppression

Time domain suppression can be designed to cause minimal reduction in slowly varying targets, including aircraft.

There is also evidence that successive blade flashes are sufficiently similar that an appropriate matched filter will provide further and sufficient suppression.

8.4.3 Frequency domain suppression

Figure 8.26A and B show spectrum amplitudes for samples centred on a model turbine blade flash with an approaching target, (A) before and (B) after FD suppression.

Wind turbine returns have been found to be repetitive and consistent. In the return spectrum in Figure 8.26, the modelled response is substantially suppressed on the basis of the previous same-blade return and the target aircraft may be effectively isolated at lower levels than this example.

The vector histogram for the encoded, turbine-suppressed spectrum is shown in Figure 8.27, with the main target and spurs indicated.

Suppression here may operate on the modular spectrum, using data segments that contain an advancing leading edge and a receding trailing edge to achieve this form of turbine suppression.

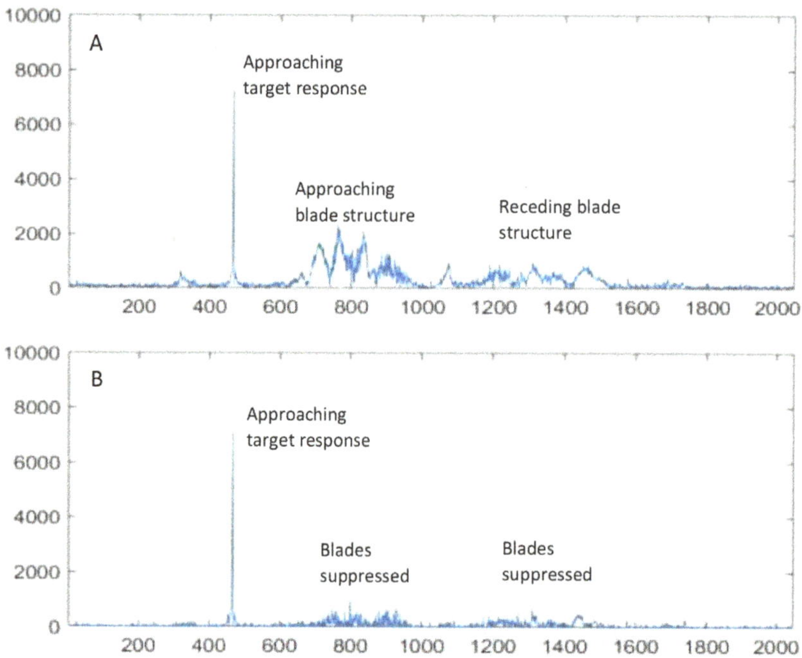

Figure 8.26 A and B. Target and turbine spectra before and after frequency-domain suppression

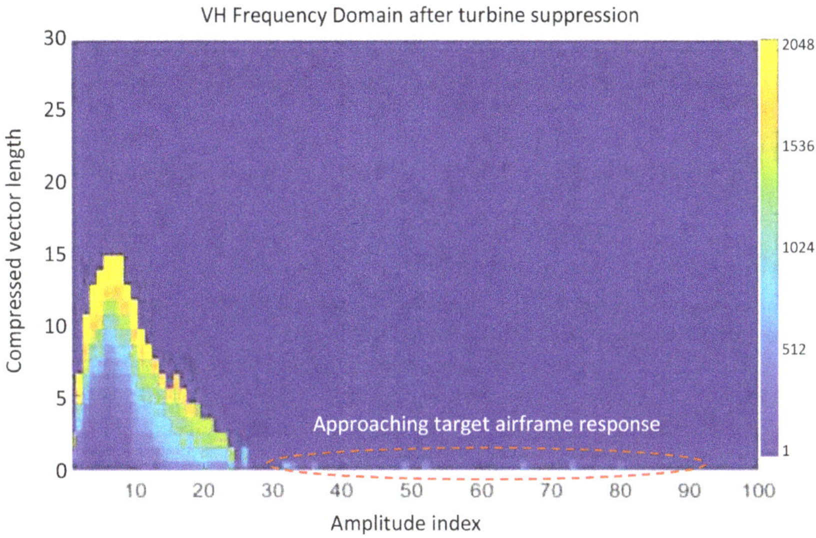

Figure 8.27 *VH format for post-FD suppression target capture. Blue series represents the airframe.*

8.4.4 Turbine shadowing and ghosting

There is a further effect of wind turbines on air surveillance that differs from that of wind turbine radar clutter on the rate or probability of false reports. In many cases during trials of surveillance in the presence of wind turbines, the detection and tracking of aircraft has been found degraded when the aircraft's trajectory is at low altitude, beyond the wind farm, resulting in the wind turbines interposing between aircraft and radar.

In principle, wind turbines, especially their towers, have two effects on radio and radar propagation.

8.4.4.1 Diffraction and direct shadowing

The first, directly linked to shadowing, is that signal energy is diverted from the forward propagation path by these large structures as a result of wave diffraction. At typical radar frequencies, especially lower bands such as L band, because of diffraction around the turbine tower, the 'optical shadow' in the Fresnel zone behind the tower exists but will not extend further than several hundred metres to a kilometre. Beyond that distance, energy is scattered out of the forward path, but the shadow will be diffused and the irradiating amplitude reduction is proportional to the square root of the obscured area fraction.

This is an active area of research. The effect varies not only with the size and density of the turbines and the wind farm but also with the frequency of operation of the radar. To maintain wind-energy efficiency, turbines are placed hundreds of metres or kilometres apart, while their towers and blades have width dimensions of metres. For example, with turbine towers of diameter 10 m and separated (as seen through each rank of wind turbines) by 500–1 500 m, within the Fresnel diffraction zones – i.e., within a distance of several hundred metres – a target at an elevation within that of the turbine will be severely shadowed by a single turbine, but the effect at higher altitude or longer range will be small. The loss per rank of turbines should then be in the region of 1% (for 10 m diameter towers, beyond 1 km, at L band).

For more distant targets (kilometres) beyond the turbines, both forward and target-scattered waves will be reduced by not more than 0.5 dB, depending on the size and density of turbines, and resulting in a total loss of about 1 dB at altitudes up to a few times the height of the turbines but not higher.

The EUNIT makes no distinction between objects that are directly in view of the radar and those that may be obscured behind other objects. The effects of all such objects are subject to the Theorem and in principle should be present in the observed signals, though at a low level. The necessary distinction is whether, granted that such signals exist, they can be resolved given the power budget and resolution of the radar. There will indeed be cases where the total losses incurred on reflection, by absorption, or by diffraction and scattering lead to too slight a signal-to-noise ratio for discrimination, classification or reporting. These will then be imperceptible and at this level, turbine towers will contribute to the limit on sensitivity.

8.4.4.2 Track confusion by 'satellite' targets

The second effect is that of multipath, introduced in Chapter 3. In addition to ground reflections, signals propagating either from transmitter to target or from target to receiver will be reflected by turbine towers and may have detectable amplitude if the tower is close enough either to the radar or to the target. Such 'satellite' reflections occur either in the direction of the target, but at extended apparent range, or else in the direction of the reflecting tower, also at extended range. They are termed 'satellites' because the additional reflection reduces their intensity and they tend only to be reported when close to either. Such satellite returns have been observed, and their recognition and suppression are discussed in detail in Chapter 9.

8.5 Longer-term retrospective surveillance

Beyond the exploitation of signal sequences of seconds to extract target and clutter features of interest, there are also surveillance functions of longer-term significance that may arise from this ability to retain continuous, raw radar signal information.

BSRs derive track reports from intermittent position estimates, whereas HSR acquires a persistent sequence of pulse returns for a chosen period. Occupied cells can be discovered from cell signal power and spectral distribution, and resolved in fine Doppler detail as well as position. As noted, cell returns can be evaluated and re-evaluated when concatenated with earlier or subsequent periods. The EUNIT recognises that, except in extreme and destructive circumstances, targets of interest do not actually disappear. They fade, for the purposes of a momentary observation, as a result of intra-signature interference, but extended observation allows targets within the CVoR to emerge from any physically consistent fading pattern. In principle, HSR can measure positions, motions and other characteristics of all objects in that cell and in neighbouring cells as they migrate.

A surveillance radar that can exploit that characteristic is better-placed to maintain contact with the target than one whose operation is based on only intermittent observation of each target.

The most likely scenarios here are (1) target fading, as referred to in Section 8.3.2, and (2) multipath fading, referred to in Chapters 3, 7 and 9. Low-observable aircraft are a special case; the same issues of fading still obtain, but these aircraft are designed using geometries and materials that render the radar cross-section itself much smaller than for conventional aircraft, at certain look directions. These are beyond the scope of this chapter but are relevant to Chapter 10.

8.5.1 Clutter imaging (NCD)

A conceivable area of weakness in the EUNIT may be seen in intermittent effects of static clutter, by which signals from moving objects are reflected to the receiver as plane waves.

With respect to measuring the shape and effects of static clutter, two solutions are available. One is to extend the receiving aperture until clutter features can be resolved. Assuming that the aperture is fixed, as with an ATC sensor, the second solution is a long-term radar application based on archived target responses that are subject to multipath reflections and diffraction from each clutter item (as all are, at some level). This results in either fading or satellite targets, and the correlation between direct target responses and those propagated via reflecting or diffracting clutter is entirely determined by the target and the locations, size and shape of the clutter. These can be solved subject to noise, the power budget of the system and the length of the CPI. We shall return to this subject in Chapter 9.

8.5.2 Multipath fading

Reduced target responses may result from interference between signals propagated via different paths. This effect has been referred to in Chapters 3 and 7 and will also resurface in Chapter 9 in relation to the study and assessment of clutter within the CVoR, in the case of non-interfering multipath. Clutter imaging and multipath fading are in fact different constructive views of the same phenomenon, which constitutes multiple 'looks' at each scattering source.

8.5.3 Target accounting (CST, CDT, RDT)

Targets of interest are either assets or threats that need tracking, not as a way of associating and interpolating between intermittent detections but as a security procedure.

A target of interest may come in many forms, and having been observed by a HSR will have been classified in some way, in terms of location, speed, size (in terms of radar cross-section), size (in terms of fading characteristics) and capability (in terms of dynamics). It may also be recorded in terms of apparent objective and possible intent.

Targets arrive in the CVoR and leave it later. For air traffic, targets may take off or land at known field locations. Qualified targets that disappear after the forms of search described above may indicate cause for concern.

While providing the basis for HSR to offer full performance as a surveillance radar, these methods are also the starting-point for deriving a more complete air surveillance picture. Here we seek to describe non-cooperative contributions to such a picture, based on the known and demonstrated, or anticipated, capabilities of HSR sensors.

A key aspect of the data processing power appropriate to a HSR is its ability not to discard signal information once detection decisions have been made but to maintain it in extended storage. It is carried forward on the assumption that information, especially information initially buried within a noise distribution (as most HSR channel information will be), will remain of continuing value provided it is retained as linearly derived, fully sampled and coherent signal data received on a known aperture.

When such information is carried forward, it can be concatenated with prior or subsequently acquired signal data, providing opportunities for continuing evaluation of targets or clutter sources (both of which are assumed to exist over extended periods). The intent is that these are then classified appropriately to clarify the air picture.

8.5.3.1 Targets and airspace

Returning to the items in the set of applications in Chapter 4, radar signature features that contribute to or distort such a clear air picture can arise from various types of scatterer (T1–5) that may be considered for reporting:

T1: Aircraft in flight (CDT) that support identification as:

- fixed wing propeller aircraft
- jet aircraft
- single or twin rotor helicopter, etc.

T2: Airspace affected by wind turbines (RDT)

T3: Air targets that support identification as UASs but not birds (CST)

T4: Natural air targets such as birds, at heights, speeds and with Doppler signatures that satisfy avian models (CST)

T5: 'Satellite' objects, derived from air targets by reflection and diffraction (CMT)

Type T1 represents the standard set of targets for air surveillance. These are expected to be detectable quickly, and to follow trajectories that yield smoothly varying and well-defined Doppler signatures that may be matched with secondary radar or cooperative multilateration measurements. In these cases, time concatenation may be used to maintain target contact to measure airframe or propulsion characteristics, speeds and blade lengths or to evaluate pilots' manoeuvring and responsiveness to controller requests. Recent signals in storage and their concatenation can be used in response to the detection by a controller of unplanned manoeuvres and events.

Type T2 represents airspace compromised by wind turbines. These large, moving structures have presented a major challenge to surveillance radars over recent years. Radar manufacturers have made strenuous efforts to remove the effects of these targets from the surveillance reports of their radars, either by hiding the radar by terrain, by increasing the bandwidth and reducing the beamwidth of the radar to minimise the affected area, or by other signal processing algorithms.

The real-time processes we have discussed are designed to meet the defined needs of, in this case, the air traffic management function, in the presence of aircraft and clutter in the normal course of operations.

Type T3 refers to drones. Small, slow drones may well have radar cross-sections similar to those of birds, and it will be an essential aspect of surveillance of drones that they should not be mistaken for birds and vice versa. The key capability offered by staring surveillance radar is that once a Doppler return is discovered (of a representative speed and cross-section), a parallel process deploying extended dwell time may be used so that, with continued detection of the drone's airframe, reduced thresholds may be used to discover the mirror symmetry in nearby spurs that move with the airframe. These spurs may be 10, 20 or 30 dB below that of the airframe; however, they are so directly indicative of the nature of the drone that such discrimination may be required only at intervals along the trajectory to maintain high confidence in the classification.

Type 4 includes birds, which cannot be expected to follow known transport routes but may take more or less predictable trajectories in feeding, predating or migrating. Type 4 may also include turbulent air movements that follow identifiable sources such as aircraft on take-off or large trucks. Differences in the behaviours of birds have been observed between night and day, which may represent the difference between feeding and migrating bird trajectories. In the latter case, they are unlikely to be close to the ground. The range of targets T4 require detailed analysis for classification.

Type 5 refers to satellite targets that arise from multipath propagation with respect to Type 1 or 3 targets. These signals are distinguished by being correlated in several ways with other targets, particularly being associated in range and Doppler, and with known or derived sources of reflections.

8.5.3.2 Air accidents

In the event of an air accident or incident, the likelihood that its subject is directly observed by a scanning radar at the key moment is about 1%.

In the case where at a time between scan N and scan $N+1$ of a BSR surveillance radar, contact is lost with the target of interest, there is then a further delay between the first loss of contact and the subsequent recognition by the controller or the system that a significant break has occurred. In that case, it is known that for between 4 and 12 seconds following the last plot, the target will not have been observed at all as the radar completes its scan.

HSR surveillance allows for subsequent determinations of behaviour, other target approaches or dismount events, vibrations, engine failures or airframe disintegration, millisecond by millisecond.

In cases where it becomes clear that such an event has occurred – e.g., where an aircraft has been lost from the traffic controller's screen over too long a period – then radar that acquires and retains continuing signal data provides the necessary coverage for remote but real-time event analysis.

Information about all objects and their behaviours within the CVoR is acquired and, given appropriate signal data storage, will remain available. Since the stored data must cover the loss of continuing contact, it is then possible to retrace the action sequence, possibly using target models different from those normally considered for air traffic surveillance.

An example of a target model that lies outside those considered above would be that the aircraft has been subject to unusual dynamics, such as entering a steep dive without otherwise changing in its appearance (cross-section, fading behaviour, etc.).

In this case where signal data is retained long enough to retrace from the point of loss of contact, and to extend with increasing CPI, the situation can be re-evaluated to determine whether either of these models fits the signals observed. In either case, communication with the pilot may have been lost. The ability to detect either anomalous behaviour of the aircraft itself or the approach of a hostile, fast-moving and uncorrelated object offers the opportunity for more definitive event analysis than is available from present-day surveillance.

8.5.4 Aircraft classification

8.5.4.1 Analytical track solutions

As a method of communicating directly with the airframes of targets, what is different about staring radar in comparison with other surveillance methods, and what questions may be answered?

In current air traffic management systems, primary radar surveillance provides reports of objects of sufficient size, apparently travelling in a way consistent with aircraft speeds and dynamics, through track association and filtering.

Holographic staring radar is about measuring all available aspects of aircraft, to be complemented by cooperative systems where possible, but to provide for the unexpected in both military and civil aviation. It maintains continued contact with aircraft in Newtonian trajectories, measuring their evolving position, speed, direction, acceleration and their externally moving parts.

In principle, continuous analytical solutions optimise the information available from the target. Filtering to obtain the most accurate trajectory and most robust classification (Target vs. False alarm) is an integral part of such solutions; therefore, a separate non-cooperative, statistical tracking procedure will have no information to add.

Table 8.2 Aircraft type discrimination

Type	Speed	Altitude	RCS	Props	Rotors	JEM	Route	Group	Agility	Precision Doppler
Civil airliner	Sub-sonic	High, steady	High	None or 2	None	Turbines	Corridor planned	Single	Low	Consistent
Business jet	Sub-sonic	Variable steady	Medium	None or 2	None	Turbines	Free flight	Single	Low	Consistent
General aviation	Low	Low, variable	Medium	1 or 2	None	None	Variable	Single	Variable	Consistent
Altitude bomber	Sub-sonic	Strato-sphere	High	None or multiple	None	None or multiple	Threat	Sparse	Low	Consistent
Ground attack	Sub-sonic	Very low	Medium	None	None	Single, twin	Threat	Group	High	Variable
Jet fighter	High	Agile	Medium	None	None	Single, twin	Agile	Group	High	Variable
Transport	Sub-sonic	Steady	High	None or multiple	None	None or multiple	Planned	Variable	Low	Consistent
Helicopter	Low	Low	High	None	1 or 2, #blades, speeds	Turbines	Variable	Variable	High	Spread, variable
UAV	Low	Low	Low	None or 1	1 to 8, #blades, speed	None or 1	Variable	Single > swarm	High	Variable

Table 8.3 Flight characteristics suggestive of purpose

Type	Engine speeds	Acceleration and diversion	Responses	Evasion/ manoeuvre	Direction	Approach
Altitude bomber	Steady	Small	Large scale diversion	Climb	To target	Maintain altitude
Ground attack	Variable	High	Aggressive diversion	Lateral divert	To target	Reduce altitude
Jet fighter	Variable	High	Persistent engagement	Engage	To threat	Converge
Transport	Steady	Low	Avoidance	Protect	To operation	Risk averse
Take-off	Variable	High	Disciplined	Climb	Runway	Control
Landing	Variable	High	Disciplined	Descent	Runway	Control
Helicopter – operational	Steady	Agile	Persistent engagement	Low altitude/ landing	To operation	Converge
UAV	Variable	Agile	Persistent engagement	Risk resilience	To operation	Converge

8.5.4.2 What kind of aircraft are present?

Secondary radar transmits a message requesting responses from aircraft equipped with the appropriate transponder, through which the target can be identified. Other cooperative tracking systems exist, including satellite navigation and Wide Area Multilateration systems, whose distributed ground stations elicit responses from aircraft, again appropriately equipped.

At various times, the lower cost and higher message information content of cooperative systems has appeared ready to supplant the need for non-cooperative air surveillance; however, unexpected circumstances continue to arise and to emphasise the necessity for independent interrogation of airspace.

With key exceptions regarding unique low-observable aircraft, distinctions can be drawn based on military or civil ATC target histories, as shown in Table 8.2.

In many cases, aircraft routes and grouping provide indications of their participation in civil air traffic or their potential as a threat. However, boxes in blue in Table 8.2 represent aspects of targets that are measured directly in the course of HSR surveillance and may aid air traffic controllers, military commanders or flight investigators.

8.5.4.3 Measurements of intent

It is possible to use particular forms of target information available to a staring radar as on-the-spot analytical, rather than statistical, intelligence.

Evolving measures of acceleration and vector Doppler (available from a HSR network; see Chapter 10) will be valuable in assessing incomers' intentions and responses, illustrated in Table 8.3.

8.6 Conclusions on CPI concatenation and target histories

The potential offered by staring radar for continuing coherent target processing is supported in this chapter by a number of examples linearly modelled on scattering processes under the EM Uniqueness Theorem.

These include aspects of the known CVoR environment, the extension of target analysis based on concatenation of CPIs, the extension of clutter analysis on the same basis for repetitive and persistent sources and the opportunity for gap-free, analytic airspace surveillance, target classification and tracking.

These examples illustrate that when target observation is persistent, and (again) provided appropriate computing resources, the promise of the EM Uniqueness Theorem that adequate information will be communicated to the radar to allow unique solutions to scattering challenges, set by complex and manoeuvring targets, is supported in theory and by numerical analysis.

Experience with early HSR has demonstrated that such information is present in the scattered target returns, and it is regrettable when practitioners in the radar field adhere to the view that the physical effects of such information are limited to momentary detections created within the radar, rather than to continuing integration with target scatter information.

Successful tests have been carried out in various contexts by Thales-Aveillant, and radars in continuing operation have, especially in the case of wind turbine mitigation, demonstrated effective use of holographic staring radar as practised by the company. These generate data that, when paralleled by modelling, confirm the EUNIT and can be developed to yield further and detailed target interpretation.

Chapter 9
Multilook mapping and multipath suppression

Staring radar surveillance incurs multipath propagation for radar targets near large ground structures. This can result in the appearance of 'satellite' target artefacts in the azimuth plane that may threaten the one-to-one association of reports with RToIs and potentially the utility of HSR.

This chapter provides, in some detail, the physical basis on which HSR satellite returns are identifiable and how the information they convey can be decoded, and describes why that information provides for their suppression. For radar practitioners whose constant and justified focus is on minimising false alarms, satellites generated by scanning radar are anathema, but are relatively rare. For HSR they occur more frequently, but persistent, coherent observation makes them identifiable as features of propagation in the CVoR whose origin can be found and the effects suppressed.

A Huygens model is used to help the reader to visualise multipath propagation along different paths and in simple examples in support of subsequent descriptions of the effect in the field. All the processes are linear, and more congested clutter situations can be treated as linear superpositions of simple examples.

This level of detail may be excessive for readers on the first introduction to the HSR radar mode. In summary, the EUNIT and the reciprocity theorem provide that satellites occur in a specific pattern for each combination of HSR, RToI and scatter source. This pattern is directly recognisable, and the satellite signal information can be captured, identified and suppressed, as described below. For those readers, and subject to this conclusion, the following detail may be reserved for later reading.

9.1 Sources and effects of multipath

The presence of clutter is an occupational hazard for surveillance radar. Efforts to minimise it are based on safeguarding regulations and site selection; it remains a performance-limiting factor, but static clutter responds to Doppler filtering. High-performance radars often use a clutter map to respond appropriately to signals, especially with low Doppler shifts, received from known, fixed clutter. The map may be developed on installation and commissioning of the radar and may be updated as new, persistent clutter targets evolve.

Clutter with high radar cross sections such as some terrain forms, buildings and structures within the CVoR can also cause secondary multipath reflection and scattering from targets, all of which may interfere with direct signals or may cause false target reports. Single-look, scanning radars are affected in this way when a clutter item reflects the beam both towards and back from a target, with a reflection loss in both directions that may preclude detection. For HSR, clutter due to azimuth multipath has been observed as a result of scattering from wind turbine towers and large buildings and may contain non-zero Doppler components, but the precise relationships between those signals and the targets themselves need to be explored.

Radar that transmits and receives in all azimuth directions will experience more instances of multipath propagation than will a narrow-beam radar. The EUNIT applies to all the contents of the volume of regard in reflecting and scattering the irradiating field to the receiving aperture. A first approximation limits the solution to direct line-of-sight propagation, but a staring radar will also observe the presence of indirect or secondary multipath propagation. There is then a basic concern that for a staring radar the increase in instances of multipath will create many unwanted targets, thereby severely stressing the processes of detection, discrimination and track filtering.

A clutter map may provide a first indication of the location of sources of scattering, but the actual effect of multipath depends on the size, materials, position and shape of the structures. Robust solutions for HSR surveillance must seek to accommodate these signals. Multiple scattering, in the context of the EUNIT, reduces the intensity of signals in secondary and higher order multipaths but may leave residuals that emulate interesting targets. Not only does the presence of multipath generate unwanted returns for a staring radar at different directions and ranges; these aspects of clutter also acquire non-zero Doppler shifts that will defeat simple low-Doppler clutter rejection.

Under the EUNIT, there are no 'false returns'. All radar returns contain coherent information about the CVoR; the question is whether the information can be resolved, processed and used to report RToIs only. This chapter focuses on that issue and its implications both for performance and cost of surveillance. If the process is valid, then the information encoded in the returns will permit the multipath components to be identified, suppressed and discarded from reports and may also be exploited in forming a target's trajectory. Functions of detection, discrimination and tracking may also be subsumed into the signal analysis.

If the radar signal process is based on the assumption of only one look into the VoR, then the radar returns arising from azimuth multipath (referred to here as 'satellites' since they tend to occur close to targets and to scatter sources) inevitably lead to false reports. As indicated in Chapter 7, they present what could be the most cogent objection to staring radar. To overcome that objection, it must be shown that the 'one look' solution can be replaced with a multilook form that (a) generates a fast, robust and correct solution, and (b) is affordable. Such a solution would identify, associate, interpret and suppress these satellites: it will become a necessary part of a general staring radar solution and may also offer broader benefits in surveillance.

In this chapter, we indicate what the form of such a solution may be, where the necessary information may exist in the signals and how it can be decoded among the known signal variables.

For a staring radar, multipath propagation routes constitute several 'looks' at a target, and the direct, first arrival route encodes the actual target position. Associated secondary routes will have longer delays, and some will encode the direction of the scatter source as opposed to that of the target.

9.1.1 Interfering surface multipath

In the absence of multipath and assuming far-field conditions, a target return should be a plane wave at a constant amplitude throughout the receiving array. If the amplitude varies either across or down the array, that is a direct indication of the occurrence of interfering multipath propagation on reception. Measures to process and recover its effects have been introduced in Chapter 7 in relation to the variation of received field strength with a position in the vertical dimension on the receiving array, at a given range and Doppler frequency, arising from surface multipath (SMP).

In summary, and for completeness here, the vertical interference pattern formed on the array is a direct indicator of the vertical direction of arrival of the target signal and any reflection in the earth or ocean surface. Its depth from constructive to destructive interference is also a direct indicator of the reflecting characteristics of the surface at the range of the surface reflection (which can be inferred from the height of the radar and the topography of the surface) and the look direction after receive beamforming.

The reflection coefficient will vary with weather conditions and also with the state of use of the surface (with agricultural uses in particular), but the topography may be expected to be steady over time. A map may be developed from these inferred reflection characteristics, provided that target fading itself is taken into account – that is, by averaging over observed fading patterns.

9.1.2 Shadowing and absorption by buildings

Shadowing has been discussed in outline in Chapter 8 in relation to further effects of wind turbines. Other structures and buildings may also have such effects and should be considered in relation to possible multipath propagation.

The field strength distribution to be expected from a radar target is calculated from the radar equation in free space. Intervening material that causes absorption, reflection or scattering along the propagation path will modify received field strengths. Under the EUNIT, the target will then be illuminated by diffraction around the obstruction, and scattered signals that intersect with the receiving aperture will be modified in each direction. This may introduce blind zones for any practical radar, which must be minimised through safeguarding rules and procedures, according to which the area surrounding a radar must be protected from shadowing buildings and structures.

The effects of large structures depend on their size, materials and shape, the distances to and directions of structure and target, and the operating wavelength. The Fresnel diffraction zone (the 'near field' for diffraction purposes) extends beyond the shadowing structure and will dominate the irradiation and reception signal amplitude up to a distance known as the Fresnel diffraction zone beyond the obstruction, and bounded by:

$$\mathbf{R}'_{\mathbf{Fresnel}} <\sim 2.\mathbf{D}^2_{\mathbf{obs}}/\lambda, \tag{9.1}$$

where $\mathbf{D}_{\mathbf{obs}}$ is the diameter of the obstruction.

9.1.2.1 ATC shadowing

Large structures and buildings may well cause the loss of targets flying at low altitude and should be avoided by appropriate safeguarding. For targets within the shadow of large structures, their effect under the EUNIT is that of diffraction around the structure edges, both in transmission and reception. It depends on the details of the obstruction and will typically yield attenuated returns.

The region affected by shadowing is that directly behind the structure where the range differentials caused by near-direct diffraction are large, and angles steep. These effects are best considered in terms of local loss of sensitivity rather than as examples of multipath.

9.1.2.2 SRC shadowing

For SRC at S-band and a building 40 m wide, the shadow effect would extend 32 km beyond the radar range of the building. Its effects, especially for HSR if not appropriately positioned and adequately safeguarded, may be significant and in extreme cases (as suggested above) are likely to constrain the effective coverage area of the radar. In this case safeguarding rules are a necessary defence.

9.1.3 Non-interfering azimuth multipath (NIMP)

Safeguarding rules under which radar sensors are deployed should ensure that surveillance radar is not adversely affected by buildings and structures within a limited part of the CVoR. However, scattering in azimuth can result in unwanted detections, especially for targets at low altitudes.

There are cases in which azimuth-scattered signals are received within the resolution bounds of the radar and cause interference. Unless the position and the radar cross section of the scattering source incur only a small loss factor (see Section 9.3), the effect will be relatively minor; it will occur within a narrow azimuth range and will have the limited effect of low-reflectivity interfering multipath.

Effects of azimuth multipath are mainly observed where the secondary scattering routes lead to delay differences such that satellite returns are delayed by more than the pulse duration and can be detected as smaller but separate targets. In this case, instead of interfering with the direct target reflection, range-resolved multipath returns give rise to satellite returns, as indicated in Chapter 7, Sections 7.3.3 and

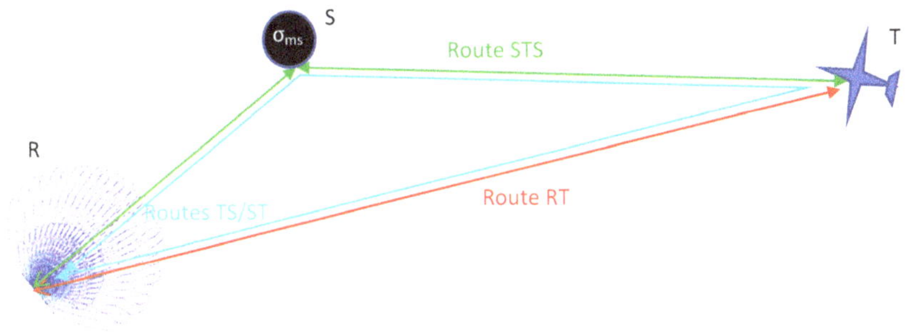

Figure 9.1 *Alternate propagation and scattering routes define ranges* $\mathbf{R}'_{RT}$, $\mathbf{R}'_{ST}$, $\mathbf{R}'_{TS}$ *and* $\mathbf{R}'_{STS}$

7.3.4. These are consistent with the EUNIT, and the multiple propagation routes (provided the coherence and persistence of HSR) provide additional target and trajectory information that can be exploited first to suppress unwanted reports. Below are outlined methods for recovering and potentially enhancing performance under these conditions.

We have seen in Chapter 3 that the radar range equation includes a loss term reducing the received power by the fourth power of the range. This arises from the product of two area ratios, in propagation to the target and in the echo propagating as a scattered signal to the array. Each is inversely proportional to $\mathbf{R}'^2$**.** The term $\mathbf{R}'$ is used to refer to range values that are not measured in geographical terms but are inferred from delayed times of arrival of radar signals, multiplied by $\mathbf{V}_c$.

Multipath propagation via an intermediate object takes the form of scattering or reflection, as illustrated in Figure 9.1, in which the different propagation routes are labelled. It depends on the form of the scatter source within the first Fresnel zone (**fFz**) around the secondary propagation route (**TS/ST** in Figure 9.1). For a highly regular and specular reflector extending beyond the **fFz**, the signal intensity received is reduced below that from the direct route by the reflection loss in the surface and by the inverse square of the fractional increase in the total path length for each reflection (offset by associated changes in the **ST** or **RT** path lengths). Each such route is a straightforward case of multipath, in which the presence of the multipath return will be clear, and its effect stable and localised. It can be considered and treated under 'optical' propagation rules, and once again should be provided for under safeguarding and commissioning rules and activities. The received power includes a third loss term for the scatterer cross-section and its distance from target or radar.

More common are multipath returns via intermediate scattering objects or surfaces (i.e., not large and specular reflectors). These may be buildings or other structures. A particular case is that of observing targets beyond a wind farm, whose turbines stand on tall, cylindrical towers, each of which can be a source of multipath satellites. This is a known and difficult case and deserves detailed attention. We shall

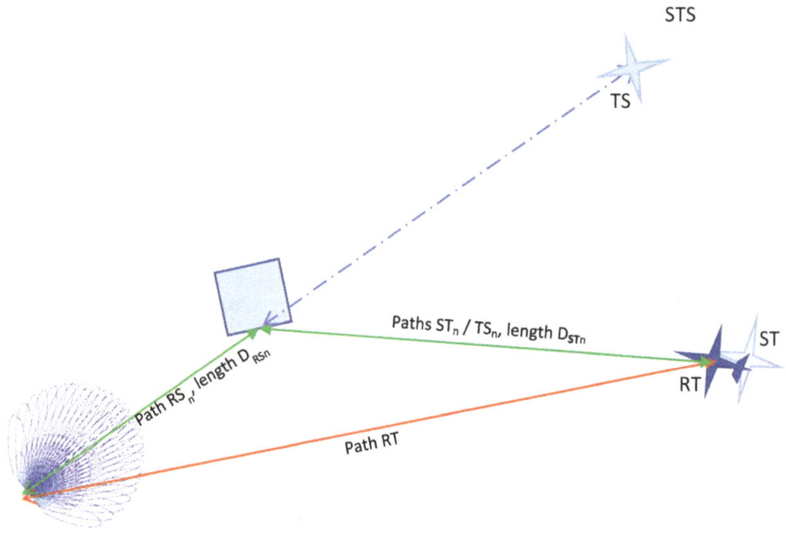

Figure 9.2 Outline for azimuth multipath with point-to-point propagation paths (from Figure 7.5)

focus on this scattering condition, as illustrated in Figure 9.1, assuming large scattering cross sections in each case.

In Figure 9.1, this notation distinguishes the multiple return routes available, based on propagation paths **RT**, **RS**, **ST** and **TS** and generating target **RT** and satellites **TS**, **ST** and **STS,** as in Figure 9.2:

1. **RT**: Radar–Target–Radar;
2. **TS**: Transmit to the target; receive via the scatter source; Radar–Target–Scatterer –Radar;
3. **ST**: Transmit via the scatter source; receive from target; Radar–Scatterer–Target –Radar;
4. **STS**: Transmit/receive via the scatter source; Radar–Scatterer–Target–Scatterer –Radar;

Where the path lengths differ by more than the radar range resolution, typically for multipath in the azimuth plane, it results in resolved satellite targets as in Figure 9.2. This chapter focuses on their decoding.

HSR experiences more complex multipath effects than a BSR in the same situation. For BSR, TS and ST satellites are only observable via sidelobes, with the accompanying losses. STS satellites are as for HSR. Figure 9.2 reflects Figure 7.5, illustrating the HSR case where, in addition to satellites arising at ranges $\mathbf{R'_{TR}} = \mathbf{D_{RT}}$ and $\mathbf{R'_{STS}} = \mathbf{D_{RS}} + \mathbf{D_{ST}}$, as would be observed with a narrow beam, satellite returns can also occur via Paths **TS** and **ST**, both at apparent ranges

$\mathbf{R'_{ST}} = \mathbf{R'_{TS}} = (\mathbf{D_{RT}} + \mathbf{D_{TS}} + \mathbf{D_{RS}}) / \mathbf{2}$ for satellite ST in the direction of the target and satellite TS in the direction of the reflector.

In this chapter, we explore the interpretation of resolved multipath satellites, recognising that from the perspective of the EUNIT these additional propagation routes are legitimate components of the CVoR. They can be solved, and they provide 'second looks' at the target.

From the perspective of target motion and Doppler frequencies, satellites are not restricted to reasonable Newtonian dynamics – for example, if the target passes immediately by the scatter source, the ST and TS satellites will swap from incoming to outgoing Doppler in a very short space of time, with a short and highly dynamic apparent turning or accelerating trajectory. Satellites also represent targets that appear and disappear within the CVoR as the target approaches and recedes from the scatter source. They, therefore, violate the EUNIT if interpreted as independent targets. Because of the additional losses they tend to be observable only within a small radius of each scattering source. The reporting process within a clutter-congested area may exploit the measured target position (the highest amplitude) to predict where satellites are likely to occur and use the actual measurements first to suppress the satellites but potentially to update the scatter source characteristics and increase the accuracy of the target position.

9.2 NIMP scattering and measurement

Chapter 7 (Section 7.3) introduced how the effects of surface multipath may be recovered or resolved as part of the surveillance operation, without compromising the manageable range of target characteristics and without violating any physical laws. A basis and a method for NIMP satellite processing is also required: here it is given in some detail to explore its probable effectiveness.

We have referred to the meaning of measurements made by the radar and to the physical reality of the CVoR, the radar, the targets and the signals that travel between them. The symbol **R′** is range inferred by the radar on the basis of a 'there-and-back' assumption.

9.2.1 Secondary satellites

Secondary satellites (**ST** and **TS** as above), again, are formed with single multipath scatter or reflection events. They share exactly the same range increment and the same Doppler shift but yield reduced amplitudes and appear at different azimuth directions. Their equal Doppler shifts depend on the direction of target motion with reference to the paths joining target and radar and target and scatter source. The necessary information to identify these returns will be encoded in each return received at the array.

9.2.2 Tertiary satellites

Tertiary satellites (Route **STS**) form when the same multipath route is repeated on transmission to and reflection from the target. They exhibit twice the range increment as the secondaries, twice the loss in amplitude and twice the Doppler shift, which also depends on the azimuth direction of travel. They are subject to twice the loss factor (in dB) of the secondaries compared with the primary return.

Higher orders of multipath exist under the EUNIT, but because of the geometry of objects on the scale of tens of metres distributed within a CVoR on the scale of kilometres, each scattering tends to reduce the signal level by in the region of 40 dB. The first step is to find a solution for secondary satellites, arising from a single reflection.

9.2.3 *Measured positions, Doppler and time*

To be consistent with the reality of propagation, we need to maintain a separation between physical distances and radar range measurements. Physical distances ($\mathbf{D}_{XY}$) are what would be measured by a ruler between two physical points, or the objects located there. Radar ranges (**R'**) are measurements made by the radar in which the delay at which a return is received is interpreted in terms of distance to a target by multiplying by one-half light speed. The assumption that the signal travels the same distance twice is true for the direct route but not for multipath, so that confusion can result between distance and radar range.

Here, **D** indicates distance on the ground. $\mathbf{R'}_{RT}$, $\mathbf{R'}_{ST}$, $\mathbf{R'}_{ST}$ and $\mathbf{R'}_{STS}$ denote the apparent ranges, including scatter paths, inferred from the return delays τ_{RT}, τ_{ST}, τ_{TS} and τ_{STS}. $\mathbf{P}_L$ is the total path length: $\tau = \mathbf{P}_L / \mathbf{V}_C$ and $\mathbf{R'} = \mathbf{P}_L / \mathbf{2}$. The differential ranges for satellites are $\Delta\mathbf{R'}_{ST} = \Delta\mathbf{R'}_{TS}$ and $\mathbf{R'}_{STS}$.

Doppler shifts are measured for target and satellites as $\mathbf{F}_{DRT}$, $\mathbf{F}_{DST}$, $\mathbf{F}_{DTS}$ and $\mathbf{F}_{DSTS}$. They are proportional to the target speed $\mathbf{V}_{TT}$ and will allow that to be derived as a Doppler shift $\mathbf{F}_{DTT}$, but they vary differently as the target position and flight direction change.

9.2.4 *The RAED data structure in satellite suppression*

To suppress secondary and tertiary satellites, the relevant signal information is to be found primarily in the FD beam-wise segment of the **RAED** structure (Figure 6.1), but rapid access to associated time-domain segments will also be needed and available.

9.2.5 *Scatter source information*

To achieve a multilook target process, information about scatter sources within the CVoR must be available and can be entered into the HSR system in three ways: from prior and updated information, from direct clutter measurements and in the course of each time step by inference from target satellites.

9.2.5.1 Prior information

The installation of HSR sensors should, as far as practicable, be subject to safe-guarding in the same way as other surveillance radars. Where possible, there should be clear zones near the radar, with slopes of exclusion such that nearby structures provide minimal reflection paths to the radar.

As with current-generation surveillance systems, the installation needs to be carefully placed and prepared. Prior information on positions of structures can be used to provide a starting point for clutter mapping, also using detailed terrain maps and building dimensions with a computer simulation. The location, shape and orientation of steep terrain, tall buildings, towers, broadcast masts, wind turbines, etc., should be referenced in choosing the positioning of the radar but also in preparing details of the radar data process.

9.2.5.2 Direct measurement

During commissioning of an HSR, position measurements of static clutter likely to cause reflection and scattering from targets can be initiated by measuring zero-Doppler returns in each resolution cell. The scale of target returns and their persistence over a period give initial guidance on their likelihood to generate secondary scatters from nearby RToIs. However, multipath reflections tend to occur obliquely in building faces that do not provide direct clutter returns to the radar.

It may be possible to use detailed numerical, geometrical models of a built-up area or a suitably equipped UAV to estimate or determine the scattering characteristics of a CVoR segment. However, this would be difficult to make fully representative of the clutter and scattering effects to be expected for the HSR.

9.2.5.3 Inference from 'satellites'

Signals scattered from a secondary source themselves encode information about the CVoR contents, including their source. This chapter focuses primarily on that information and its use in Sections 9.3 to 9.5, also using a numerical model that simulates the propagation of EM waves to, between and back from objects within the CVoR. The model applies the bounds of physical constraints (Chapter 3) to specific propagation geometries.

9.3 Reflection and scattering geometries and satellite ranges

The radar range equation (3.1) embodies two-loss factors containing the ratios of the target RCS to the area product $\mathbf{4\pi D^2}$ and of the receiver effective area to the same product:

$\mathbf{P_r = P_t.G_t.G_r.\sigma Tg.(\lambda^2/4\pi) / (4\pi D^2)^2}$ (leaving out processing gains $\mathbf{G_{coh}}$ and $\mathbf{G_{inc}}$ from (3.1)

Secondary scattering includes a third loss factor on the scale of the ratio of the scatter source cross section $\sigma_{\mathbf{TSR}}$ to the term $\mathbf{4\pi D^2_{ST}}$, where $\mathbf{D_{ST}}$ is the distance from the scatter source to the target or to the radar, whichever is smaller, for routes **TS** or **ST**. Multipath is most significant for scatter sources closer either to the radar or to the target than the target is to the radar. Tertiary scattering (in both transmit and receive directions) includes a fourth loss term at a similar scale for route **STS**.

Below, in describing the propagation geometry, distances ($\mathbf{D_{XY}}$) are one-way scalar values where X and Y indicate the end points (**R, T** or **S**).

Ranges, as measured by the radar, are $\mathbf{R'_{RT}}, \mathbf{R'_{TS}}$ $(= \mathbf{R'_{ST}})$, or $\mathbf{R'_{STS}}$ for the four complete routes.

The azimuth directions from radar to the target and scatter source are α_T and α_S and the azimuth offset between target and scatterer, $\boldsymbol{\Delta\alpha_{TS=\alpha_T-\alpha_S}}$.

Object scatter cross sections vary with the directions of irradiation and scattering. The target is irradiated from directions **R** and **S**, and observed after scattering in directions **S** and **R**, with different scatter cross sections for propagation routes **TS**, **ST** and **STS**.

9.3.1 *Sensitivity, range and Doppler for satellite propagation*

Propagation for a target **Tg** and scatter source **Sc1** is observed by a radar **R** via three paths of lengths $\mathbf{D_{RT}}$, $\mathbf{D_{TS}}$ and $\mathbf{D_{RS}}$. The propagation routes, starting and ending at point **R** are labelled **RT**, **ST** (radar return from radar), **TS** (radar return from scatter source) and **STS** (both directions via scatter source).

The radar cross section $\boldsymbol{\sigma_{TSR}}$ of the scatter source for the alternate paths **ST** and **TS**, by the reciprocity theorem, is equal. For either asymmetric path (**ST** or **TS**), the received signal amplitude differs only in relation to the array beamforming at the correct receive direction.

By the direct route, the measured range $\mathbf{R'_{RT}} = \mathbf{D_{RT}}$. By routes **ST** and **TS**, the measured range is $\mathbf{R'_{ST}} = \mathbf{R'_{ST}} = (\mathbf{D_{RT}} + \mathbf{D_{TS}} + \mathbf{D_{RS}})/2$, and via route **STS**, $\mathbf{R'_{STS}} = \mathbf{D_{TS}} + \mathbf{D_{RS}}$.

Multipath propagation for target **Tg**, illuminated from the direction of the radar along path **RT** at a distance $\mathbf{D_{RT}}$ and azimuth $\boldsymbol{\alpha_T}$, is seen with cross section $\boldsymbol{\sigma_{TRS}}$, via a secondary scatterer at a distance $\mathbf{D_{RS}}$ and with effective cross section $\boldsymbol{\sigma_{Sc1}}$ at azimuth $\boldsymbol{\alpha_S}$, along route **TS** at a delay corresponding to range $\mathbf{R'_{TS}}$ (equal to $\mathbf{R'_{ST}}$; Eq. 9.4).

The staring radar equation (3.6) for a transmitter with gain $\mathbf{G_t}$ and receiver with effective area $\mathbf{Ae(\alpha_{S,T}) = G_r(\alpha_{S,T}).(\lambda^2/4\pi)}$, via either route **ST** or **TS**, becomes:

$$\mathbf{P_{rST,TS}} = \mathbf{P_t} \,.\, \mathbf{G_t} \,.\, \mathbf{Ae}(\alpha_{S,T}) \,.\, \alpha_{TRS} \,.\, \alpha_{Sc1} \,/\, ((\mathbf{R'_{RT}}^2 + \mathbf{D^2_{RS}} - \mathbf{2.R'_{RT}} \,.\, \mathbf{D_{RS}} \,.\, \mathbf{cos}(\alpha_{TS})).\, \mathbf{R'_{RT}}^2 \,.\, \mathbf{D^2_{RS}} \,.\, (\mathbf{4}\pi)^3) \tag{9.2}$$

For double scattering (i.e., via route STS), the loss, with the target cross section $\boldsymbol{\sigma_{TSS}}$ for both views from and back to the scatter source, has the form:

$$\mathbf{P_{rSTS}} = \mathbf{P_t} \,.\, \mathbf{G_t} \,.\, \mathbf{Ae}(\alpha_S) \,.\, \alpha_{TSS}.\alpha_{Sc1^2}/((\mathbf{R'_{RT^2}} + \mathbf{D_{RS^2}} - \mathbf{2.R'_{RT}} \,.\, \mathbf{D_{RS}} \,.\, \mathbf{cos}(\alpha_S))^2 \,.\, \mathbf{D_{RS^4}} \,.\, (\mathbf{4}\pi)^4) \tag{9.3}$$

The **STS** and higher order satellites are reduced in amplitude by additional scattering loss terms compared with the **TS** or **ST** routes and are not essential in this discussion of satellite radar returns.

The multipath satellites formed by a staring radar are directly correlated with the target, at extended range and modified Doppler frequency shift and phase, and their apparent positions are determined by the target and the reflector. These multipath

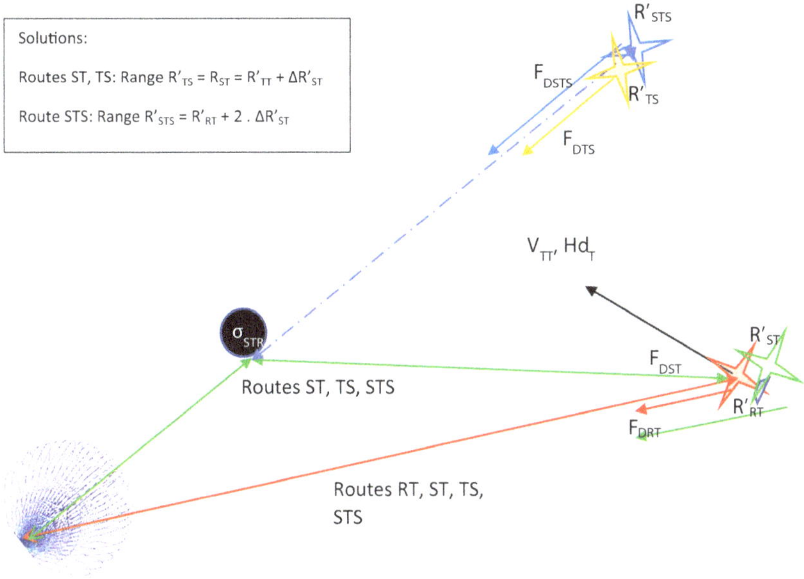

Figure 9.3 Relative locations of target and satellites

returns, for HSR, are coherently related and encode the relation between target and scatter source.

The return via a scatter source relatively close to radar or target is reduced below that of the direct return by approximately the loss ratio:

$$\mathbf{L_{ST}} \approx (4\pi\ .\ (\mathbf{R'_{ST}}2 + \mathbf{D_{RS}}2 - \mathbf{2.D_{RS}}\ .\ \mathbf{R'_{ST}}\ .\ \mathbf{cos}(\alpha_S - \alpha_T)))\ .\ (\alpha_{STR}/\alpha_{TRR}) \tag{9.4}$$

For a scattering cross section $\boldsymbol{\sigma}_{\mathbf{STR}}$ = +40 dBsm and a target 30 dB above the minimum, the scattering distance beyond which the scatter signals fall below the minimum is:

$$\mathbf{D_{STMax}} \approx \mathrm{sqrt}(10^7/4\pi) \approx 1\,000\ \mathrm{m} \tag{9.5}$$

At shorter ranges or for a target with large cross section, resolved satellites are likely to be captured unless the reflection route is transitory. The presence and position of secondary scatter sources are, therefore, significant items within the CVoR, and they require explicit, analytic or logical suppression.

Figure 9.3 illustrates the apparent positions and motions of the target and satellites. The satellites are always more apparently distant than their target. These relations are detailed in (9.6–9.10). Their Doppler shifts are related to that of the target but are also strongly dependent on its direction of travel in relation to the scatter source. In this case, the target velocity vector is closer to the path towards the scatter source than to that towards the radar, so scattered Doppler $\mathbf{F_{DSTS}}$ will be less than $\mathbf{F_{DTT}}$, as would correspond to the actual target velocity $\mathbf{V_{TT}}$, but will exceed $\mathbf{F_{DST}}$ and $\mathbf{F_{DTS}}$, which will also exceed the direct radar-target Doppler $\mathbf{F_{DRT}}$.

Routes **ST** and **TS** are reduced in sensitivity, as in (9.2), by a scattering ratio estimated in (9.4), and route STS will be reduced by the square of that ratio as in(9.3). Routes ST and TS, if within the distance D_{STMax}, will result in two, correlated, lower-intensity satellites, and in false reports unless suppressed or otherwise excluded.

If satellite propagation is built into the process, it will be possible, in the presence of NIMP from a known scatterer, directly to derive vector Doppler information for the primary target prior to track filtering.

9.3.2 Modelling satellite propagation

Radar propagation in the presence of substantial ground structures and clutter is complex to predict and difficult to test under accurate, repeatable conditions. Numerical modelling, based on a Huygens decomposition of the structures and waveforms, is an appropriate way to explore and test the effectiveness of staring radar under these conditions, and a model is used here that takes a first step towards bringing NIMP into a manageable framework.

From the standpoint of scanning radar, whose method naturally reduces the occurrence of secondary, but not tertiary satellites, they appear to threaten the singular association of returns with targets. Each target–scatter source pairing generates at least two satellites. The presence of many scatter sources appears to multiply this threat to the extent that tracking and reporting performance will be degraded. The question is whether the satellite returns can be quantified and identified prior to submitting a target plot to the tracking and reporting process.

For satellites observed from different directions at different times (as with a narrow beam), there is no coherent correlation between any **ST**, **TS** and **STS** satellites, and no potential for treating them quantitatively as associated returns arise. Coherence between satellites is provided by HSR and supports a method of satellite classification and suppression.

With HSR, satellite returns have been observed in proximity to substantial scatter sources such as wind turbines but in too few cases to provide a satisfactory account of the general case. Those observations confirm that satellites have tended to occur relatively close to the target, in cases where the flight has been near the position and height of the of scatter source.

To validate the promise of the EUNIT that HSR will provide for clarification of azimuth satellite returns from the radar display and also for improved trajectory information, a simulation of the effect, based on a Huygens propagation model, offers more opportunity to investigate its resilience in this respect. Such a numerical model can place bounds on the character of the satellites, to find whether they can be reliably identified and can therefore be suppressed.

An important aspect of such a model is the necessity to maintain distinct variables in terms of (a) distances as measured on the ground and (b) ranges as derived at the radar by interpreting propagation delay. The apparent equivalence of delay and range is an outcome of the narrow-beam operation and breaks down in the staring case.

In the model, a transmitter generates a radio signal, typically an amplitude-modulated pulse that propagates at light speed into the CVoR, its field strength decreasing proportionately to distance.

The field impinges on all scattering objects, including targets, which become the source for fields scattered in all directions. The intensity varies with direction because real targets are many wavelengths in extent (scale of metres or tens of metres) compared with wavelengths of small fractions of a metre, and their cross section will vary radically with the look directions both from the transmitter and towards a receiver or towards a scattering centre.

The fields scattered by either the target or the scattering source will propagate outward from each, and for secondary scattering will then impinge on the alternate objects (scatter source or target). The secondary scatters, either **ST** or **TS**, propagate outward to each element of the receiving array.

The model calculates all these components. The initial form uses a description of a single scatter source in terms of a large, azimuth-symmetrical scatterer and of the target also as an azimuth-symmetric object. However, a simple (but software-extensive) linear superposition of Huygens scatterers allows for exploration of both complex targets and a CVoR containing multiple multipath scatter sources.

Staring radar in the case of NIMP operates subject to and with the symmetry provided by the EM reciprocity theorem. The RT provides that for signals passing between two antennas via any propagating space (the CVoR and its contents), the signal received will be indistinguishable whichever antenna is used to transmit and which to receive. In the secondary multipath case, this means that the direction of travel between **ST** and **TS** satellite routes will be simultaneous and at the same precise range and Doppler shift.

This symmetry provides the basis whereby the scattered signals can be identified. Once identified they can be suppressed or exploited; the simplest exploitation being first to suppress reports that might arise from them and secondly to locate the scatter source for each secondary satellite pair.

Figure 9.4 illustrates a scenario in which an aircraft (◂) flies a distance of 120 m at a low altitude near a high-cross section item of clutter (O), yielding a direct target return and satellite return at the radar ( ) as outlined in Section 9.2. The target is at (7 000, 100) m with respect to the radar, approaching at 60 m/s. The scatter source is at (5 300, 1 000) m. To achieve visually clear multipath returns with the model at reasonable noise level and distance scales, the scattering cross section is set here at 2 000 000 m^2; +63 dBsm [1]. Typical scatter targets are more likely in the range 1 000–100 000 m^2 or +30–50 dBsm [1]. This would be an extreme case in reality but illustrates satellite signal components that are visually easy to compare. The **ST** and **TS** satellites are equal in range, frequency and amplitude, and they occur at ranges about 150 m greater than the **RT** return.

In associating measurements based on target returns, a **RAED** data structure is helpful, which makes signal characteristics (amplitude and phase) available versus seven dimensions: Range (**Rg**), Elevation (**ε**) and Azimuth (**α**), in array element positions in the vertical (**V**) and horizontal (**H**) directions, and in both the time (**TD**) and frequency (**FD**) domains. This appears to be a large structure, but in reality it

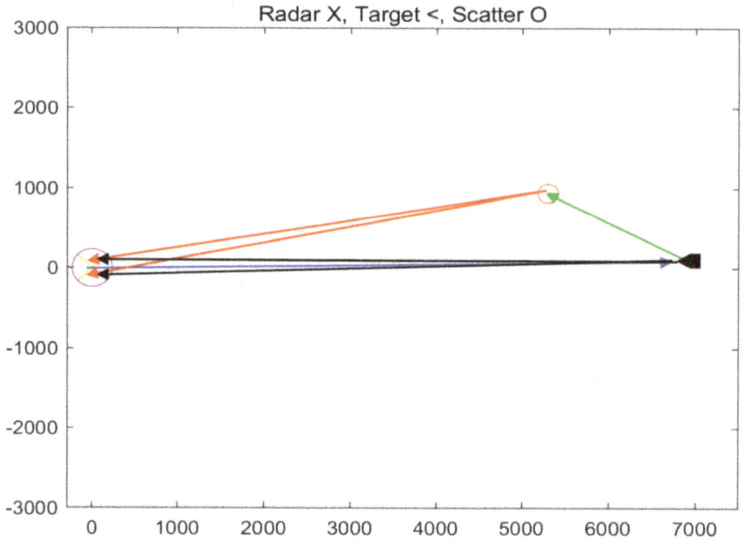

Figure 9.4 *Geometry of scattering, showing positions of radar, target and scatter source (plan view)*

will contain a manageable dataset that can be accessed in the necessary ways after parallel data processing. The processing capacity for ATC is described in Chapter 7. For SRC, the total memory requirement for the RAED is less than 20 GB, which at this date is well within affordable capacity.

Analysis of resolved azimuth multipath as discussed in this chapter may be restricted to **FD**. Others (e.g., the analysis of interfering multipath discussed in Chapter 7) need access to rows of elements prior to forming elevation beams, so the data structure and access to it (illustrated in Figure 3.3) require adequate capacity, access flexibility and speed.

9.3.2.1 Time-domain satellite responses

Time-domain sequences for successive range gates model the amplitudes of direct return (**RT**) and **ST**, **TS** and **STS** satellites, illustrating the range offset between the satellites and the direct return.

Figure 9.5 illustrates time domain passes of the RT, ST, TS and STS returns through resolved range sampling windows, for the scenario of Figure 9.4. The model confirms that ST and TS satellites are precisely superposed. Here, to provide a clear illustration of the progression of range in the time domain, return amplitudes are shown in three successive range gates at RGs 52, 53 and 54 (ranges close to 6 930, 7 070 and 7 210 m) for pulses 1–2 048. The modelled target is travelling at −60 m/s (inward), starting at a distance of 7 km (illustrative of SRC configuration).

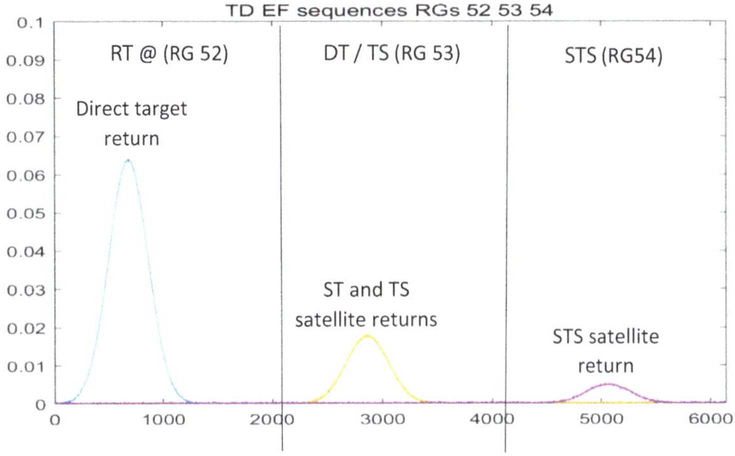

Figure 9.5 *RT (blue), ST (red), TS (orange) and STS (violet) envelope for the same time interval at successive range gates (RGs 52, 53 and 54) for the same time interval*

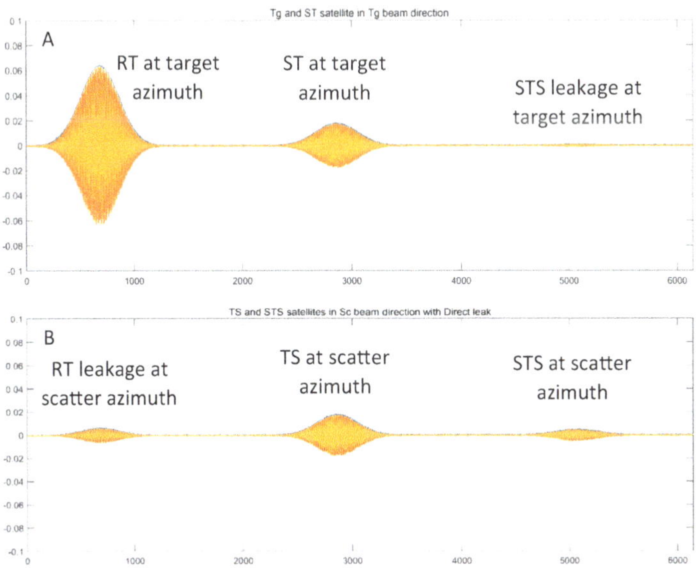

Figure 9.6 *(A and B) Beam-weighted returns for **RT**, **ST**, **TS** and **STS** in the target and scatter directions respectively*

The model returns **RT**, **ST/TS** and **STS** satellites with accurately equal range increments. Noise in each receive channel will of course be uncorrelated. With this geometry, range gate setting, speed and direction, and for the purpose of illustration, the time-domain returns occur near the centre of each time period at each range gate.

The time-domain returns are separated in their azimuth measurements. Relative phase measurements at the receiving array provide direction measurements to the target (**RT** and **ST**) near 0.8° and to the scatterer (**TS** and **STS**) of 10°. These measurements (with a relatively low noise level) compare with set positions of 0.82° and 10.7°.

When resolved into azimuth beam directions, with finite (~ −20 dB) cross-beam leakage, the time-domain returns at the target azimuth and the measured scatter azimuth are shown in Figure 9.6A and B.

As expected, **ST** and **TS** signals are the same in amplitude and range, and their phases are matched in the appropriate beam forms.

9.3.2.2 Frequency-domain satellite responses

In the frequency domain, the returns are illustrated in Figure 9.7. The satellites appear in the different range gates at different Doppler frequencies and reveal different route ranges. Figure 9.7A looks in the target direction; Figure 9.7B looks in the direction of the scatterer.

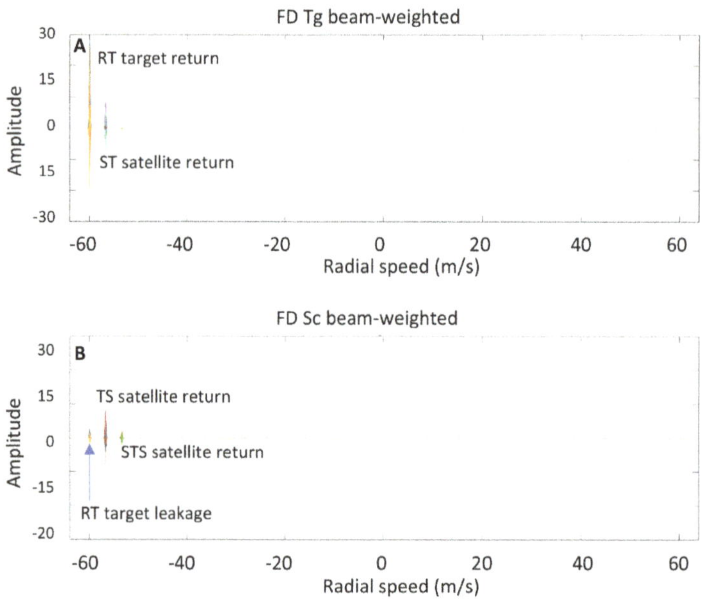

Figure 9.7 *(A and B) Beam-weighted frequency-domain responses* ***RT, ST, TS*** *and* ***STS*** *in the Figure 9.4 scenario*

The model illustrates leakage between the beam forms, but the **ST** and **TS** satellites appear at the same amplitude and frequency at their shared range.

These waveforms and spectra include noise in all signal channels set at a level to permit visual perception of signals in the time domain (see Figure 9.14 for realistic noise in the time domain). Low noise here illustrates the information content but not the limits in sensitivity.

9.3.3 Summary of satellite returns

The characteristics of NIMP satellite returns are:

a. There are two secondary satellites for each target and scatter source.
b. They are at the same, longer measured range than the target itself.
c. Their amplitude is substantially reduced below that of their associated direct target.

- Only scatter sources close either to radar or target can generate satellites of significant amplitude.

d. One single-scattered satellite is in the same azimuth direction as the target.
e. The other is at as the scatter source, yielding its direction.

- These paired, single-scattered satellites are directly identifiable with an identifiable source.

f. A third, dual-scattered satellite arises for each target and scatter source.
g. Its amplitude is further substantially reduced below the single-scattered satellites.
h. If detectable, it is also found in the direction of the scatter source.
i. The dual-scattered satellite is at doubly increased range.

- Resolved satellite Doppler shifts will differ from that of direct targets in each case.

The differences in measured range determine the position of the scatter source, and the different Doppler shifts determine the vector Doppler of the target.

Known or accurately repeated positions for large scatter sources allow for the relative azimuth position of the target to be measured more precisely than from in-process beam positions, especially at longer range, and for its tangential speed to be determined from a single, coherent dwell period.

9.4 Scatter analysis and treatment

To realise the information encoded in NIMP satellites, several process steps are necessary.

9.4.1 Satellite identification

Following signal discovery according to Chapter 6, target candidates are identified. Each is associated with a resolution cell.

Resolution cells are matched with neighbours in range, azimuth, elevation and Doppler shift to derive position estimates for all captured targets. This will imply an increase in the process burden associated with cell matching.

Some of the matching cells will represent direct (**RT**) targets, but some will be nearby **ST/TS** satellite pairs and **STS** satellites, always associated with one **RT** target. Satellite amplitudes will be less than that of the associated **RT** and their ranges will be greater (they are typically weaker in proportion to the scatter cross section as a fraction of the area factor $4\pi \mathbf{D}_{TS}2$, where $\mathbf{D}_{TS}$ is the distance from target to scatter source).

For each target capture, **ST** satellite candidates are sought within an upper range window $\mathbf{R}'_{ST}$ + at the same measured azimuth direction α_T as the target and with amplitude between the target and the capture threshold.

At the ranges of **ST** candidates, **TS** matches are sought at beam directions near the target azimuth. Signals matched accurately in Doppler shifts ($\mathbf{F}_{DST}$, $\mathbf{F}_{DTS}$) and range ($\mathbf{R}'_{ST}$, $\mathbf{R}'_{TS}$) at azimuths α_T and α_S are identified as an **ST/TS** pair, and the associated, higher amplitude signal is identified as a target at $\mathbf{R}'_{RT}$, α_T with multipath satellites.

The **STS** satellite can then be sought at twice the range offset and in the scatter source direction, but may not be captured due to the double scattering loss.

These match searches will be well suited to data organised in a RAED structure.

9.4.2 Satellite exclusion and report clarification

Identified **ST**, **TS** pairs and any **STS** returns are labelled as satellites and excluded from lists of targets for reporting. **ST** candidates not meeting match criteria proceed to tracking as target candidates.

This method of exclusion operates on every single CPI for each resolution cell and promises to remove the risk of radar display contamination with multipath satellites. The physics of the situation prescribes that **ST** and **TS** satellite signals will match in this way when observed by a staring HSR and assures that clarification is achievable.

This process is linear with respect to other targets, and their probability of detection is not compromised except in extreme and coincidental circumstances.

9.4.3 Scatter source inference from multipath satellites

Scatter source positions can be found as follows, and can then be used as part of the clarification process, or for updating scatter maps.

Measurements of satellite range increments and the scatter azimuth allow the position of the scatter source to be derived and reported or stored for future application.

Approaches for this process are given in Section 9.4.4. Once these are established, the Doppler shift measurements can also be exploited.

The satellite Doppler shifts and differences are related to the scatter source position and the direction of travel of the target.

In the scenario of Figure 9.4, the Doppler differences between **RT** and **ST/TS** and between **ST/TS** and **STS** are equal. They are equal to zero when the trajectory bisects the **RT** and **ST** directions.

The processes being outlined here require flexible access to signal data and a RAED data structure described in Chapter 6 (Figure 6.1). Depending on the actual data processing architecture and memory structure, the aim is to allow the identification of data blocks in both domains and azimuth and range order so that a single memory reference will result in the delivery of complex data in time sequences or spectrum series, channel series or cell volumes.

Appropriate processing to bring the suppression or exploitation of satellite returns within the normal operation of an HSR first requires the steps of target candidate capture, satellite identification and target clarification, followed by scatter source location and Doppler exploitation described above.

The discussion above, and the basis for the numerical model used in developing the figures and arguments above, is derived from the variables and relationships described below.

The model is based on what the EUNIT provides and properly represents the known circumstances under which multipath propagation occurs. These effects have been observed in reality in a small minority of circumstances; they may not occur at all in many surveillance cases and situations. However they might also be regarded as potentially disqualifying the case for staring radar and must be assessed in recognition of their potential real effects on radar operators.

Actual scattering sources can only be properly determined, either during a commissioning period or during the operation of the radar, by observing RToIs in flight and analysing their radar returns to measure the effects of scatter sources and satellites. Test flights are needed during commissioning that will provide data about scattering in the CVoR.

In the interim, the modelling approach is appropriate in assessing the nature and likelihood of the threat. Using a numerical Huygens model of the multipath situation, satellite returns can be identified and excluded by means of their matching information and derived information about scatter sources can be maintained and updated on that basis.

We have seen that **ST** and **TS** satellites form pairs that are matched in range and in Doppler (and also closely in amplitude) but not in their azimuth measurement. The association of **ST** and **TS** satellites with their target and with a specific scatter source is uniquely determined by the geometrical progression of the target in the frame of each scatter source and the radar. Satellites will, over time and except in extreme circumstances, demonstrate a more dynamic trajectory than the target.

Staring radar can discriminate satellite returns arising from these multipath events, distinguishing them from real or non-correlated false targets by a series of criteria:

1. They occur in time correlation with the target itself.
2. The **RT** route will be at the shortest range of the group, R'_{RT}, and yielding Doppler shift, F_{DRT}.
3. Satellites **ST** and **TS** are at reduced amplitude and at ranges $\mathbf{R'_{ST} = R'_{TS} = R'_{RT} + \Delta R'_{ST}}$.
4. Satellite **STS** occurs at range $\mathbf{R'_{STS} = R'_{RT} + 2.\Delta R'_{ST}}$.
5. Satellites **ST** and **TS** occur with Doppler shift $F_{DST} = F_{DTS}$.
6. Satellite **STS** occurs with Doppler shift F_{DSTS}.
7. If the range differences are within the range resolution of the radar, they will interfere, with target amplitude beating and positional oscillations.
8. **RT** and **ST** routes yield satellites in the azimuth direction towards the target, and **TS** and **STS** routes yield satellites lying towards the scatter source.
9. R'_{RT} and R'_{STS} will differ from $R'_{ST}(= R'_{TS})$ equally in opposite directions.
10. F_{DST} and F_{DTS} will be equal but will differ from F_{DRT} depending on the target position and direction of motion in relation to the scatter source position.

The relationship between measured ranges and Doppler shifts for **TT**, **ST**, **TS** and **STS** satellites can be used to calculate the vector motion of the target, provided that the reflector position is known or has been estimated.

If the multipath occurs such that neither the ranges, the Doppler shifts, nor the directions of the multiple routes can be resolved, then the positions and motions cannot be calculated separately.

9.4.4 Scatter source position

A target T is captured, in a resolution cell known to be affected by scatter sources, and measured at range R'_{RT}, azimuth α_T and elevation ε_T. For the following discussion using a propagation model, only events at low elevation are considered for azimuth multiple scattering events and elevation terms are excluded on the basis that $\cos(\varepsilon) \approx 1$ in all significant cases.

As described in Section 9.4.1, the target cell is compared with all low-elevation cells at the same azimuth and nearby range to seek candidate satellite cells. Those at longer range but within the maximum such that $\Delta R'_{ST} < D_{STMax}/2$, and with amplitude less than the target, provide numbered candidate satellites labelled ST_n.

The different interpretations of the range measurement process yielding R'_{RT}, R'_{ST} and R'_{TS} are the basis of a multilook radar.

The next step is to search nearby azimuth cells for exact TS satellite matches. If there is no matching TS_n satellite, then the candidate is not a satellite and is processed as a target. If that decision were incorrect (e.g. where a TS satellite is present but not resolved in azimuth), the ST_n satellite will continue to follow the target.

The TS_n satellite (matched in range, Doppler and elevation, but not in azimuth, to the target T) will appear close to the direction of the scatter source n, α_{Sn}, revealing the source's azimuth.

The third requirement is the distance from the scatter source to the radar or to the target.

In Figure 9.2 (after Figure 7.4), routes $\mathbf{ST_n}$, $\mathbf{TS_n}$ and $\mathbf{STS_n}$ involve a large scattering structure, allowing the secondary, reflected signal to form a detectable satellite target. A scatter source such as a building near the radar (for SRC) or a wind turbine tower near the target (for ATC) (such that $\Delta \mathbf{R'_{STn}} < \mathbf{D_{STMax}/2}$) can yield a target capture at the radar.

As described, distance values assigned or known between objects are designated **D**. Source range values measured by the radar are designated $\mathbf{R'_S}$, and azimuth angles measured by the radar are $\boldsymbol{\alpha_S}$. For a scatter source n at a fixed position ($\mathbf{D_{RSn}}, \alpha_{Sn}$) and a target at measured range and azimuth ($\mathbf{R'_{TT}}, \boldsymbol{\alpha_T}$), the apparent ranges of the satellites, $\mathbf{R'_{STn}}$ (also = $\mathbf{R'_{TSn}}$), $\mathbf{R'_{STSn}}$ and $\mathbf{D_{STn}}$ are given by:

$$R'_{STn} = R'_{TSn} = (R'_{RT} + D_{RSn} + \mathrm{sqrt}(R'^{2}_{RT} + D_{RSn}^2 - 2.R'_{RT}.D_{RSn}.\cos(\alpha_{Sn} - \alpha_T)))/2; \quad (9.6)$$

$$R'_{STSn} = D_{RSn} + \mathrm{sqrt}(R'^{2}_{RT} + D_{RSn}^2 - 2.R'_{RT} \,.\, D_{RSn} \,.\, \cos(\alpha_{Sn} - \alpha_T)); \quad (9.7)$$

$$D_{STn}^2 = R'^{2}_{RT} + D_{RSn}^2 - 2.R'_{RT}.D_{RSn}.\cos(\alpha_{Sn} - \alpha_T) \quad (9.8)$$

The value of $\mathbf{D_{RSn}}$ is obtained from the relations:

$$X_{RS} = (R'^{2}_{RT} - R'^{2}_{ST}/(2.R'_{RT} - (R'_{RT} + R'_{ST})/(2.\cos(\alpha_{Sn})) \text{ and } Y_{RS} = X_{RS} \,.\, \tan(\alpha_{Sn}), \quad (9.9)$$

$$\text{So that } D_{RSn} = \mathrm{sqrt}(X_{RS}^2 + Y_{RS}^2) \quad (9.10)$$

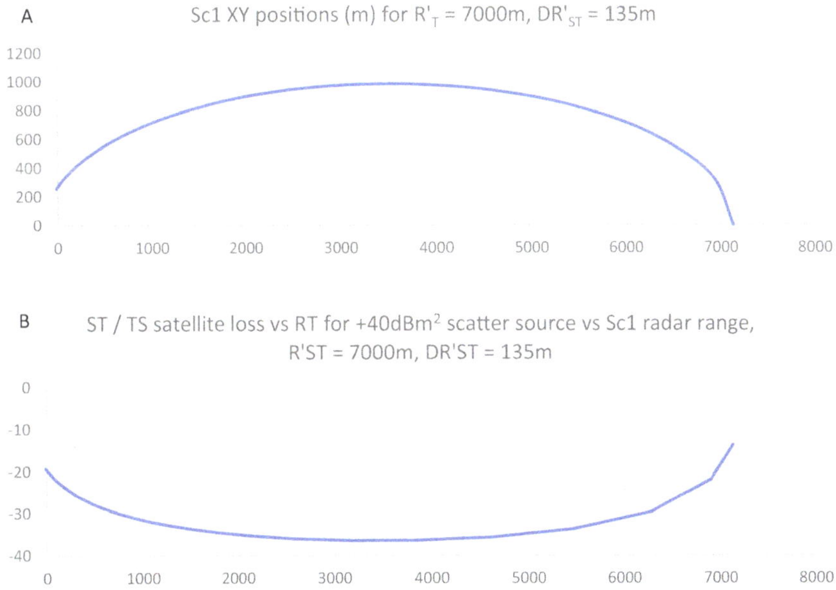

Figure 9.8 (A) Scatter source positions for target and a given satellite range increment $\Delta R'_{STn}$, (B) Power loss for satellite compared with the ***RT*** *target.*

Equations (9.6) or (9.7) can be solved anew for the unknown $\mathbf{D_{RS}}$ in each case, but the measured differential ranges $\Delta\mathbf{R}'_{STn}$ and $\Delta\mathbf{R}'_{TSn}$ and direction offset $\Delta\alpha_{TSn}$ can be used with functions illustrated in Figure 9.8 to infer the distance $\mathbf{D_{RSn}}$ to the scatter source for those satellites, as follows:

A target captured at $(\mathbf{R}'_{RT}, \alpha_T, \mathbf{F_{DRT}})$ with ST satellite at $(\mathbf{R}'_{STn}, \alpha_T, \mathbf{F_{DSTn}})$, $\mathbf{R}'_{RT} < \mathbf{R}'_{STn} < \mathbf{R}'_{RT} + 2\mathbf{D_{STMax}}$, and a return at $\mathbf{R}'_{TSn} = \mathbf{R}'_{STn}$ and $\mathbf{F_{DTSn}} = \mathbf{F_{DSTn}}$ with closely similar amplitude in a cell at azimuth direction $\alpha_{Sn} = \alpha_T, \pm 2\alpha_{res}$ yields the distance from the target to the scatter source from Figure 9.8. As an example from the figure, a satellite with range difference $\Delta\mathbf{R}'_{STn}$ sufficient for resolution (say 150 m for SRC) but within $\mathbf{DST_{max}}$, and at an azimuth offset $\Delta\alpha_{TSn}$ of 20° indicates a secondary scatter source at $\mathbf{D_{RSn}}$ (here ≈ 2 km) and at the satellite azimuth α_S.

The locus of points of constant $\Delta\mathbf{R}'_{STn}$ is an ellipse illustrated (for positive *Y* values) in Figure 9.8A, and the power loss with a scatter source of cross section +40 dBm² in those positions, compared with the target intensity, is illustrated in Figure 9.8B.

Measuring target position and scatter position from the model range and azimuth, with a noise level 20 dB greater than that illustrated in Figure 9.6, gives the results illustrated in Figure 9.9.

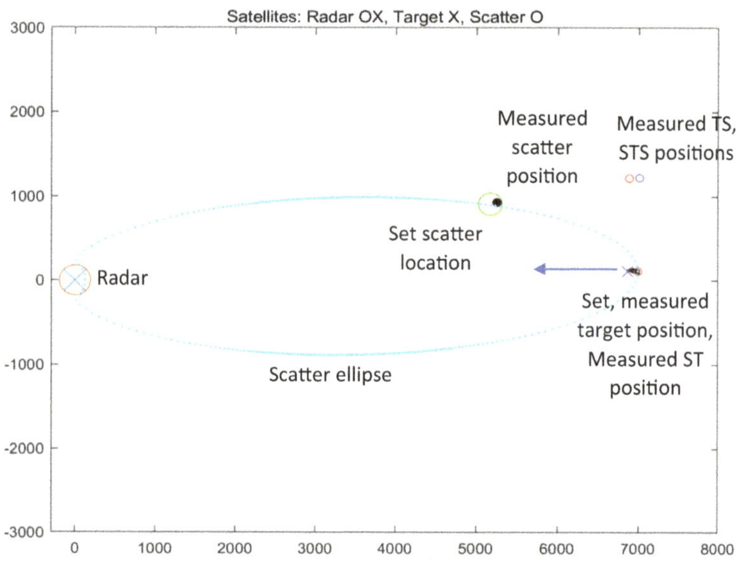

Figure 9.9 Results prior to satellite identification and suppression and scatter source location. Cyan plot is the ellipse matching the scatter source with radar and target as foci.

A modelled ST/TS satellite ellipse is shown in the figure; the scatter source is on the ellipse at the TS azimuth, with a small range error. The symbol X represents the measured (modelled) target position and • being the scatter source position derived within the model. The red O is used both for the ST and TS satellites, one appearing behind the target, the other in the direction of the scatter source, as expected. The blue O represents the STS satellite. The target for Figures 9.8 and 9.9 is heading directly towards the radar.

9.4.4.1 Satellite suppression: radial trajectory

The match between the range and Doppler frequency of the **ST** and **TS** satellites is precise. The **STS** satellite is located near the same azimuth (with somewhat greater error due to its low signal-to-noise ratio) with exactly twice the Doppler increment and twice the range increment. The delay increments measured from the **RT** target to the **TS** and **ST** satellites, if the same, define the ellipse on which the scatter source must lie. The radar is at one focus and the target at the other. The **TS** satellite azimuth determines where on the ellipse the scatter source lies.

This accurate match is not a statistical feature but is a necessary consequence of the EUNIT; the laws of physics would be severely challenged in the absence of these features while in the presence of a large scatter source. When a test for equality between these measurements returns True, the satellites are rejected and the output

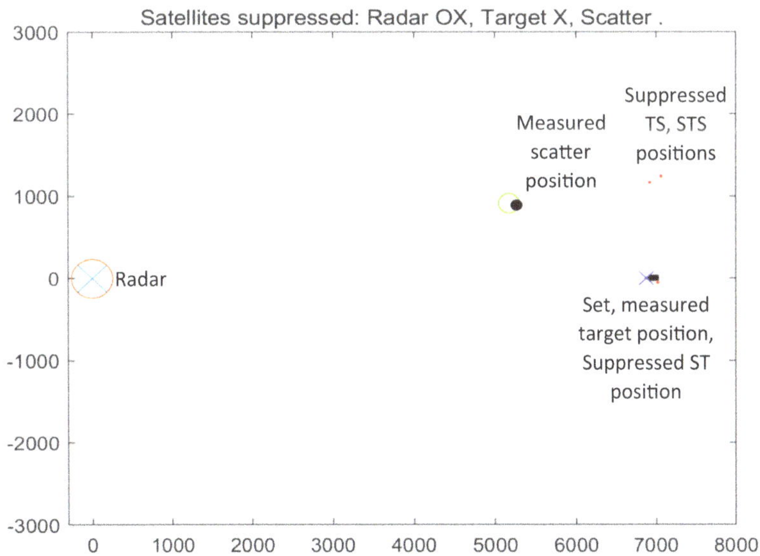

*Figure 9.10 Plan of results after satellite suppression using the **ST/TS** match. Dots indicate the suppressed locations.*

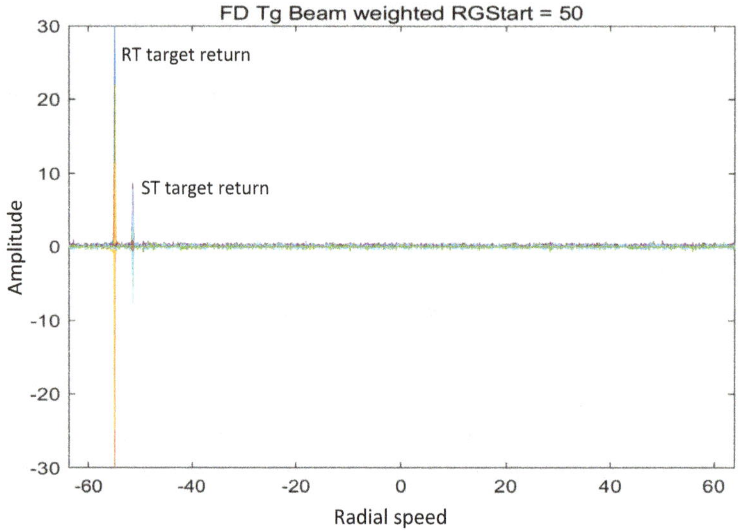

Figure 9.11 *Figure illustrates the Doppler spectrum in a beam weighted towards the target*

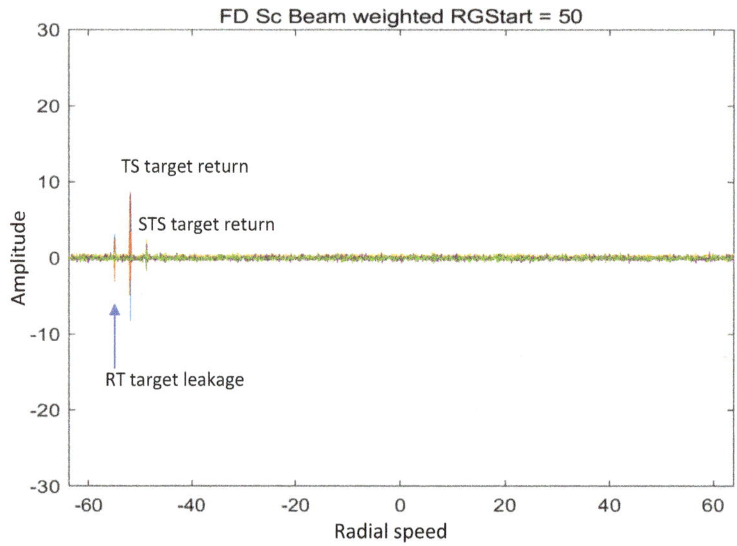

Figure 9.12 *Figure illustrates the Doppler spectrum in a beam weighted towards the scatter source*

plot becomes as shown in Figure 9.10. The Doppler spectra under these conditions are shown in Figures 9.11 and 9.12.

9.4.4.2 Satellite suppression: oblique trajectories

The model, represented in Figure 9.13, now represents a trajectory in which the target moves along the bisector between the directions towards the radar and towards the scatter source. In that condition, the Doppler shift for the ST and TS satellites are equal, and the same as the RT target Doppler, but the ranges remain separate from the target. The model uses a low signal-to-noise ratio for the satellites to illustrate position errors, as in Figure 9.14. Different trajectories are used to illustrate the model catering to different realities.

When heading in this direction, the Doppler spectra converge precisely, as shown in Figure 9.15, where the **RT**, **ST**, **TS** and **STS** outputs are superposed at the same frequency. **ST** and **TS** are at separate target and scatter source azimuths, where **STS** is at twice the range increment.

Figure 9.15 illustrates the **RT**, **ST**, **TS** and **STS** Doppler superposed when the target moves along the bisector between radar and scatter source.

In the frequency domain, the Doppler returns **RT**, **ST** and **TS** are illustrated with increased noise in Figure 9.15. In the time domain, Figure 9.16A and B, A shows the RT target return with the **ST** satellite, while B gives the **RT** leakage with the **TS** satellite and the **STS** return.

This accurate match is a statistical feature in the presence of noise but is a necessary consequence of the EUNIT; the laws of physics would be severely challenged

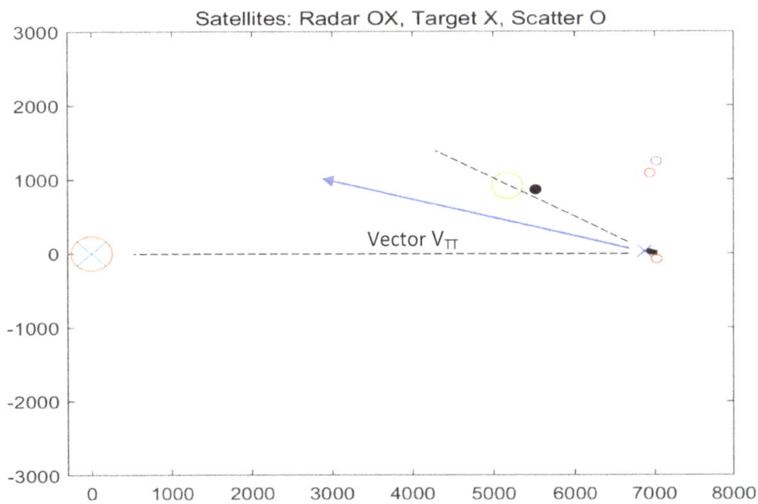

Figure 9.13 *The target trajectory bisects the directions towards radar and scatter source*

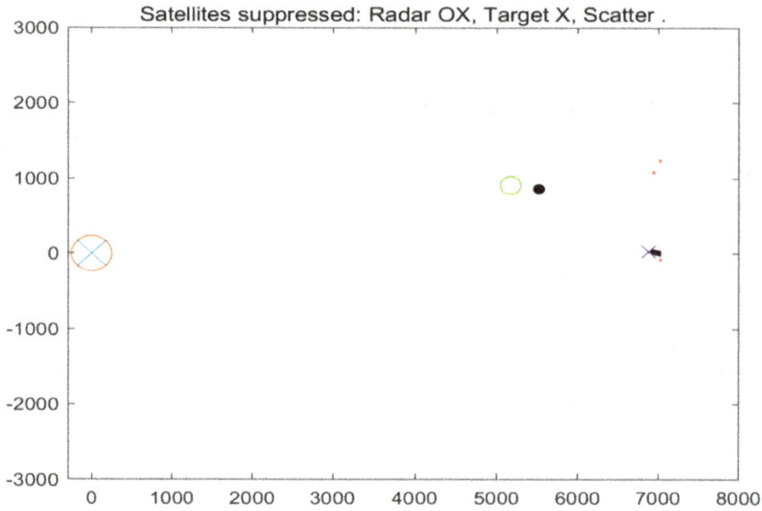

Figure 9.14 *The satellites (.) are matched and suppressed despite a shift in scatter source estimate*

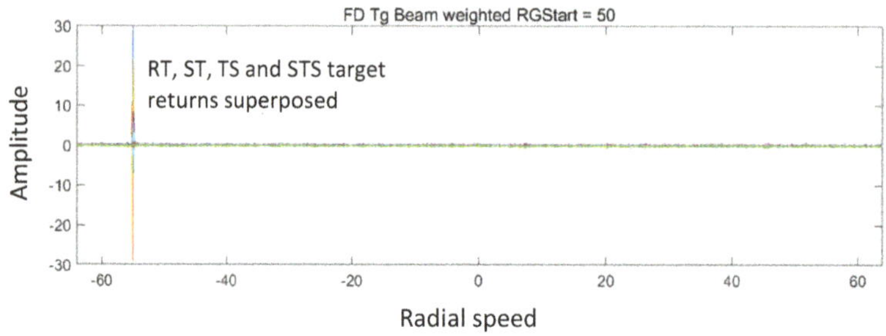

Figure 9.15 *In this bisecting case, the Doppler frequency for the target and all three satellites are the same*

in the absence of these features while in the presence of a large scatter source. When these measurements are equal, the satellites are identified and the output plot becomes, as shown in Figure 9.14.

This demonstrates that the separation of satellite and direct Doppler frequencies is related to the direction of travel of the target. When the scatter source lies between the radar and the target, there is no significant effect arising from multipath, and no information is available to measure it. Significant multipath and satellites arise when the scatter source lies away from the direction to the target and from its direction of travel.

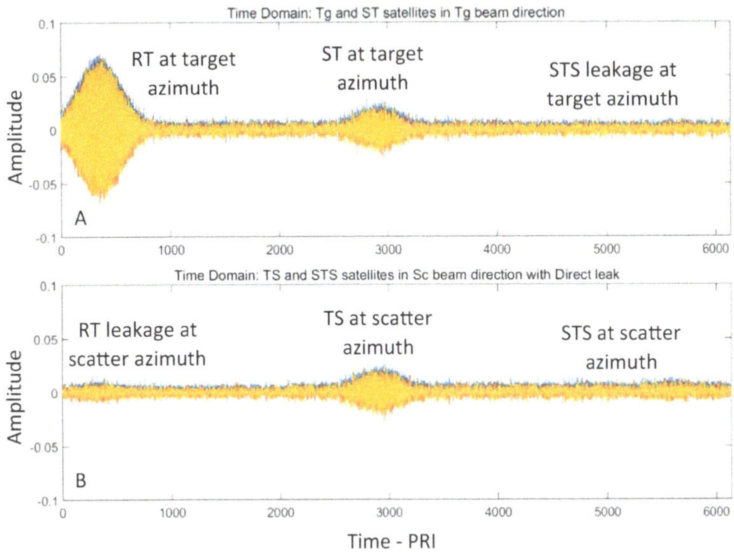

Figure 9.16 *(A and B) Range measurements in the model are performed in the time domain. Low signal-to-noise ratios in that domain yield some errors.*

When the target bypasses a scatter source, the EUNIT ensures that the presence of significant satellite signals can be recognised and in fact allows the vector motion of the target to be measured.

The next model example shows the target moving near tangentially across the bisector between the radar and the scatter source. The Doppler returns are shown at greater frequency scale in Figures 9.17 and 9.18 for beam weightings in the target and scatter source directions, and the plot results in (with the target trajectory) in Figure 9.19, and (after suppression) in Figure 9.20.

It is notable that the **ST** and **TS** satellite components are distinctly spread in Doppler shift. This is because the distance from target to scatter source is relatively short, and in this trajectory, there is an appreciable variation of radial speed during the target's 120 m travel, perceived here as a deceleration.

The use of staring radar may be seen to be prone to distraction by satellite targets that may stress association and tracking processes. The model shows that the signals acquired by HSR satellites, if present, will be recognisable, and associated not as time-separated statistically independent events but as direct, coherent expressions of the EUNIT and the reciprocity theorem. It should be noted that in physics a theorem is a fully proven element of overall theory, in contrast with a hypothesis. Satellite returns can be identified and removed from the reporting function of the radar.

Their use as indicators of the presence of scatter sources and their identifiability as satellite returns is a necessary characteristic of staring radar data, in eliminating

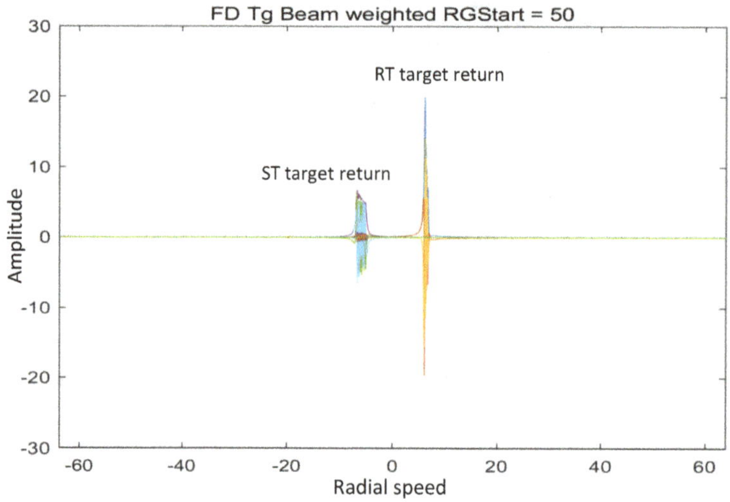

Figure 9.17 A target moving across the radar/scatter bisector with Doppler components of opposite sign, weighted in the target direction

satellites as false targets. This model has shown its functionality in the presence of single targets and scatter sources, and the ability to suppress satellite returns under those conditions means that the apparent vulnerability of staring radar to azimuth multipath degradation is fully recoverable when the appropriate signal information

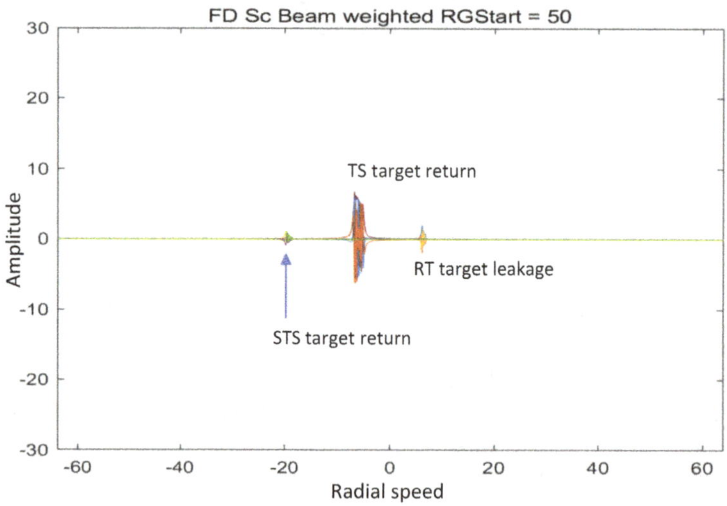

Figure 9.18 As shown in Figure 9.17, weighted in the scatter source direction

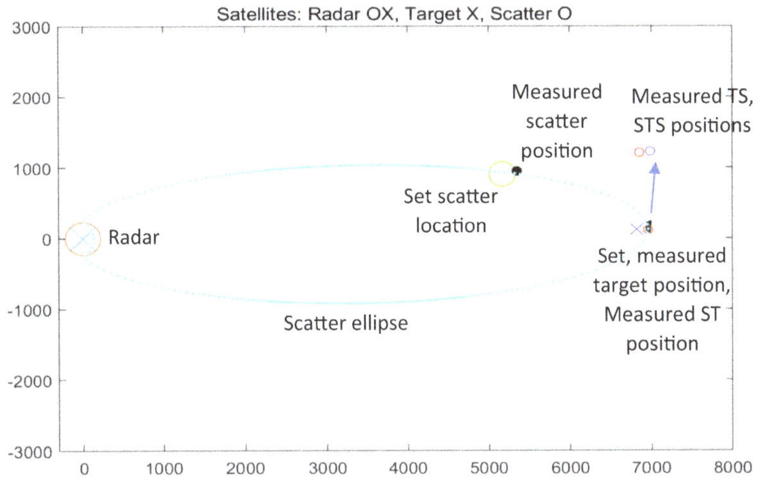

Figure 9.19 *Geometry: Target crossing the radar/scatter bisector with plotted satellites*

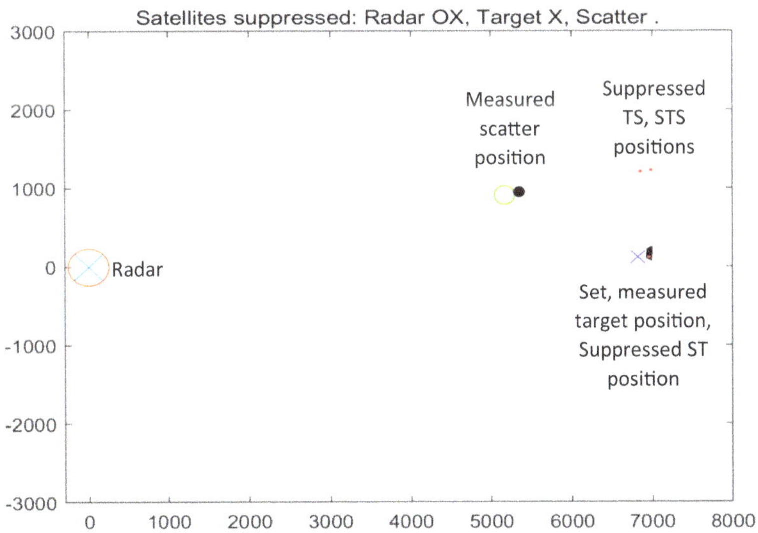

Figure 9.20 *Geometry, as shown in Figure 9.19, with satellites identified and suppressed*

content is extracted and applied. The expected effect in the presence of multiple scatter sources is outlined in Section 9.5.

Because satellites are a contingent occurrence, dependent on the location of scatterers and the presence and motion of the target, they should not be seen as a regular source of target or scatter information.

However, when flight occurs within a built-up area, the added information encoded in satellites may indeed be valuable for close target monitoring. Below is illustrated the method for vector Doppler measurement using satellite returns. The potential for vector measurement is more robust and becomes a significant opportunity in staring radar surveillance, when used with a network of coherent staring radars, as described in Chapter 10.

9.4.4.3 Vector Doppler

The vector velocity of the target can be found from the satellite Dopplers and directions. Propagation is illustrated in Figure 9.21.

The scatter source position and the speed and heading of the target are determined using the ranges, azimuths and Dopplers of the target and the ST/TS satellite pair, R'_{RT}, α_T, α_{Sn}, R'_{TSn}, R'_{STn}, ΔR_{TSn}, F_{DRT}, F_{DSTn}, F_{DTSn} and the derived distances D_{RS} and D_{STn}, all of which are measured or (for D_{RS}) may be known. The existence of a unique, consistent solution is the expression of the EUNIT in this case.

The position of the scatter source and the position, speed and heading of the target have all now been estimated from the known measurements. The EUNIT describes the reality of EM wave propagation, generating unique responses at the receiving aperture, and the model confirms that satellite signals encode the information to exclude satellite returns prior application of tracking filters.

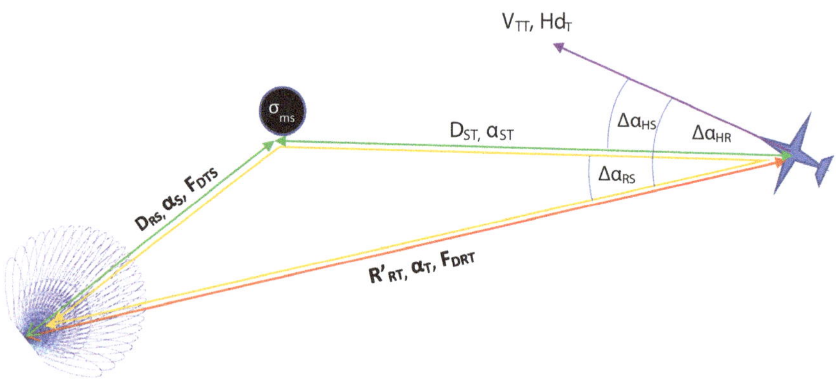

Figure 9.21 Propagation paths, ranges, directions and target velocity vector for NIMP

This possible application of multilook surveillance is contingent on the presence of NIMP satellites, and Chapter 10 will describe its formal incorporation using a coherent radar network.

9.5 Clutter-congested airspace

Illustrating the challenge of staring azimuth multipath, we may use the additional information provided by the staring radar about targets and clutter in the CVoR to prevent congestion of the display, provided continuing signal data and sufficient computing capacity for the analysis.

For a complex CVoR, containing many structures, each of which may result in two or up to three satellite returns per target, we need to assess the challenge in terms of the computing burden. The operation of an HSR with a CVoR near a built-up area or observing aircraft flying in the neighbourhood of a large offshore wind farm will require confirmation that the solution is calculable.

We consider a CVoR segment area of 20 km^2, containing one relatively tall structure per square km, with a radar cross section of 10 000 m^2. This allows an assessment relevant to either the ATR or SRC circumstances mentioned above. The structures can scatter in any azimuth direction.

9.5.1 Satellite discrimination and assignment

A set of range cells beyond and at the same azimuth as a captured target $(\mathbf{R_{RT}}, \boldsymbol{\alpha_T})$ is tested as outlined in Section 9.4 for the presence of up to **n** candidate ST satellites. In Figures 9.22 and 9.23, these are labelled SaST1n and are aligned in azimuth with the target. Associated azimuth cells are then tested for up to *n* satellites matching accurately in range and Doppler to be labelled SaTS1n.

In Figure 9.22, the black dots represent secondary scatter sources. Those circled in red are within the first Fresnel zone; their scattering effect will be IAMP rather than NIMP, as discussed in Chapter 7.

In Figure 9.23, there are three structures within a NIMP maximum distance of 2000 m, each of which will generate **ST/TS** satellite returns; a ratio of six satellites per target, with a loss factor of 30–40 dB. The remaining S numbered structures in Figure 9.22 are those for which the loss factor $\mathbf{L_{ST}}$ will be greater than 40 dB. For NIMP satellites, we are concerned with the satellite amplitude being sufficient to introduce a new captured target.

The geometric characteristics of the satellites are as follows in Table 9.1.

Assuming that the target is detectable at azimuth $\boldsymbol{\alpha_T}$, satellite **SaST1** from scatter source **S1** would be found in the direction of the target but with a range increment that depends uniquely on the distance to the scatter source and its azimuth offset from the target. In this case, the apparent range increment will be in the order of several hundred metres or a few range cells. For scatter sources such as IS1 the range increment will be less and low-amplitude interference will result.

In either case, the satellite due to route **ST1** will appear beyond the target, in a range sequence depending on the arrangement of structures. The satellite due to

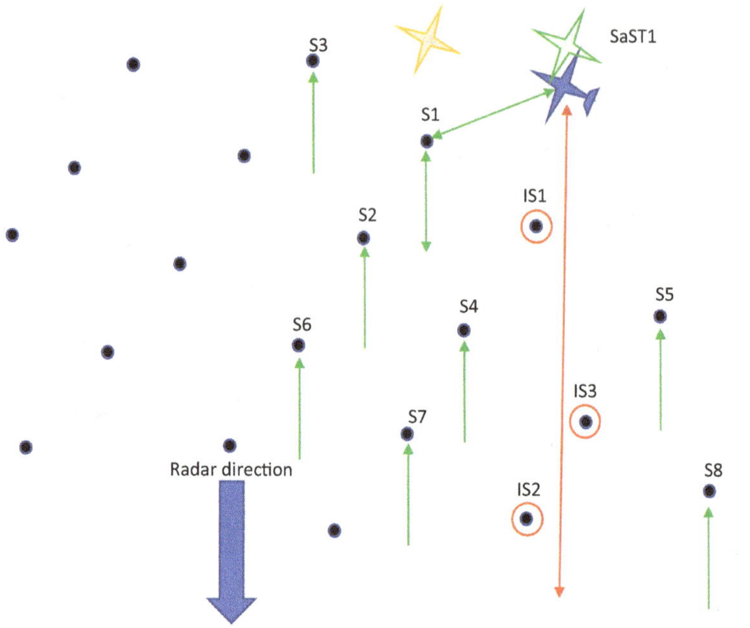

Figure 9.22 A congested CVoR segment with a single target flying at the far edge of the segment

route **TS1** will be located beyond structure 1, at the same range as **ST1**, with the same Doppler frequency and amplitude. If it is resolved in azimuth by the radar, it will be in a neighbouring azimuth cell. If not, it will interfere with **ST1** satellite; according to the reciprocity theorem, the interference should be constructive.

The effect of this assembly of scatter sources is to create satellites in a halo near the target, increasing in amplitude as they approach and decreasing as they recede. This is illustrated in Figure 9.24, later in this target's imagined trajectory, and represents in some detail a stringent test for a staring radar. Note that it occurs over a period of many seconds, not all at once.

In Figure 9.24 and Table 9.1, the angles $\Delta\alpha_{\mathbf{HR}}$ and $\Delta\alpha_{\mathbf{HSn}}$ are those between the target velocity vector and the directions to the radar and to scatter source *n*. Given the range and Doppler resolution of the **HSR**, for NIMP satellites, the **SaST/TS** pairs for each scatter source are expected to appear in different Doppler bins, and many in separate range bins. The **SaST/SaTS** members of each pair will only be resolvable in azimuth.

SaTSsatellites will follow beyond the target, as illustrated in Figure 9.24, approaching and then receding in range with apparent Doppler appropriate for the motion relative to the radar.

Following the aircraft's path in Figure 9.24, satellite amplitude increases as the target approaches each relevant scatter source. Doppler for each $\mathbf{SaST_n}$ satellite

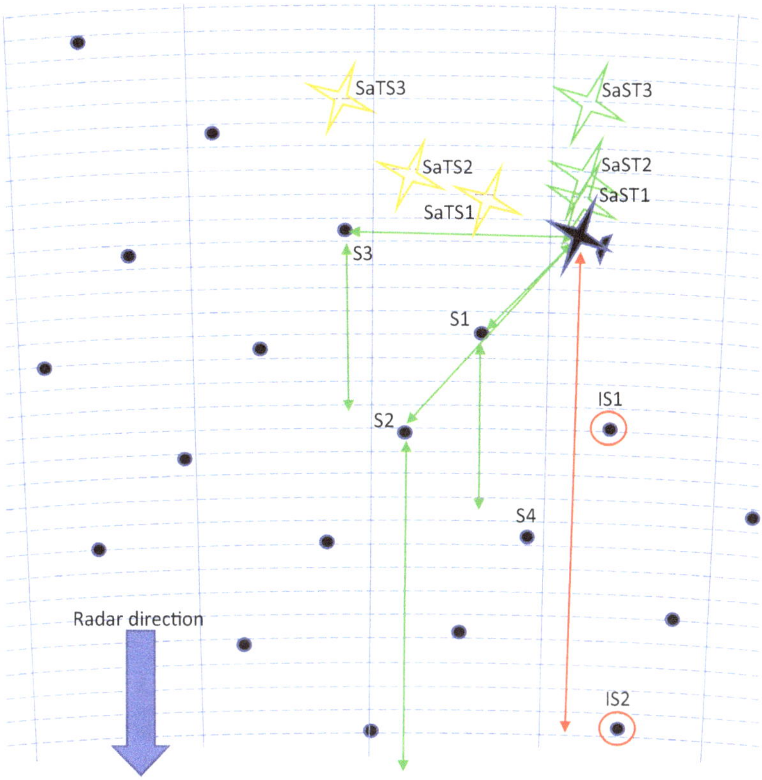

Figure 9.23 *The cumulative effect of NIMP secondary satellites in a congested CVoR near large scattering structures, where the maximum source-target separation is 2 000 m*

Table 9.1 *Secondary and tertiary satellite geometries*

	Range	**Azimuth**	**Dopplers**	**Resolution**
SaST1-3	R_{ST}1-3 – Eq.(9.6)	α_T	VT $(\cos(\Delta\alpha_{HR})+\cos(\Delta\alpha_{HS}1\text{-}3))$. FC/c	SRC/ATC
SaTS1-3	R_{TS}1-3 – Eq.(9.6)	α_S1-3	VT $(\cos(\Delta\alpha_{HR})+\cos(\Delta\alpha_{HS}1\text{-}3))$. FC/c	SRC/ATC
SaSTS1-3	R_{STS}1-3 – Eq.(9.7)	α_S1-3	2VT $\cos(\Delta\alpha_{HS}1\text{-}8)$. FC/c	SRC/ATC

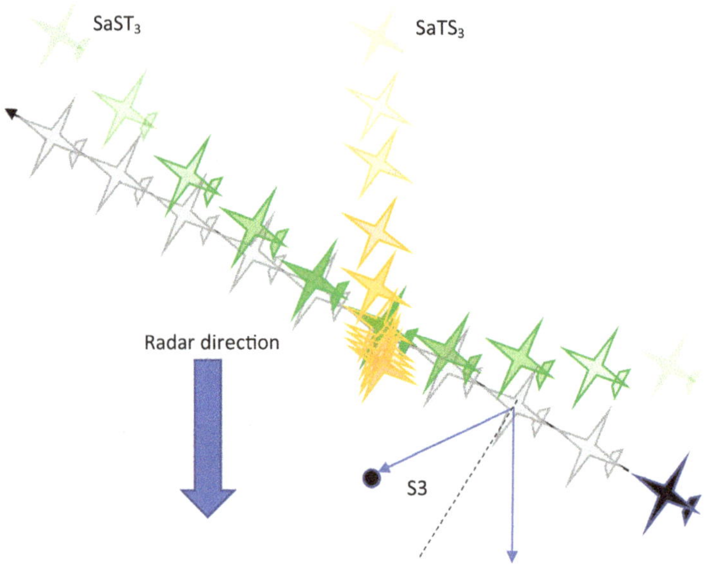

Figure 9.24 *Successive positions of a target in transit past a scatterer and associated* ***ST/TS*** *satellites*

decreases, matching that for the **SaTS$_n$** satellite, as the target approaches the bisector between the vectors towards the radar and each scatter source. The reciprocity theorem provides that **SaTS** satellites will appear and disappear when the target approaches and leaves the neighbourhood of their scatter source, at the same, constant azimuth as the source, and at the same range and with the same Doppler frequency as the **SaST** satellites. In such a CVoR segment, and with the positions of scatter sources known, the process may use the Doppler components $\boldsymbol{F_{DRT}}$ and $\boldsymbol{F_{DSTn}}$ and the scatter azimuth offset $(\alpha_{S-\alpha_T})$ to derive the target's velocity vector. For each scatter source, as the target reaches the bisector, and the satellite Doppler reaches zero, the **SaST/TS** satellite ranges reach their minimum. As the target reaches the line of the scatterer azimuth (from the radar), the **SaST** and **SaTS** satellites coincide and also with the target itself. The parameter linking these variables is the tangential speed of the target.

The scatter source positions are derived with azimuth accuracy related to that of the target, but at a lower signal level. This may allow them to be associated with accurately known positions of CVoR surface features. If measured over an extended period using different targets of opportunity and provided they are themselves separate, fixed structures, their positions might be refined with increasing precision, but this function would be contingent on targets' trajectory relative to each scatter source.

The target's azimuth position can then be refined relative to established scatter sources, at long range if the scatter source positions are precise, and this updated measurement may be more accurate than that provided based on the radar's single-target resolution and accuracy.

Recounting the conclusion of Section 9.4, the information used to complete the calculation is:

1. positions of captured targets and satellites $(\mathbf{R_{RT}}, \alpha_T)_n$, $(\mathbf{R_{STR}}, \alpha_T)_n$, $(\mathbf{R_{TSR}}, \alpha_S)_n$
2. doppler frequencies of target and satellites $(\mathbf{F_{DRT}}, \mathbf{F_{DTS(n)}}, \mathbf{F_{DST(n)}})$
3. the position of the radar and the derived position of each scatter source

When the returns are not resolved either in range or Doppler (i.e., when the reflecting angle is so close to grazing from the reflector surface that the range and Doppler overlap and result in slow interference), these distinct measurements would merge to form a sinusoidally varying resultant that requires through-time analysis to resolve and is discussed in Section 9.7.

9.5.2 Target assignment and the computing burden

The feasibility of using this method of analysing the scatter sources and CVoR clutter depends on the extent of the processing burden it implies.

Once the associations have been confirmed between the **n TS** / **ST** satellite pairs, the associated, shortest-range and most intense target at $\mathbf{R_{RT}}$ is confirmed as the target and goes on to support tracking and reporting.

For a congested CVoR segment, there may be a significant number of satellites within a region of scale about 1 nautical mile of each target, which is consistent with ESASSP and CAP 670 aircraft separation regulations. We will explore the number of operations required, per target, per radar time step, to carry out the positioning, discrimination and assignment of targets and satellites.

Returning to Figure 9.23, in the congested segment illustrating a wind farm layout, the target and six satellites are spread over about 15 resolution cells. Between them there will be 3 **ST/TS** Doppler frequencies, some with significant apparent dynamics as the target passes tangentially near a scatter source.

An outline calculation of the computing burden is as follows. For each cell containing a target or satellite, at ranges extending to $\mathbf{D_{TSMax}}$, totalling less than 100 cells within a congested, say 10 km^2 CVoR zone (such as a large wind farm) there are at most 12 algebraic calculations, each requiring an average of six additions, multiplies, cosine or square root assignments, and totalling less than 100 operations. Here, there are 3 **ST/TS** range and Doppler comparisons for each of three satellite pairs and the target, totalling 19 and a grand total of less than 100 × 100 × 19, or less than 1 M operations per captured target per cell for the discrimination operation, per CPI (more than 1 s). A large target may be captured in several cells, say 16, each of which needs this analysis plus amplitude comparisons, either by single or multi-thresholds, and the total per zone appears to be up to 20 M operations per second per target.

Assuming that this density of scatter and traffic extends over all the range/azimuth area of the two lowest elevation beams, the affected CVoR segment for a wind farm, for an HSR such as ATC, with maximum range of 110 km, will cover up to 1 000 km^2 and about 10 000 resolution cells. For the wind farm, the rate of calculation approaches 20 GFlops overall. If the scatter source density is maintained not only by wind farms but also by urban areas, there will be more than one such segment. This process will contain conditional calculations likely to be unsuitable for highly parallel processing, and the treatment of scatter-congested CVoR segments is likely to require multiple-core CPUs, with fast access to a RAED-structured memory.

This calculation is a loose approximation, but it does not threaten to be beyond the capacity of available machines at a small cost in relation to surveillance radar. It is well within the expected computing capacity of surveillance HSRs such as SRC or ATC, and should not be seen as barrier to successful satellite suppression.

9.6 Solution maintenance

Having established which signals represent targets and which are satellites, the satellites can be suppressed explicitly to clarify reports but should also be used to aid in maintaining the CVoR clutter status. They provide a significant component of the information available under the EUNIT.

Once the position of a scatter source capable of generating satellite returns has been established by observation and evaluation of satellites, it can be incorporated in a map of clutter and scatter sources.

Significant parts of the HSR reporting process may build on prior knowledge of the positions of substantial scatter sources, which may first be used during modelling work to determine the expected positions of scatter sources as seen by the radar. In the course of HSR commissioning, fixed clutter returns can then be assessed against these expected positions. Commissioning flights may be designed and planned to test the effects of actual scatter sources, and then to test and update the satellite exploitation and suppression process.

During operation, the additional information that can be extracted from the target and satellite trajectories will be available to maintain and update the clutter map and the suppression process.

9.7 Interfering multipath

The most common multipath condition for radar is surface multipath, since any ground or sea surface is a strong reflector at low elevation angles. It occurs with respect to RT signal routes low over the surface and routes **TS**, **ST** and **STS** via reflection in the surface itself. As propagation nears the grazing angle in reflection, for both conducting and dielectric surfaces, the reflection coefficient approaches unity with a phase reversal. The increase in satellite range is also reduced and interference begins.

Asymmetric, unresolved routes occur for low-elevation targets as described in Chapter 7. As with any form of radar, surface multipath may affect detection and accuracy in elevation measurement but in most cases can be identified and countered over time.

9.7.1 Transmit surface interference

A radio wave grazing the ground or sea surface will arrive at the target with at least one phase reversal compared with the direct transmission, which can lead to a null condition along the surface. For propagation at small but increasing grazing angles, the two waves will arrive with differing phases, and the intensity with which the target is illuminated is a sinusoidal function of the target's elevation angle. This effect is common between HSR and BSR radars, and has been dealt with in Chapter 7.

9.7.2 Reception surface interference

The effect of and recovery from surface multipath on reception have also been described in detail in Chapter 7.

In summary, the exact condition for a zero on the receiving array occurs analytically at only one vertical location on the array; the intensity for rows above or below the zero will differ from zero to an extent that varies with the target's elevation.

The scalar sum of intensity on the array will be non-zero, and the elevation of a target can be recovered by analysis of the pattern of intensity, which at the lowest elevations tends to a parabolic locus around a near-zero. Elevation may be derived from this pattern and the speed at which zeros or maxima of the power distribution travel across the array. This is not a 'cure' for surface multipath effects but a useful limit on their effect.

9.7.3 Interfering azimuth multipath (IAMP)

Unresolved and interfering azimuth multipath (from scatter sources IS1, IS2 and IS3 in Figure 9.22) will occur under the EUNIT. The interference effect will be negligible (amplitude varying by less than ±10%) unless the scattering loss $\mathbf{L}_{\mathbf{ST}}$ is less than 20 dB, which will occur only when the scatter source (assumed to have a scattering cross section of at least thousands of square metres) is within about 300 m of the target or the radar. It may be recommended not to fly aircraft within that distance of a structure taller than its flight altitude.

At such short distances and properly safeguarded, the significant effect will be that of shadowing, outlined in Sections 8.4.4, 9.1.2 and 9.4.4, and may cause intermittent loss of signal as a trajectory progress.

9.8 Conclusions on multipath scatter suppression and exploitation

In the operation of HSR, if the discovery process assumes a single look at each target, clutter and multipath propagation have the potential to confuse target data and reporting.

The EUNIT and the reciprocity theorem provide that signals associated with these propagation routes in fact encode necessary information rather than random or inconsistent interference. The Huygens model here demonstrates that it is possible to identify and suppress reports of scattered 'satellite' radar returns, clarifying the radar display, including in the case of the tertiary satellites that affect BSR displays.

These signals can also be explored to reveal more about targets and their motions within the CVoR. They can extend the function and effectiveness of radar surveillance in mapping the presence of real terrain and ground structures; however, further benefits such as the ability to calculate vector velocities by exploiting satellite target returns may depend on too many coinciding conditions to make a reliable contribution to the surveillance function.

For ATC applications, the numbers found for the computing burden suggest that a broad range of flexibility and clutter resilience will be affordable for a HSR.

Shorter-range HSRs should also have this capability, but the effects require research for use under yet more complex urban clutter conditions.

Readers who have penetrated this far in Chapter 9 will have obtained a perspective on the balance between the value of full CVoR signal information and the possible computing challenge inherent in multipath analysis.

Chapter 10
Spectrum efficiency and HSR networks

Earlier chapters have shown that persistent observation by radar can support effective air surveillance, achieve target capture that is tolerant with dynamics, extend coherent target and trajectory analysis, and provide azimuth multipath and clutter suppression.

Smaller scales of HSR have been illustrated with configurations designed for tracking of drones, for close-in wind turbine suppression, and discussed for automotive radar, and for some military and security applications. Ranges to 40 nautical miles have been successfully tested but further than that remain to be explored. Larger-scale solutions under new surveillance requirements and constraints, including resilience to wind turbines, have been proposed in some depth and configurations at both SRC and ATC scales have been outlined in Chapter 5.

Radar, including both SRC and ATC configurations of HSR, requires the use of signals of significant RF bandwidth to resolve air targets. A number of spectrum bands are allocated for radiodetermination, but there is pressure to reassign, share or otherwise reduce occupied bandwidth. This chapter explores ways whereby better use can be made of the allocated bands, and what other aspects will arise under those conditions.

Chapter 10 is an enquiry into constructive spectrum re-use within a wide-area air surveillance network, under conditions where congestion, re-assignment or sharing of the radio spectrum with other services might threaten aspects of performance.

The primary issues in this more speculative chapter are as follows:

- spectrum demands for operation
- mutual interference of surveillance radars
- operation, target identity and measurement
- resilience to external interference.

The discussion is based on the physical theory and numerical modelling. As in Chapter 3, the physical theory provides the essential bedrock and the structure within which new challenges to radar performance can be met and solutions found. Numerical modelling provides robust tools to evaluate highly linear physical processes, including electromagnetic propagation.

It is difficult to predict the time scale within which better spectrum efficiency and the surveillance capabilities that arise will be specified in requirements and implemented. In some respects, that time has already arrived, but the realities of product life cycles and those of engineering practice necessarily constrain the adoption of improvements arising out of science. No assumption is made here about that time scale, and whereas earlier chapters have dealt with presently required capabilities, this chapter enquires about directions for long-term and larger-scale developments.

The enquiry addresses the possibility, in the context of HSR, for wide-area surveillance using a single, common carrier frequency, in the context of a network of holographic staring radars (NHR).

10.1 Spectrum requirements

Present-day wide-area radar surveillance requires a network of sensors both because the achievable sensitivity and accuracy degrade with distance, and also because the Earth's curvature prevents each station from observing low-altitude airspace at distance. While achieving full cover for the smallest targets at full range, it is inevitable that the VoR for any single radar will overlap with those of neighbouring sensors over much of the total airspace, so that mutual interference must be taken into account.

Where multiple stations are in use, many targets may be observed by more than one sensor at the same time. It becomes necessary to be able to identify each radar return with a specific transmitter, and the sensors are therefore operated on different frequencies and with different and unsynchronised pulse repetition rates. Each is assigned more than one selectable operating frequency, and each of these typically occupies several megahertz of spectral bandwidth.

This arrangement has served the air traffic control (ATC) and defence sectors effectively in terms of both performance and spectrum occupancy, during a time through which operators and regulators have had access to extensive spectrum allocations, and the density of traffic and threats has grown, but within the capacity of the surveillance system.

In addition to expected further growth in traffic, the present ATC architecture, in which extended flights are routed into relatively narrow corridors, may be supplanted by greater freedom of flight. That will entail a less dense but more complex air picture, placing greater stress not only on controllers but also on surveillance technology and systems.

Airspace contains different and more complex occupants and events, both as recognised users and as clutter or potential threats. Further demands on the content of air surveillance information, beyond detection, tracking and reporting of single, isolated targets of interest, may also enforce a higher level of performance on surveillance infrastructure, including improved track continuity, and capabilities and confidence measures in target analysis and classification.

These demands are in conflict with pressures on the radio spectrum. Substantial growth in demand from other users of the radio spectrum, especially data services, is expected in the twenty-first century, and the spectrum is indeed becoming increasingly congested. Improved spectrum efficiency has therefore acquired a new urgency, and in recent years, the value of access to spectrum has been emphasised by a number of spectrum auctions carried out by governments including the United Kingdom and the United States. Values are measured in many billions per megahertz, sufficient to justify changes in approach.

This chapter aims to give a less formal survey of NHR's potential to achieve high spectrum efficiency as well as providing additional capabilities.

10.1.1 Factors influencing the irradiation requirement

The EUNIT provides that where a sufficiently and coherently irradiated VoR is observed at an aperture with sufficient sensitivity and resolution, there is one and only one solution for the positions and motions of objects within the extended Volume of Regard (eVoR).

10.1.1.1 Sensitivity and range resolution

The requirement for sensitivity is supported by a coherent incident EM field throughout the VoR, with one or more receivers, capable of satisfying the radar range equation with sufficient field strength, a known waveform at a frequency that scatters effectively from air targets, and appropriate bandwidth and timing. Motion or manoeuvring metrics prescribe signal sampling and processing capabilities.

For a monostatic radar, the range is a scalar quantity and is related directly to time by the factor $\mathbf{Vc/2}$. In a multistatic network, with one or more receivers and more than one transmitter, this simplicity is lost in exchange for the multilook network function. In this case, 'time delay', $\mathbf{T_D}$ is the appropriate parameter for signal descriptions, and 'radius', $\mathbf{R_G}$ for geographical distances from the receiver. Provided that signals can be identified with transmitters, geographic positions of resolution cells are still determined by $\mathbf{T_D}$ and by azimuth and elevation directions for each source, with one value of $\mathbf{R_G}$ from the receiver for each transmitter.

10.1.1.2 Directional resolution

For monostatic radar, directional resolution prescribes a combined minimum antenna dimension and maximum operating wavelength; target features of interest place an upper limit on wavelength (feature sizes must be comparable with or greater than a half-wavelength), and atmospheric absorption or cost may impose a lower limit on wavelength. For radar networks, direction resolution may be needed for target isolation as well as for position measurement.

10.1.1.3 EM compatibility

The third requirement that has influenced the design, deployment and regulation of surveillance systems, where air targets may be interrogated by more than one radar sensor, is that they must operate at different, filter-separable frequencies, and this necessity is the first subject of this chapter.

10.1.2 Spectrum occupancy versus time resolution

The effectiveness of radar surveillance is related to its ability to resolve returns from targets in close proximity. The dimensions for achieving resolution have been range, azimuth, elevation and Doppler shift, and high range resolution is related to using a wide spectral bandwidth in the transmission. For monostatic radar, resolution in range (**Rres**) is related to resolution in $\mathbf{T_D}$(**Tres**) by

$$\mathbf{Tres} > \mathbf{1/BW_C} \text{ and } \mathbf{Rres} > \mathbf{Vc.Tres}/2 \tag{10.1}$$

Many efforts have been made to undercut the limit with signals of varying amplitude, but the margins of improvement have been narrow and signal-dependent.

Time resolution itself is indirectly related to the occupied spectral bandwidth of a signal, whose time-evolution affects Rres and BWc in different ways and whose sidelobes are inversely connected; that is, tightly defined edges in time imply extended frequency sidelobes.

10.1.2.1 Rectangular amplitude-modulated carrier pulse

In early days of radar, the achievement of maximum signal energy output, in-band, within a hard limit on peak power from a single vacuum device, was a high priority. At that time to minimise spectrum occupation was of lower importance, and a rectangular pulse was the preferred form even though it implies wide spectral sidelobes.

With the shortest possible rising and falling edges, a rectangular pulse is associated with frequency sidelobes defined as:

$$\mathbf{A(f)} = \mathbf{Ap} .\sin(x)/x \tag{10.2}$$

where $\mathbf{x} = (\mathbf{f}-\mathbf{fc}).\mathbf{tp}.\boldsymbol{\pi}$, **fc** is the carrier frequency, **Ap** is the rectangular pulse amplitude, and **tp** is the pulse duration (here, definitions of amplitude and duration differ between the pulse and bandwidth forms). The sidelobes reduce only gradually in power as the inverse square of the sidelobe order.

In the twenty-first century, increasingly demanding spectrum limits prevent a rectangular AM pulse from achieving compliance with regulations, with sidelobes exceeding −40 dBc up to 20× its nominal bandwidth.

10.1.2.2 Gaussian pulse envelope

The reality of spectrum congestion and the constraints of:

- European Telecommunications Standards Institute (ETSI),
- European Conference of Postal and Telecommunications Administrations (CEPT),
- Federal Communications Commission (FCC),
- the National Telecommunications and Information Administration (NTIA) and
- the International Telecommunications Union (ITU) regulations

and recommendations now focus attention on spectrum conservation, at the cost of some reduction in efficiency in the use of transmitter peak power.

From the perspective of achieving a specific delay resolution with the minimum spectral width, the ideal pulse shape is Gaussian, since it also yields a rapidly diminishing spectrum:

$$\mathbf{A(T)} = \mathbf{A}_{GT}.\exp(-\mathbf{T}^2/2\tau^2),\ \text{and}\ \mathbf{BW_C} = \mathbf{1/\tau};\ \mathbf{A_G(F)} = \mathbf{A}_{GF}.\exp(-(\mathbf{F}^2/2\mathbf{BW}_C^2)) \qquad (10.3)$$

where $\mathbf{A_G}$ is the Gaussian pulse amplitude and τ is its characteristic width.

A Gaussian pulse yields −40 dB sidelobes at approximately 2.6× its nominal bandwidth.

10.1.2.3 Frequency- and phase-modulated waveforms

Spectrum occupancy in the future will become an unyielding constraint, and radar spectra can now be defined with increasing precision: ITU recommendations for FM 'chirp' waveforms aim to minimise spectrum width at −40 dBc. As the cost of signal processing effort reduces, this with various phase and frequency profiles will yield increasing benefits.

Transmitter design to meet out-of-band spectrum constraints may involve some compromise in transmitter power dissipation. The preferred waveform tends towards a frequency-modulated 'chirp' with a carefully designed amplitude and dispersive frequency profile to achieve efficiency, whose spectrum should approximate a Gaussian, minimising the −40 dB bandwidth.

10.1.3 Passive radar

As described in Chapter 4, in the last several decades there have been many developments towards spectrum-efficient radar including several in the form of 'passive radar' – that is, radar systems without their own transmitter. Detection and positioning are achieved using signals transmitted by other broadcasters such as radio or television stations. Indeed, the first radar experiment in the UK was carried out near Daventry, UK, using signals transmitted by a local radio station and received by a passive station both directly from the station (although shielded by the local terrain) and after scattering from an RAF aircraft flying nearby.

In this and other cases, reception at several antennas has allowed both the direction and the delay of scattered signals to be measured. Therefore aircraft can be detected, tracked and reported without additional radio transmissions. Efforts continue to achieve a level of surveillance performance consistent with air traffic regulations, and systems exist that may provide that. However, they will be subject to constraints on the reliability and consistency of ambient signals used, the density of targets and the number of stations required to provide uninterrupted cover.

Following extensive development and trials activity in several countries, the capabilities of passive radar are increasingly well-defined. Well-developed approaches are Thales' MSPSR and Hensoldt's TwinVis product.

The limitations of passive radar are also well understood. These include the potential for mismatch between the volume of irradiated airspace required by the surveillance function (up to a high altitude regardless of geographical position) compared with that required for terrestrial broadcast (close to the ground over similar or smaller areas); the lack of control or uncertainty of availability of the ambient signals; and the suitability of the ambient signals in frequency, bandwidth, coherence and power for the targets and trajectories of interest.

The benefits of passive radar in avoiding the need to occupy spectrum with a dedicated transmission may justify these limitations in performance. However, if it were possible to retain the benefits of active radar while occupying a much smaller region of the spectrum, that would carry the benefit of maintaining the regulatory basis for the safe operation of ATC.

10.1.4 Sharing with other services

Separation of frequency bands between different, managed user communities has been the method of achieving low levels of interference in each application. However, the EUNIT does not specify that each 'solver' of the EM propagation problem must operate separately in frequency, nor that each solution in transmission will be adequate for all applications.

The following discussion will be limited to spectrum sharing or control between different stations within the air surveillance ('radiodetermination') service and with comparable functions.

10.1.5 Common-spectrum surveillance

A networked holographic radar configuration, **NHR**, is outlined with the objective that spectrum occupancy by a whole surveillance system is minimised while continuing to yield improvements or extensions of surveillance function. **NHR** is a multistatic radar system and is intended as a means of exploring additional capabilities without expanding the spectrum.

The EUNIT assumes a known irradiating field throughout the eVoR. It provides that an EM solution is unique but does not require that the field originates at a single position. Rather than conceive of surveillance sensors as isolated and independent units, the **NHR** network maintains interrogation of an extended airspace volume ('**eVoR**') with the transmission of identical signals and reception at a number of positions within or at the edge of the volume to measure signal scattering from within it and to decode the target information they contain.

To achieve an integrated NHR-based air picture, a set of sensors are positioned and operated as a coherent network. The aim is to construct unique solutions for accessible targets, with high dimensionality, resolution and precision, and with the minimum occupation of radio spectrum.

The **NHR** concept will be used to enquire into the feasibility of assembling a system of HSR stations using a common irradiation spectrum to provide effective wide-area surveillance. The enquiry addresses the key challenges introduced by this concept: mutual interference and transmitter identification.

10.2 Mutual interference

This concept runs counter to almost all radar surveillance practices, except that of passive radar. However, in this case the several transmissions used, instead of being unknown, are defined precisely for the network and all its transmitters.

The first challenges to this proposition are that direct transmission may saturate more than one receiver and that the transmission origin of each received signal needs to be identified without the conventional unique frequency association.

10.2.1 Common frequency in a BSR network

In Figure 10.1, two scanning radars observe a target at the same time. Each receives its 'own' scatter, but also since the transmissions are unsynchronised, a randomly timed scatter arising from the other Tx. The crosstalk is occasional but occurs

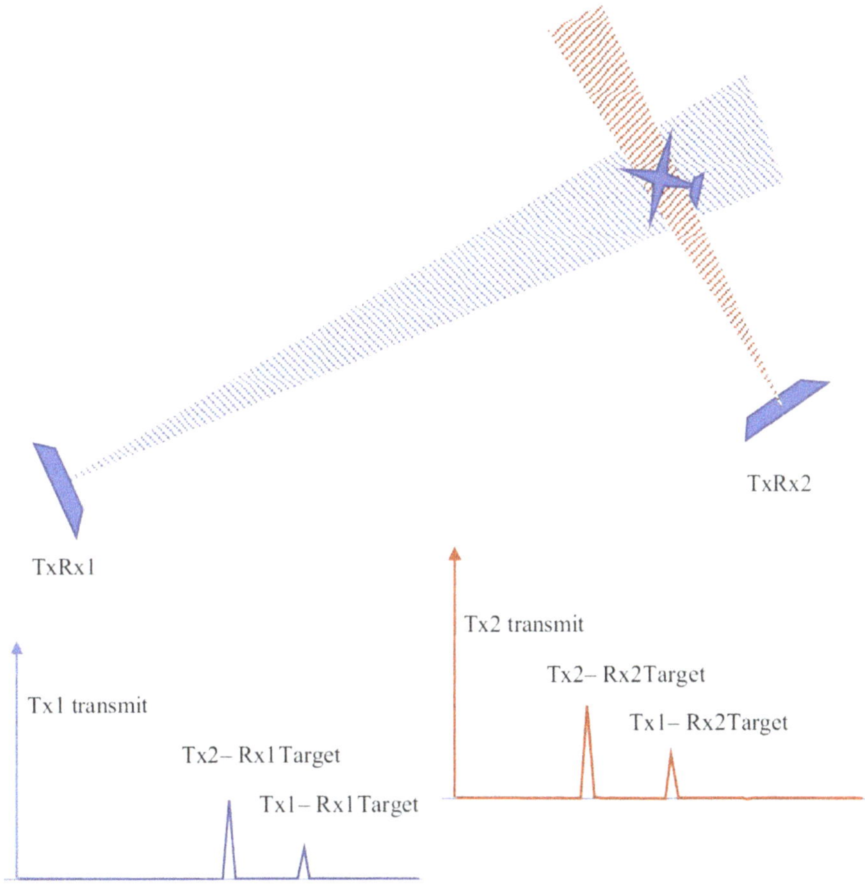

Figure 10.1 Use of a common frequency in a wide-area scanning radar network

unpredictably for all targets, including clutter targets, and can only be rectified by matching locations in each pair of reports, which also will not be synchronous.

The solution for a BSR network has been to operate with frequency diversity, occupy additional spectrum and maintain independent reports.

Any potential for spectrum sharing within a staring radar network will reduce the factor by which the necessary sensor bandwidth is multiplied to achieve network compatibility. This chapter is an effort to identify that potential.

10.2.2 Coherent, synchronous, persistent transmissions for NHR

The EUNIT requirement is for known irradiation. It does not exclude more than one simultaneous source, at different places in the VoR, nor does it state a requirement for separate frequencies.

The NHR configuration suggests a method of achieving broad coverage and high surveillance performance with reduced overall use of spectrum, based on distributed, coherent signal sources.

The use of a common frequency $\mathbf{F_c}$ again means that the receiver is sensitive to scatters from the target arising from both transmitters, and common timing will allow the two to be correlated.

A simple concept of NHR ($\mathbf{NHR_0}$) will be modelled to explore the possible vulnerabilities or failings of such a network. $\mathbf{NHR_0}$ contains a single, 360-degree receiver and two, separate 360-degree transmitters.

Two issues again occur that might preclude this approach:

1. inevitable interference between the transmitters
2. the need for resolution and identification of returns from each Tx.

10.2.2.1 Interference between NHR transmitters

Any interference between two separate but synchronised and coherent transmitters will form an interference pattern, which will affect the irradiation of a target and might compromise sensitivity overall.

For $\mathbf{NHR_0}$, each segment of the eVoR is interrogated by both remote transmitters and the receiving array, each operating at geographical locations separated by distances on the scale of the maximum range, at a common frequency, and transmitting the same repeated waveform at the same instants. The first aim of this chapter is to determine whether such a simple network, with transmitters and receivers properly controlled, breaks down in sensitivity or resolution as a result of arising interference.

For $\mathbf{NHR_0}$, locations occur at which the interrogating field from one transmitter will interfere coherently but destructively with that from another, forming 'null zones'. A common single frequency will be ruled out unless the network geometry effectively prevents the formation of null zones in which a RToI can be lost.

In the $\mathbf{NHR_0}$ model, synchronous pulse transmissions occur at the common frequency with matched or slowly varying phase (within the performance constraints of timing and waveform control circuits), and the receiving site is synchronised by

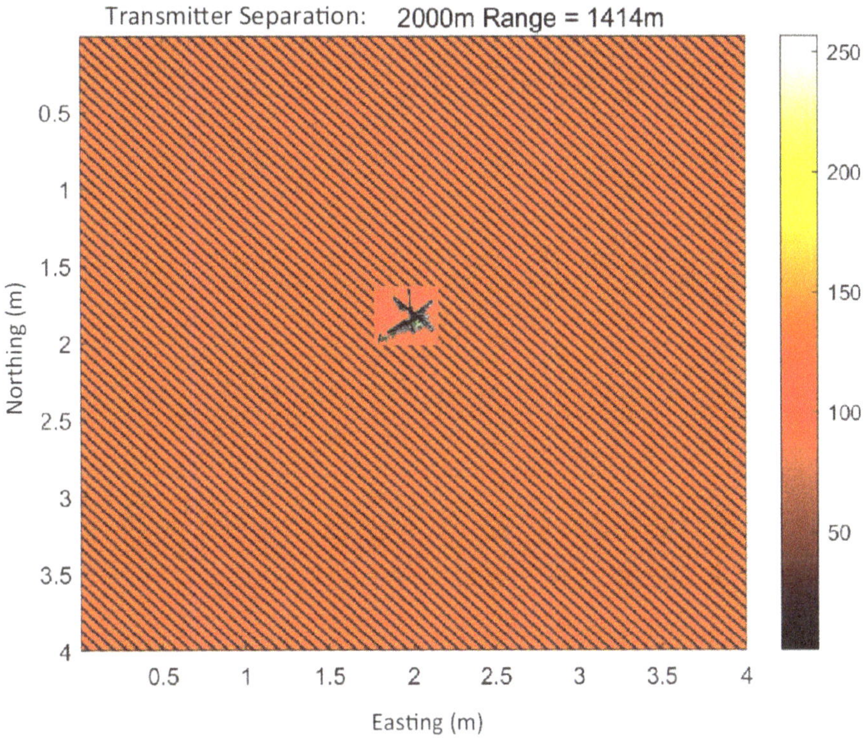

Figure 10.2 *Interference (scales in metres) at S-band: null loci for equal target radii less than 1 km, with a scaled, small drone image*

the same timing control system. Such a system needs to be disciplined either by a GPS-like timing source, by a single terrestrial atomic clock such as MSF, or by quantum-entangled oscillators.

The two coherent transmitters are separated to a reasonable fraction of the maximum radius. They are omnidirectional in azimuth and are separated from the receiver by similar but different radii and azimuth directions. Interference occurs between them, still with the risk that it will impair sensitivity.

Using an **NHR$_0$** model based on SRC, operating at S-band, the interference pattern within a small area when the transmitters are distant from each other is illustrated in Figures 10.2 and 10.3, with images of a very small drone target superposed to approximate scale. The null and peak zones are illustrated at 1 and 4 km, respectively, for **NHR$_0$** under coherent timing control with transmitters separated by 2 km. They are calculated in small areas (4 m square) and at positions equidistant from the transmitters. The interference patterns vary smoothly in scale between these radii, and only occur at transmitter-equidistance, where the two pulses overlap spatially and have similar amplitudes.

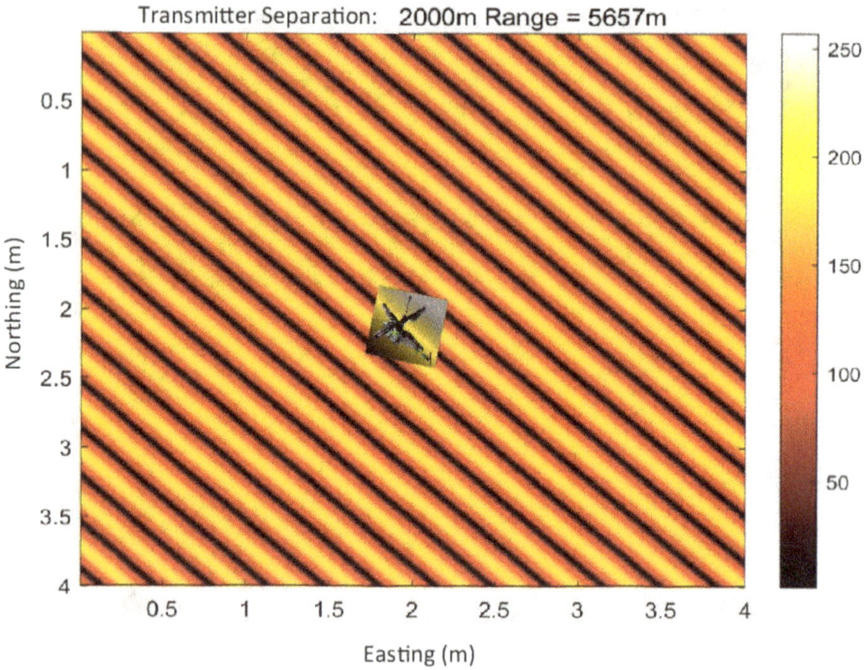

Figure 10.3 *Interference at less than 4 km for 2 km Tx separation and coincident propagation delay*

The scale of transmitter separation is very much greater than the wavelength, and the peaks and nulls in Figure 10.2 are separated by approximately one quarter-wavelength (25 mm), expanding to four quarter-wavelengths (100 mm) as radius increases to twice the separation in Figure 10.3; this is less than or comparable with the scale of target features, and sensitivity is maintained under this worst-case condition. We note such interference only occurs near the surface bisecting the line connecting Tx1-Tx2, as in Figure 10.4.

In these plots, a point scattering source moving exactly along a fixed null zone would produce no return, but any target of interest (with radar cross-section say 0.01 m^2 and dimensions of more than 100 mm) will not interact as a point scatterer. It will extend well into the neighbouring 'peak zones' of constructive interference, only 25–100 mm away. Null zones still exist in this case of remote transmitters, but cannot preclude meeting minimum cross-section requirement specifications.

For trajectories that do not follow exactly a null trajectory, scattering occurs at different Doppler shifts for the two transmitter sources. In this case, the scattered signals associated with the two transmitters are frequency-orthogonal and do not interfere once resolved in the frequency domain.

Similar plots can be derived for an L-band radar operating at much longer separations and radii – say separation of 50 km and cover to 100 km, yielding null-peak separations of 6 cm up to 25 cm. Then, an object of scale over 0.5 m with its centre

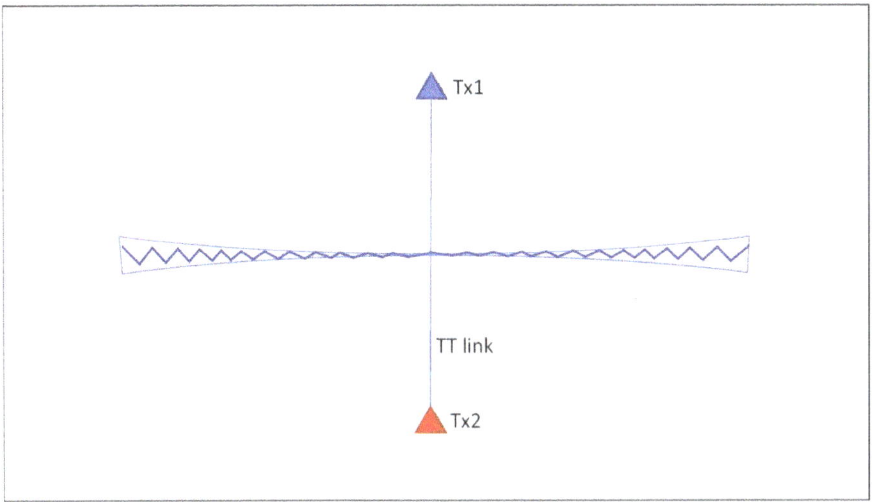

Figure 10.4 Zone of synchronous pulse irradiation, subject to cm-scale null zones

near a null will also have components in each of the neighbouring peaks in the pattern. The scattering cross section of targets of interest will vary with angles of incidence and observation and will be subject to fading in a similar way to Swerling Case 1.

For a correctly configured NHR, from the perspectives of interference and sensitivity, interference effects for targets of interest do not appear to preclude the operation of more than one transmitter at a single frequency. Provided that transmitters are separated by distances comparable with the maximum radius, these effects are excluded by the VoR geometry combined with the scale and scattering characteristics of targets of interest.

10.2.2.2 Positional constraint on transmission interference

Destructive interference between the separate, remote but synchronous pulse transmitters is only significant when propagation time differences are within the pulse duration, allowing the interrogating pulses to overlap spatially, and with similar amplitudes.

The wavelength-scale null zones occur near the thin, radially oriented surface perpendicular to the line joining transmitters, where the pulses overlap. For a pulse waveform occupying 2 MHz, with resolution of 75 m and transmitters separated by 2 km, this only occurs at all, including both null and peak zones, in a narrow region across the Tx-Tx link, growing from about 75 m thickness near the link into a hyperbola-bounded zone near the Tx-Tx bisector surface (as illustrated in Figure 10.4) in which the null zones remain of centimetre scale.

The cases illustrated are related to a version of $\mathbf{NHR_0}$ operating at S-band for UAV targets. For ATC, the frequency will be at L-band; null zones will be twice the width of those calculated here. However, targets will be many times that size and sensitivity will similarly be maintained for the L-band $\mathbf{NHR_0}$.

10.2.2.3 Target fading

The return from a null-following 'point scatterer' (an isotropic scatterer, small at the wavelength scale) will be degraded while in that part of that exact trajectory.

At aviation radar frequencies, aerial targets that qualify as point scatterers do not qualify as targets of interest. Except for a point scatterer, the incident field from one transmitter direction will result in different scattered field strength in the direction of the radar receiver from that resulting from the other, and scatter interference is similar to Swerling Case 1 fading.

For Swerling 1 fading, nulls occur when a few scatterers in a target sum to a near-zero complex amplitude for the field scattered in the direction of the receiver (illustrated in modelled form in Figure 10.5). Losses greater than 10 dB due to target fading are expected during a small proportion of the time (up to 10 per cent).

All targets of interest, for SRC or ATC configurations, qualify as multi-point targets and exceed the dimensions of **NHR** interference nulls, as we have seen. Their returns will follow Swerling Case 1, and for a multistatic radar network there is no consistent solution for null formation in both directions of incidence.

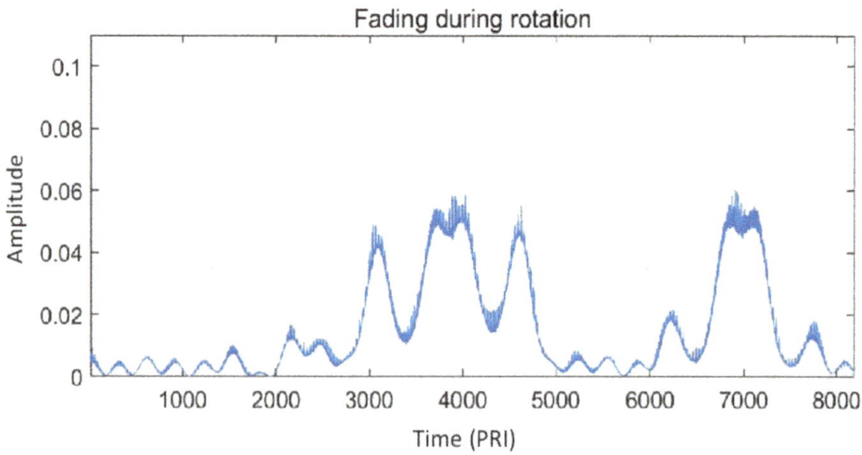

Figure 10.5 Modelled fading sequence for a simple airframe shape of scale 5–10 m, propeller driven, in a turn over 16 s

10.2.3 Mutual interference – conclusion

The requirements for spectrum to enable sufficient range resolution are clear, and the effects of operating more than one transmitter at a common frequency, within the eVoR, with different geometries of propagation, have been modelled.

The **NHR0** model shows that while interference effects occur, as expected, within this geometry and with targets of interest either for SRC or ATC they do not constitute a threat to the sensitivity requirement, and the probability of target capture will be maintained.

We conclude that the requirement for separate operating frequencies noted in Section 10.1 refers to a particular method of achieving mutual compatibility and that NHR methods may be capable of networked surveillance within a single, resolution-constrained radio frequency band.

The next step is to explore how such a system would operate, whether it will meet accuracy requirements, and whether it offers additional target information.

10.3 NHR operation, transmitter ID and target measurements

The second challenge to a common-frequency radar network is whether ambiguities that may be expected between target radar returns can be resolved unconditionally.

10.3.1 NHR_0 station siting

For every position in the eVoR, there are advantages in coverage and cost in co-locating single sources of transmission with receivers. Multiple receive/transmit sources increase the dimensionality and accuracy of target information within a single-frequency network. However, non-cooperative radar requires high-intensity transmissions and sensitive reception, and co-located receivers tend to be desensitised during transmission. Separation between transmitters and receivers may lead to coarser radius resolution but can avoid saturation and also reduce the order of dependence of signal strength on radial distance. For monostatic transmitters and receivers, target sensitivity reduces with the fourth power of the target range throughout the VoR, but in the separate case sensitivity is less steeply variable.

We start with the **NHR0** case, using two remote transmitters, to explore the effects of NHR geometry under a mix of simpler and more demanding conditions.

10.3.2 NHR_0 geometry

When transmitter and receiver are separated, the geometry of surveillance changes. In place of a fixed relationship between target distances and signal propagation delays, the link between position and delay is different for each transmitter, and spherical surfaces of constant delay are replaced by ellipsoids, each focused on the receiver and one of the transmitters.

As a result, a resolution cell for a specific receiver is defined by its azimuth and elevation directions and its sample timing, and refers to two different radial distances, one with respect to each transmitter. Each is a function of the single azimuth direction. This is clearly more complex than the simple delay-range equivalence in monostatic radar, but for each transmitter location these are fixed parameters for each time delay and each azimuth and elevation direction.

A second effect is that the ellipsoids pass through a particular resolution cell at different angles. Doppler shifts depend not on the radial speed towards the receiver but on the components parallel to the normal to each ellipsoid. The combination encodes the trajectory vector. A third effect introduces the potential for trilateration of targets to achieve higher positional accuracy.

A target lying on the line connecting a transmitter to a receiver yields no trajectory information, other than the line itself, and must rely on the second transmitter for positioning. In the case of NHR_0, the model provides that transmitters and receivers are non-collinear and able to interrogate every position in the eVoR. On each T-R link, **NHR0** geometry provides accurate position measurement from the alternate transmitter.

Elliptical geometry introduces complexity in both positional and Doppler algebra. For **NHR0**, the solutions are in the form of static, pre-calculated and accessible function references for target processing.

The siting of remote transmitters in the **NHR0** network is such as not to saturate the receiver, but transmissions are expected at amplitudes capable of accurate phase and delay measurement.

Transmitter locations should not be collinear nor symmetric with respect to receivers. Position-dependent angles of intersection between the ellipsoids yield independent Doppler shifts that resolve trajectory ambiguities, and, correctly located, there is no case where both delays are the same (allowing pulse interference) and the Doppler shifts are also the same (risking extended destructive interference).

The geometry for a target within a simple coherent network is illustrated in Figure 10.6, and the sequence of times-of-arrival of transmitter and target returns is shown in Figure 10.7. With interrogation by two transmitters, a single target yields paired returns at the receiver, each with a specific range and Doppler shift.

A resolution cell defined in an **NHR0** network provides azimuth and elevation parameters but instead of measuring range directly it has two radius parameters, as referred to above, plus transmitter directions and curvature parameters for each delay ellipsoid on which it lies.

10.3.3 Target capture

The use of two or more transmitters will have an impact on the target capture process introduced in Chapter 6; a single target will yield captured return signals in resolution cells at more than one delay value but at the same azimuth and elevation and at matching radii. These are resolved using the reference parameters for each capture cell as listed in Table 10.1.

Present radar practice benefits from the natural, sequential pattern set by its physical scan of the VoR to order its search for targets. The scan is not present in HSR, where receiving 'beams' are formed by a number of parallel processors. For ATC, CPI data are collected over 2 seconds, and the system is scaled so that

Table 10.1 *Resolution cell positional variables for monostatic and* **NHR_0** *radar*

Monostatic			
	Azimuth	α	Phase vs. rows
	Elevation	ε	Phase vs. columns
	Range	R(τ)	Time from transmission
	Doppler	Fd	2Vc / Vrad
Networked			
	Azimuth	α	Phase vs. rows
	Elevation	ε	Phase vs. columns
	Transmit delays	T(1,2)	Direct Tx delays
	Ellipse-receiver radii (2 at each (α, ε))	R'(2-2n)	Radial metres from Rx
	Tx azimuth directions (2 at each (α, ε))	Ae(2-2n)	Offset from target azimuth
	Ellipse radii of curvature (2 at each (α, ε))	Γe(2-2n)	Relative to Rx radius

beamforming is complete within each PRI, when the four segments of the RAED or its equivalent are complete. At this point, the memory search can be conducted and is illustrated below.

A possible search approach has been outlined for ATC HSR in Chapter 6. ToIs are selected where contiguous-index or singular above-noise Doppler values are present, probably using a Vector Histogram, with energies totalling more than the minimum target return, their host cell and sidelobes are identified, and their position and radial speed are passed on for tracking.

For **NHR_0**, the process is more complex in that each cell in a patch corresponds with two radius values. A target scatter returns from each transmitter, and the first task is to find the matching radii, which will be found at different time delays. The radial difference is limited to one-half of the T-T distance, and the match must be sought when captured targets at all cells at each azimuth are available.

Matching radii at the same azimuth yield trilaterated target positions, and their different Doppler values support the calculation of their speed and direction of travel.

10.3.3.1 Resolution cell parameters

For **NHR_0** each resolution cell is defined by different parameters than monostatic radar, as illustrated in Table 10.1.

A single target is observed at two range resolution cells at time delays $\mathbf{T_{R1}}$ and $\mathbf{T_{R2}}$, illustrated in Figure 10.7. Each cell is defined by time delay, $\mathbf{T_D}$, azimuth, $\boldsymbol{\alpha}$, and elevation, $\boldsymbol{\varepsilon}$, and refers to two different cell radius values at the receiver, one for each transmitter. One pair across these cell radius values must match, as illustrated in Figure 10.6.

The matching radius parameters for the two capture cells provide the position for a target based on trilateration, where the ellipses intersect.

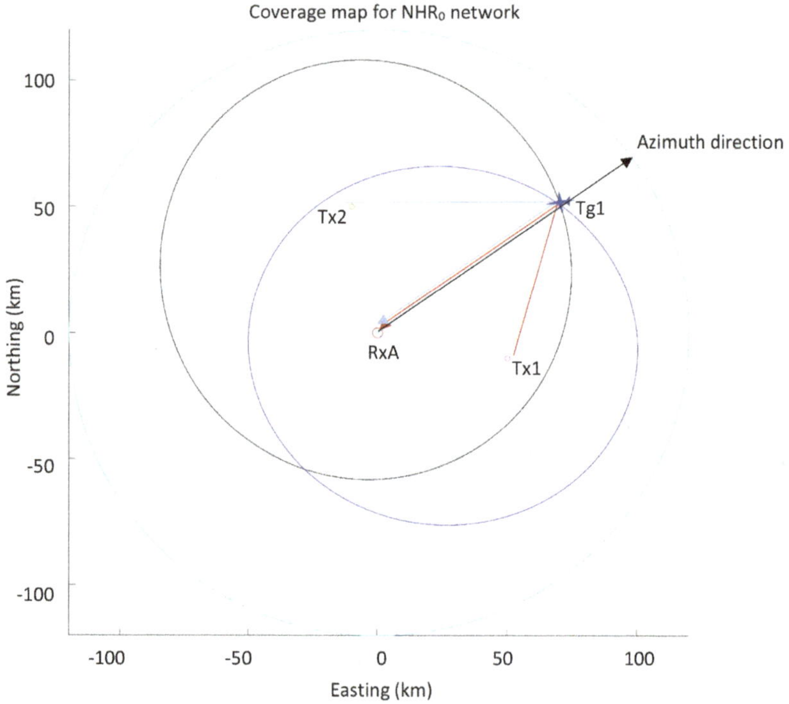

Figure 10.6 *Elliptical bistatic delay surfaces for Rx/Tx pairs coincide at the target*

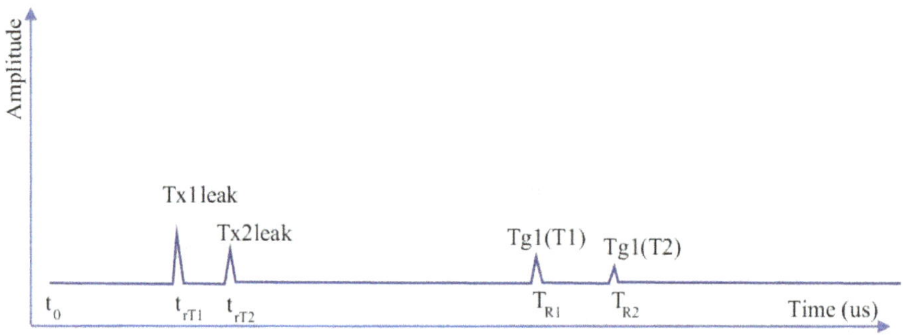

Figure 10.7 *Illustrative signal sequence at the target azimuth and elevation*

The **NHR0** geometry as a whole is shown in Figure 10.8 based on the ATC configuration.

In Figure 10.8, Tg1 is at ◯, with crossing delay surfaces. Ellipsoids become close to parallel at the position marked □, where vector Doppler takes longer to compute.

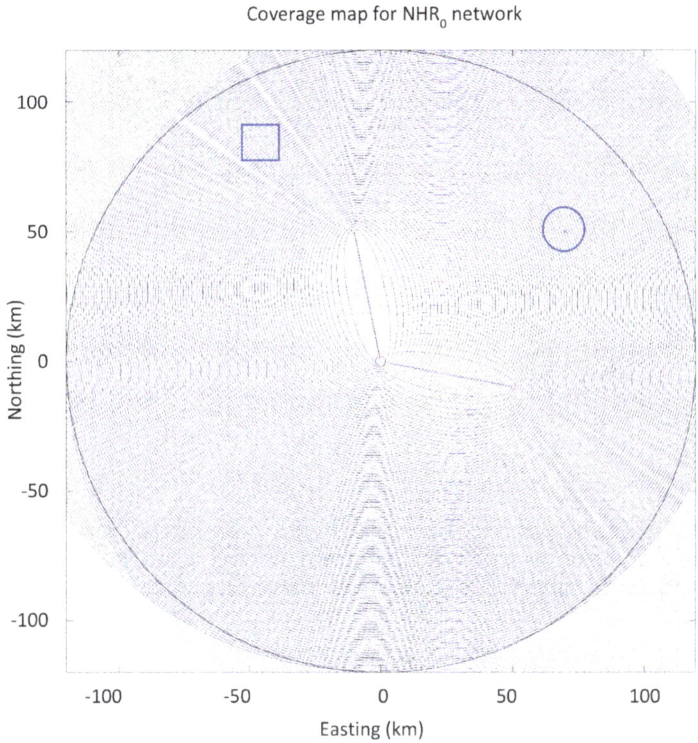

Figure 10.8 *Resolution cell cover pattern for a simple NHR$_0$ network (scale – km)*

Receivers in **NHR$_0$** are based on the ATC HSR configuration. Figure 5.10 in Chapter 5 illustrates the array, and Figure 10.8 illustrates the pattern of cover at the surface (dimensions: km). The network deploys up to 1 000 delay ellipsoids per receiver. In the figure, the radial scale has been expanded to yield a visible pattern of ellipses.

10.3.3.2 Doppler effects for NHR$_0$

A target at a specific azimuth direction generates returns at two different bistatic delays illustrated in Figure 10.7, representing different ellipsoids. Each cell registers two radius values, one for each transmitter and its ellipse at that cell. The target will be captured in two cells in that (α,ε) direction, whose two radius parameters provide a match (Figure 10.6), with two outliers that do not, as illustrated in Figures 10.9 and 10.10.

The two signals returned by the target at times $\mathbf{t_{R1}}$ and $\mathbf{t_{R2}}$ have been assigned in error to Tx2 and Tx1, respectively. They are at the wrong azimuth and will have a lower amplitude (since sidelobes are coherent and synchronous). Two positions are derived but are inconsistent in position.

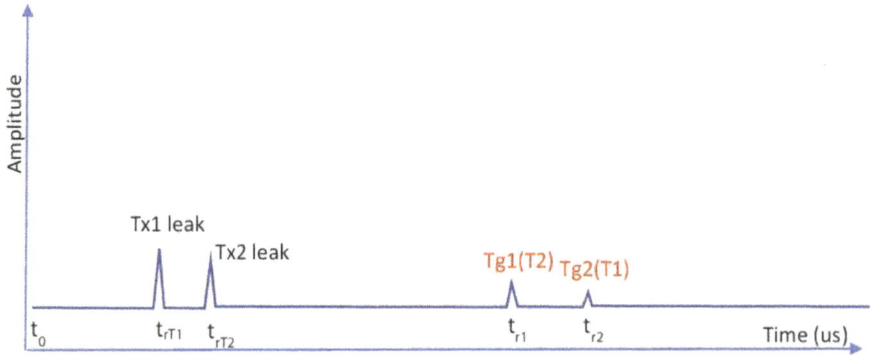

Figure 10.9 The case of incorrect transmitter assignment

The Doppler shifts for that target at each cell depend on the direction offsets to each transmitter and the receiver relative to the target's direction of travel. Phase differences across the receiving array for Doppler signals originating from each transmitter will be unaffected, and therefore directional measurements remain valid.

The time-domain returns from a target moving in the narrow region of overlap of transmissions from each source (as shown in Figure 10.4) will be amplitude modulated, representing the sum of different Doppler components generated as the target crosses between successive ellipsoids. Their frequencies depend on its speed and its direction of travel with respect to the receiver and each transmitter.

The Doppler components for a target, whether in the overlap region or otherwise, encode information about the velocity vector of the target and, over a longer period, about its shape and orientation.

10.3.3.3 Transmitter leakage

Whereas the requirement identified in Section 10.1 prescribes different operating frequencies, the EUNIT requirement for irradiation does not enforce separate frequencies for transmission and reception. What is required, under the EUNIT, is that the interrogating waveforms at each position and direction in the eVoR should be known and persistent, of sufficient intensity, of sufficiently short wavelength to scatter effectively from features of expected targets and that the $\mathbf{NHR_0}$ network geometry provides for target-scattered signals to be resolved and associated with separate transmitters. Decodable EM signals will then arrive at the receiving stations.

The transmitters deliver direct signals to the receiver at delays $\mathbf{T_{T1}}$ and $\mathbf{T_{T2}}$. $\mathbf{NHR_0}$ installations should be configured so that they do not saturate the receivers and are captured as static clutter at the appropriate azimuth and elevation. Their phase can be used as part of the CPI timing control system. They will be of high amplitude and may appear as sidelobes in other directions. They can be suppressed prior to target processing but may result in time-coincident phase noise.

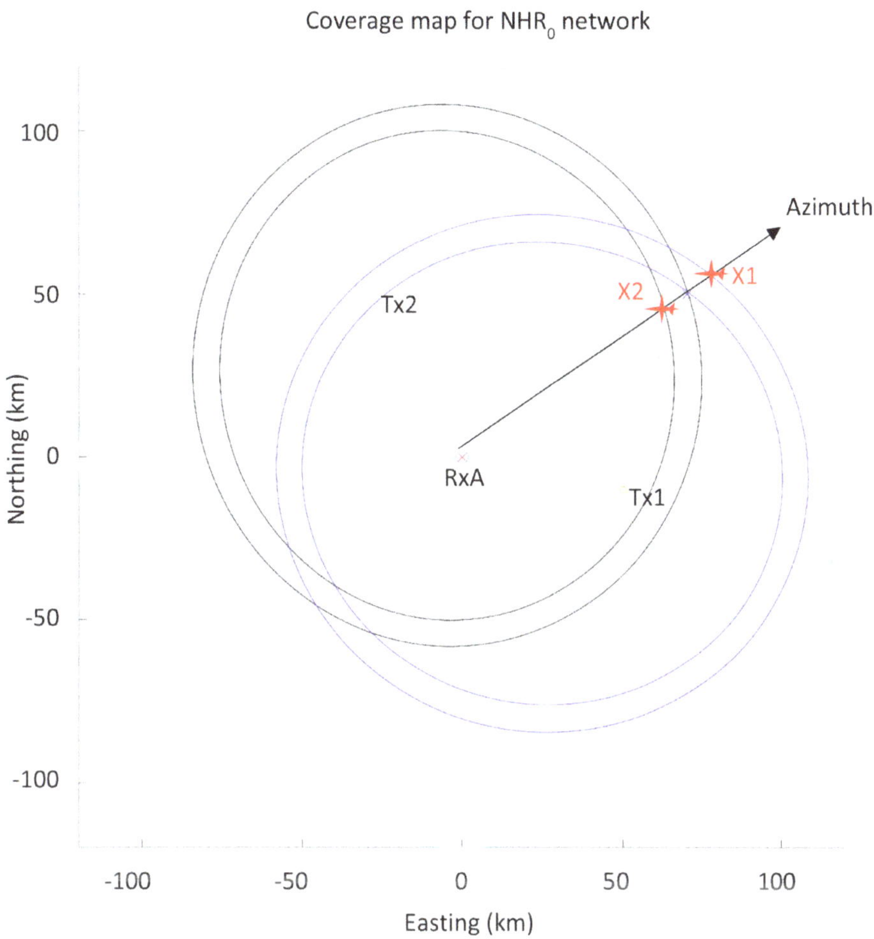

Figure 10.10 *Mis-assignment of transmitters leads to separate, unpaired target suspects*

10.3.3.4 Target–transmitter associations

A single target in the VoR, interrogated by two transmitters as illustrated in Figure 10.6, yields two target returns in addition to the target sidelobes, again as illustrated in Figure 10.7.

To obtain position measurements, the two target-received signals, based on the same carrier, must be assigned correctly to the two transmitters. In the **NHR_0** configuration the geometry allows that association to be determined.

As introduced in Section 3.2, the target lies on two distinct elliptical loci, each focused on the receiver and a respective transmitter. The ellipses (Figures 10.6 and 10.10) are determined by the time delay at the receiver and the two transmitter positions, which, again, should not be symmetric. Again, each resolution cell lies on an ellipse at the correct azimuth, and each **NHR0** resolution cell represents two receiver radius parameters, one associated with each transmitter.

Time $\mathbf{t_0}$ in the **NHR0** configuration is a clock cycle shared by transmitters and receivers across the network.

The signal envelopes in Figure 10.7 include, for illustration, leakages from the transmitters at times $\mathbf{T_{T1}}$ and $\mathbf{T_{T2}}$, via sidelobes at the target's azimuth and elevation direction.

At times $\mathbf{T_{g1}}$ and $\mathbf{T_{g2}}$ signals are received from the target $\mathbf{T_g}$, but these signals are yet to be assigned to a transmitter.

Prior to making the transmitter association, the time delays place each return on either of a pair of ellipses, focused respectively at RxA and Tx1, at delays $\mathbf{T_{g1}}$ and $\mathbf{T_{g2}}$, and at RxA and Tx2, also at delays $\mathbf{T_{g1}}$ and $\mathbf{T_{g2}}$ (Figure 10.10).

The correct ellipses and transmitters are identified by the coincidence of cell radius parameter values at the azimuth direction.

Figure 10.6 illustrates the result when the correct transmitter associations are made with the target. Two signals returned by the single target at times $\mathbf{t_{R1}}$ and $\mathbf{t_{R2}}$ have been correctly assigned to Tx1 and Tx2, respectively. After testing for matching radii of the two delay cells, these yield a single target position.

Incorrect transmitter association is indicated by two different target positions and also by the absence of a paired return in either case. If the returns at these times are mis-associated, that is, in this example, if the first is assumed to arrive from Tx2 and the second from Tx1, as illustrated in Figure 10.9, the result would be to locate two targets as in Figure 10.10. In this case, there is only one return for each apparent target, whereas two should be present. The error creates an inconsistency.

10.3.3.5 Multiple targets

When more than one target is present at the same azimuth and elevation, there will be additional pairs of signals also not distinguished by their carrier frequency but by different delays and Doppler shifts.

The assignment of signals to transmitters requires a series of tests; the presence of $2N$ returns (in pairs) indicates the presence of N targets within the beam directed at (α,ε).

The process of assignment is arithmetically straightforward but grows with the square of the number of targets present within the azimuth/elevation beam. Where N returns are captured, $2.N.(N\text{-}1)$ comparisons are made to find minimum-paired returns. A comprehensive consistency check scales in proportion to twice the square of the number of returns, $\mathbf{2.N^2}$.

Unpaired returns, where no match is found during tests, may occur when differential fading leads to a failure to capture one of a pair. Tracking logic should allow for continuity when one of a pair is maintained: lone 'suspect' returns may be located but held for subsequent matches.

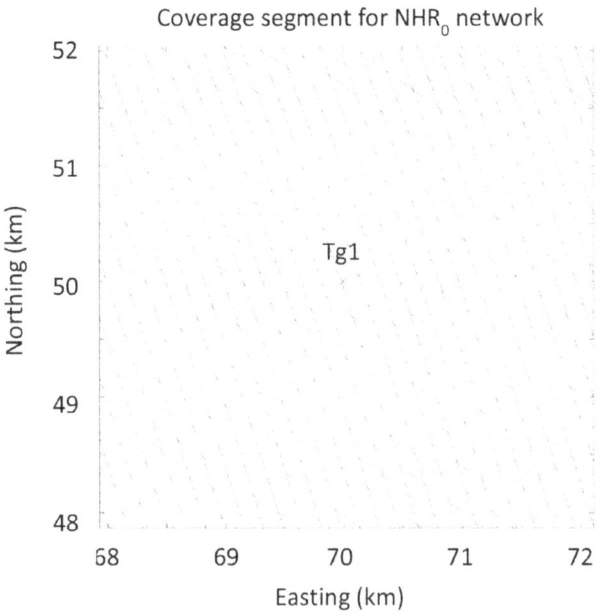

Figure 10.11 *Ellipses and azimuth beam edges for NHR0 at Tg1*

10.3.3.6 Target measurements

In Figure 10.11, the resolution cells are expanded and plotted for a 4-km-square area near the target, illustrating bistatic delay, ellipsoid tangent directions and delay/azimuth cells. Close inspection shows the two bistatic delay surfaces (red and black series) at full scale (1 ms, 1°). The azimuth beam directions are shown by the cyan radials.

This figure illustrates a 4-km square containing cross-points between the delay surfaces for each bistatic pair, near the location of this target at 70 km and 50 km, red ellipses T×1, black ellipses T×2.

Vector Doppler is available from the target's motion across these ellipsoid surfaces.

For each **NHR** station, cross-track accuracy is poorer than radial accuracy when measured by azimuth. At radii of 50 NM or 90 km, the azimuth accuracy is expected to be about 0.15°, equivalent to 250 m, across-track, at 90 km, compared with 15 m in radius (or 100 ns in delay), and is improved by trilateration.

Radius matching allows the association of received signals with transmitters in the absence of any distinction between the signal carriers.

10.3.4 Cell pair interpretation

In the presence of two transmitters, each cell corresponding to a specific delay after the matched transmit time, again, represents two distinct radii for targets in that direction – one for signals derived from T×1 and one for T×2. This results in a more complex process than that for a single sensor, but once a target has been captured and positioned automatically according to its delay/azimuth/elevation coordinates and transmitter associations, additional information becomes available.

10.3.4.1 Trilateration

With correct transmitter associations, the target azimuth position can be refined by trilateration based on the ellipse time delays from $\mathbf{t_0}$. The scales and cross-points in Figure 10.11 illustrate the positional resolution available, and the square root of the signal to noise ratio then delivers a further division to derive RMS accuracy.

Non-cooperative trilateration has been included in the model of $\mathbf{NHR_0}$ and offers more precise positioning than is available with radar networks based on sharing independent range/azimuth data. If radial accuracy is 10–100 m, the accuracy by trilateration will be similar, depending on the respective angles subtended between the transmitters and the receiver at the target.

Target positioning found by trilateration is included in Figure 10.12.

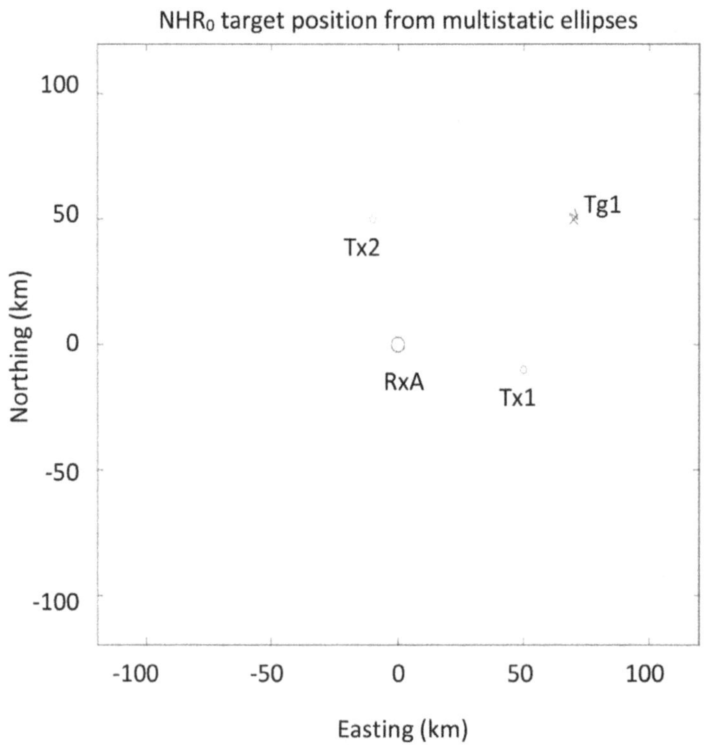

Figure 10.12 *Target Tg1 is located by delay, azimuth and trilateration by multistatic resolution cells*

10.3.4.2 Vector Doppler

The delay surfaces in Figure 10.11 are separated by many wavelengths, 650 in this case, yielding 1 300 Doppler cycles while a target crosses from one surface to the next.

The Doppler shifts of returns arising from the two transmitters are different, except for a limited set of trajectories for each cell, because the target crosses the two series of ellipses at different angles and speed components. The different frequencies act as a stereo pair and reveal the direction of travel relative to the directions to receiver and transmitters. The sum and difference are related to the radial and tangential velocity components respectively, as illustrated in Figure 10.13A,B.

Figure 10.13A shows the frequencies of the two Doppler shifts arising from an aircraft travelling at 100 m/s, for directions of travel through 360°. Sharing the same

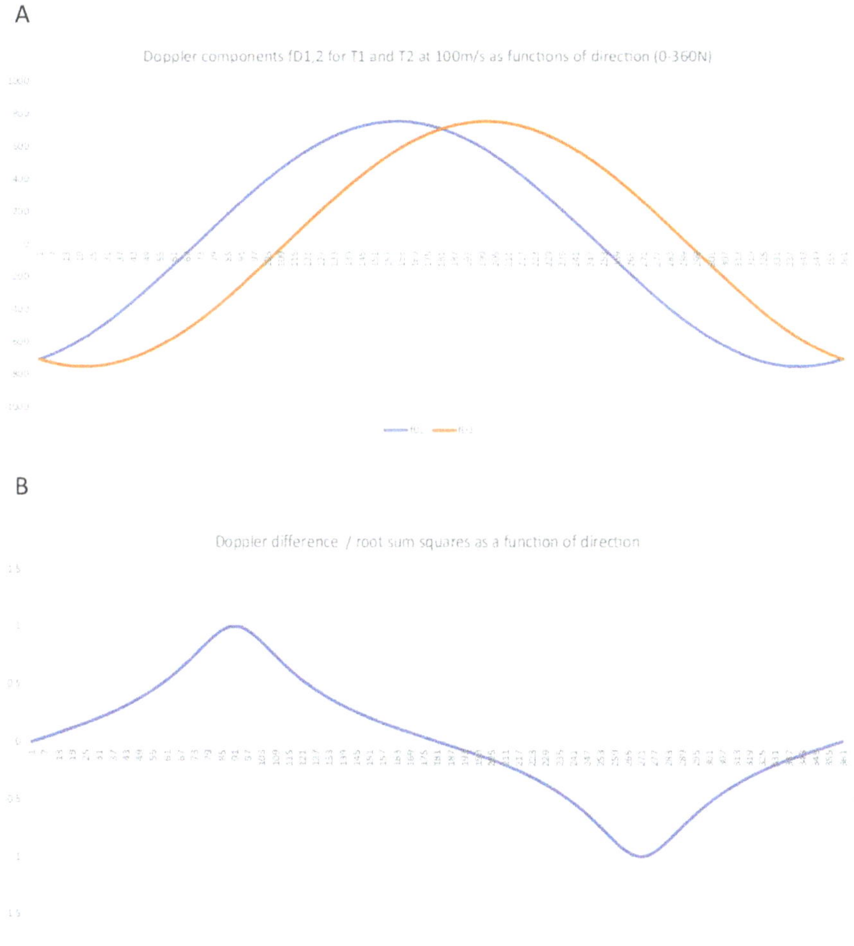

Figure 10.13 Doppler variations with direction of travel

carrier frequency and trajectory, the Doppler shifts for each return are determined by the two velocity projections normal to each ellipse.

The direction is determined from the Doppler difference, which depends on the directions to the transmitters and receiver. Each resolution cell is referenced to these direction parameters (Table 10.1). In combination, the two Doppler values measure the vector velocity of the target.

Each variable is either known as a parameter for the particular resolution cell, as shown in Table 10.1, or is measured for each target return.

Vector Doppler information is available in principle at any position, but the length of dwell required to solve for it will vary significantly in certain azimuth directions from the receiver and with the geometry of the network. Along directions in which the delay ellipses are nearly parallel, as indicated by the square box in Figures 10.6 and 10.14, high Doppler resolution will be required, by means of increased dwell time and extended or differential Fourier analysis.

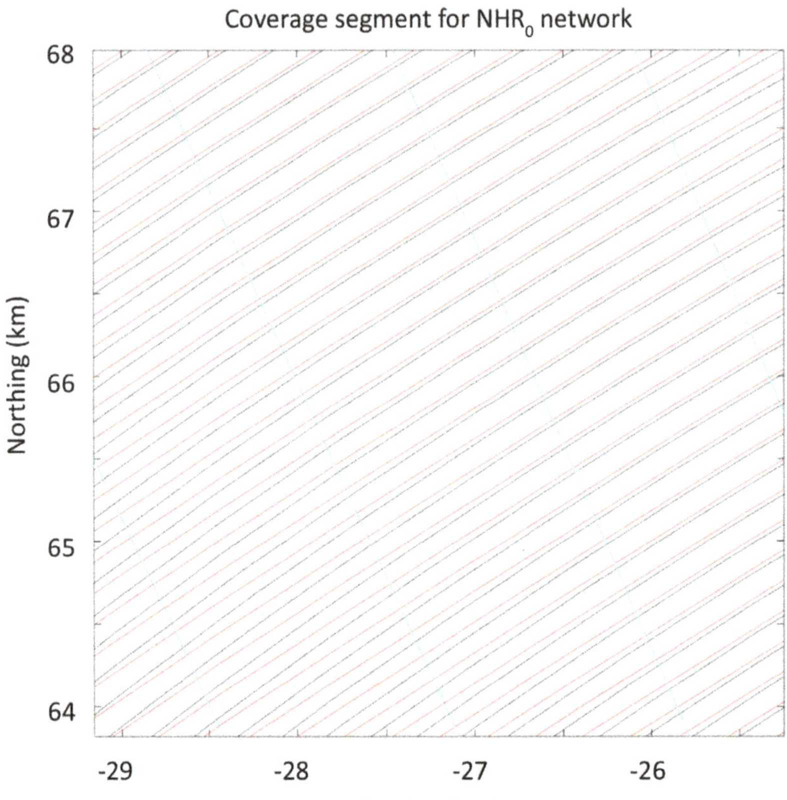

Figure 10.14 The area indicated by the square in Figure 10.8. The delay surfaces are almost parallel.

Red and black lines represent elliptical delay surfaces for transmitters T1 and T2 respectively. The Doppler vector is derived from the evolving Doppler phases $\varphi D_{1,2}$ over an extended time.

Precise cross-track positioning together with Vector Doppler shifts, both derived from NHR cell pairs, will simplify, if not replace, the task of target tracking and will improve accuracy.

Locations with parallel NHR ellipsoids are a challenge for **NHR_0** in measuring Vector Doppler. More extensive networks are more complex but will avoid low vector Doppler sensitivity in these cases.

10.3.4.3 Imaging

A single HSR acquires coherent target information that in principle contains information about an object's shape, but to interpret it satisfactorily with a single look direction to the target requires coherent target dwell through an extended, accurately reconstructed trajectory and manoeuvres.

A two-transmitter/single receiver network adds to this information by yielding multilook interrogation directions that promise to improve the resolution of target scattering; however, 'stereo' imaging will be more practical if the network includes both multiple transmitters and receivers. This is an extensive subject. It promises further dimensions in surveillance functions and deserves more detailed study than can be included here.

10.4 NHR network constraints and resilience

In designing a coherent radar network, there are constraints that may complicate operation but may also avoid errors and ambiguities.

10.4.1 Symmetries

In bistatic radar, the separation of a transmitter from its receiver introduces a zone between the two within which performance is degraded. The position of the target along a TR link is indeterminate, since forward-scatters from the target travel the same distance between transmitter and receiver. For the same reason, Doppler shifts are low and over short periods are independent of the direction of travel.

For that reason, as we have indicated, NHR receiving stations should not lie on the TT line between two transmitters, and it is preferable to set asymmetric distances between receivers and neighbouring remote transmitters.

10.4.2 Receiver saturation

Signals received directly from transmitters in a common-frequency NHR system are a significant asset contributing to timing control. However, they may represent the highest-amplitude signals at the receiving array.

A NHR transmitter may generate over 100 kW (peak effective RMS power). To avoid saturation of the receiver elements, without making provision for losses in transmission near the ground surface (which will be greater than free space losses), the remote transmitter needs to be sited beyond several kilometres, when the direct received signal at each array element may be accommodated by receivers without saturation. This supports the **NHR_0** geometry but is a less demanding constraint than that of limiting the effects of timing jitter between transmitters and receivers, especially in the presence of intense clutter returns or remote transmissions in generating phase noise.

10.4.3 Radio interference

As for HSRs in general, interference with a coherent radar network by either broadcasts or intentional jamming sources needs consideration. As for an individual HSR, spectrum agility, although not impossible, may act against the principle of continuity of observation of the CVoR.

In addition to the aspects of the interferer and signal intelligence (direction and signal-specific interception), which allow a single HSR to recover by signal-specific suppression, in an NHR system the 3-dimensional position of the interfering source can be measured accurately, and more precise suppression achieved.

As with other aspects of HSR, effective performance can be based on optimising information about the interfering signals rather than on efforts to avoid or minimise such signals by frequency agility or beam nulling.

10.4.4 Timing control

Accurate timing of transmissions and control of frequency and phase between NHR stations is a critical part of the design of a coherent radar network.

We have seen that where there is no beam rotation or e-scanning, the requirement for different transmission frequencies may be relaxed, as described in Section 10.2, provided there is accurate time and phase coordination between them. Where a number of surveillance stations operate on the same frequency, with non-colocated transmitters synchronised with a common, accurate timing source, it appears possible to identify the sources of each signal received from a particular target. The identity is based on the sequence, the azimuth, and the relative delays of receptions, and this form of identification avoids the need to separate the transmission frequencies.

Timing control has both long- and short-term constraints. In the long term (over periods of many CPIs), the timing of transmissions and reception needs to remain synchronised within about one-tenth of the time delay resolution so that distance measurement errors remain at a level comparable with the radial accuracy. Long-term control can be disciplined by reference to GPS or timing offered by MSF.

10.4.5 Control within the CPI

Within the coherent processing interval and the pulse repetition interval, timing accuracy needs to be more precise to maintain coherence and avoid noise degradation where intense target, clutter or direct transmitter signals occur. The necessary clock stability within the CPI is determined as a limit on timing jitter during conversion processes in the presence of such high-amplitude signals. Clutter may exhibit

Table 10.2 *Signal levels: clutter, jitter, Direct Tx and phase noise for SRC and ATC*

Signal source	Level: SRC (Eq. input)		Level: ATC (Eq. input)		Mitigation
Noise (with beam and Doppler BW)	−108	dBm	−111	dBm	LNA
Saturation @ Rx element	−10	dBm	−10	dBm	LNA
Saturation with beam	11.1	dBm	18	dBm	Array
Clutter; 2 km; 20 km	60	dBsm	70	dBsm	Safeguard
Clutter; RxEl	−46	dBm	−56	dBm	Safeguard
Direct Tx @ Tsep, RxEl	−29	dBm	−29	dBm	Screen
Equivalent Tx RCS	73	dBsm	93	dBsm	Screen
Minimum target; 2 km, RxEl	−122	dBm			
Minimum target; 5 km, RxEl	−138	dBm			
Minimum target 20 km			−122	dBm	
Minimum target 100 km			−150	dBm	
Dynamic range (sat/noise)	152	dB	162	dB	
Phase noise	−80	dBc/Hz	−80	dBc/Hz	
Phase noise (Clutter)	−95	dBm/bin	−98	−98	20 dB cancellable
Phase noise (Direct)	−82	dBm/bin	−71	dBm	20 dB cancellable
Dynamic range (Sat – PN, clutter, 2 km)	106	dB	116	dB	Degradation 30 dB
Dynamic range (Sat – PN, Direct Tx, 2 km)	93	dB	89	dB	Degradation 20 dB

cross sections in the region of +60 dBsm, and we shall compare the effect of these with the target returns and with direct transmitted leakage.

For reference, typical synthesised timing control systems using oven-controlled crystal oscillators and synthesizer circuits yield phase noise in the region of −70: −90 dBc/Hz, reducing towards higher offset frequencies. More stable phase is expected to be available from new quantum oscillators.

Signal levels, the effects of thermal and phase instability and the effects on the dynamic range are illustrated in Table 10.2.

Receiver saturation can be avoided by siting transmitters further than 2 (for SRC) or 20 km (for ATC) away from the receiver. Noise degradation by the direct-received signal from 1 or 10 km demands reduced system phase instability. For ATC, transmitters and receivers should be sited beyond twice the radio horizon, given approximately by:

$$\mathbf{D_h} = \text{sqrt}(10.\mathbf{K}.\mathbf{h})\ (\text{km}) \tag{10.4}$$

where **K** is the earth radius factor (typically ~1.5) due to atmospheric refraction and **h** is the height of the antennas above the surface in metres.

For transmitter and receiver sited 10 m above smooth earth, the horizon will be at about 18 km so that to meet this condition and avoid noise degradation the two should be at least 40 km apart. This is consistent with the **NHR$_0$** concept and will also have the effect of maintaining angular contrasts within the network for trilateration and vector measurements.

With respect to timing errors, synthesiser clocks based on high-quality crystal oscillators yield phase instability similar to that referred to in Table 10.2. Crystal oscillator frequencies without synthesisers to generate agile clock frequencies vary with time and temperature. Temperature can be controlled accurately, but evolution with time is a less well-controlled aspect. Typical numbers for temperature-controlled crystal oscillators indicate early aging at 1 part in 10^7 per month, falling to 1 part in 10^9 per month, or, for a 10 MHz clock, variations of $+/-10^{-8}$ Hz per second. At the carrier frequency (1–3 GHz), this converts to timing errors of up to 4 fs per CPI, and frequency shifts far within the Doppler bin resolution.

To generate the NHR timing control signals, agile synthesisers are not required. They introduce a level of phase instability that is expected with agile-PRF radar systems but here stability rather than agility is what is needed. Stability is the essence of staring.

10.4.6 Coincident and synchronised transmission and reception

There are significant advantages in operating transmission and reception at coincident sites, in terms of sensitivity and positional accuracy at short range. There is also a benefit in local resilience where control of the Tx and Rx functions is directly connected, and cost savings arise from a single site and installation.

The disadvantage that reception tends to be blocked during pulse transmissions may be circumvented by the use of a multiple-pulse sequence, by the multi-static operation of the network itself or by a solution in which the transmitter is adequately isolated (i.e., distanced and screened) from the receiving array so that the elements are in fact not saturated. In the absence of screening, the direct received signal from a transmit tower 30 m tall, and in a low-gain region of the vertical transmit beam pattern, may be of the order of milliwatts per channel. This is a lot for standard low noise amplifier components, and so requires additional isolation in the region of 20 dB in terms of screening. It suggests a tower solution that allows full eVoR coverage and clear linkage within the network, preferably with provisions for transmitter screening.

A robust NHR wide-area surveillance system can be constructed with a set of monostatic but coherent and synchronised transmit-receive stations. It would offer a non-cooperatively trilaterated air picture, with vector Doppler solutions for each target, with persistent trajectory monitoring within the track reporting system, and with the ability to retain persistent, coherent records for retrospective event tracking and imaging.

Either by a high level of screening or with the inclusion of very high-stability time and phase control, a fully unsaturated, wide-area network of either kind may be constructed in which there are no blind ranges due to transmitter saturation.

Compound NHR includes unpaired transmitters and receivers; $\mathbf{NHR_0}$ is the simplest form.

10.5 Coherent radar networks – conclusions

Radar that relies on coherence and persistence in observing a VoR can exploit the EM Uniqueness Theorem to the full.

When that persistence extends from the time domain to pervade airspace in a coherent network, it appears that the available surveillance information grows substantially to the point that it has sufficient accuracy and dimensionality to resolve signals arising from separate transmitters even when operating at the same frequency.

This then provides the ability to measure target positions precisely by coherent trilateration, and to measure directly the vector of target motion, rather than only the radial component.

It is attractive to conclude that, by using the full and growing power of digital signal and data processing, the practice of making intermittent measurements followed by credible but predictive track filters might be replaced and improved on within the linear (i.e., information-conserving) data process, including successively advancing or extending the CPI by finite intervals, thereby avoiding losses of capture while in the CVoR.

The effective limits are those of information processing, classification and storage. These capabilities will develop as the speed and capacity of computing systems continues to grow.

We have identified cases in which research is needed, both into methods and into future requirements and constraints; for example, the use of staring surveillance radar for imaging, and this possible transition from tracking based on intermittent detections to a classification and visualisation scheme that in effect maintains persistent contact with every object in the CVoR.

This chapter confirms one of the most significant strengths of radar surveillance using coherent networks of sensors, in that different components of such a network can and should operate on the same single frequency. That represents a saving of many megahertz in total operating bandwidth for the system and would enable a substantial and valuable freeing-up or leasing of radio spectrum. ECCM resilience for such systems is also expected to be effective.

Under these conditions an ATC system at the present time might use networks of several primary surveillance radars, say 10, within any area in which interference is significant, each using bandwidth of say 5 MHz. Each is then assigned to one or more different operating frequencies, and the total is at least 50 MHz and will in practice be more. If such networks were supplanted by an overall NHR system, it may operate with an overall bandwidth of just 5MHz. The saving in releasable bandwidth will then be very substantial, and the overall surveillance performance will be enhanced in the ways we have identified.

Chapter 11

Holographic staring radar – to summarise

Early in this book, some history and various concepts and applications of holographic staring radar have been introduced. Its potential for high performance in surveillance has been outlined, along with several accompanying risks. In briefest summary, using the relevant physical laws and underpinned by experience of staring radar effectiveness, numerical models have shown that surveillance information in the form generated by HSR is directly descriptive of targets and their behaviour, in addition to communicating their presence and positions.

Chapter 3 outlined how radar signals, scattered by various objects within a coherent volume of regard, encode and communicate this information, and the basis on which, under the physical laws, it can be analysed and understood remotely in terms of radar targets.

A HSR irradiates a whole CVoR and processes signals arriving in all directions, continually. It can support function-specific processes in parallel, based on coherent, complex information that it receives through EM contact with its targets throughout an extended period.

The propagation and conservation laws under which radars operate remain the same, and by interrogating the whole volume continually and coherently, the electromagnetic uniqueness theorem provides access to whole information, bounded by its resolution and power budget, about the target and its trajectory. Signal data, placed in sufficient computer memory with high speed of access and substantial process capacity, are the unique source not only for detection and tracking but also for analysis and classification of target types.

Chapters 4 and 5 have outlined suitable future applications of staring radar, and short- and long-range configurations used later to estimate performance under different circumstances.

Chapter 6 described the range of signals that can be expected at a staring radar aperture, and methods whereby targets may be captured, with other influences including clutter, noise, phase noise, multipath and interference that may influence the process, can be recognised and then suppressed.

Staring radar is by nature process-intensive, and the necessary capacity will only be manageable provided that it can use extensive parallel computing capacity efficiently. It will then support surveillance capabilities that include classification and discrimination between different target types, their behaviours and, as a result, their significance.

The method of vector histograms is introduced, as an efficient way of allowing the cell discovery process to go beyond elementary single thresholding in selecting signals of interest. It minimises conditional data processing by exploiting the many decisions already made, unconditionally, in the course of digital conversion.

In modelling staring characteristics and challenging conditions for HSR, Chapter 7 has discussed the balance between transmission gain and coherent integration, decoherence during an extended CPI, interference in surface multipath propagation, range and Doppler walk, external interference and the processing burden.

Extended coherent integration should not be conceived as a risk; it is a unique source of information. Perceived decoherence with dynamics, properly understood, translates into information about the dynamics. Surface multipath (an issue for all radars) can be measured and recovered in part by HSR. Range and Doppler walk can be accommodated by added capacity for the relevant process, and HSR has the potential for countering external or intentional interference. All these capabilities depend on extensive data processing, but none threaten to be unmanageable within the cost context of radar. All can be quantified and modelled based on the EUNIT, and all are seen to agree that persistent signal data support unique solutions.

The potential offered by staring radar for extended, continuing coherent target processing, supported by the EUNIT, is confirmed in Chapter 8 for a number of examples linearly modelled on Maxwell/Huygens scattering processes. These include aspects of the CVoR environment, the extension of target analysis based on concatenation of CPIs, the extension of clutter analysis on the same basis for repetitive sources and increasing resilience against target fading.

These examples illustrate that when target observation is persistent, adequate information will be communicated to the radar to provide unique solutions to EM scattering challenges, set by complex shapes and manoeuvring targets, and supported both in theory and here by modelling. Successful tests, and radars in continuing operation, have demonstrated the effective use of staring holographic radar as practised by Thales-Aveillant. These generate data that, when paralleled by modelling, confirm the EUNIT and can yield further and detailed target interpretation.

Chapter 9 shows that in the operation of an HSR, clutter and multipath propagation can confuse target data and reporting if the discovery process assumes (a) a single look at each target and (b) minimum computing provision.

The EUNIT and the reciprocity theorem provide that signals associated with these propagation routes, as with target dynamics, encode unique target information rather than random or inconsistent interference. They make it possible to suppress reports of scattered 'satellite' radar returns and clarify the radar display. Multipath satellite signals can also be explored to reveal more about targets and their motions within the CVoR.

For ATC applications, the numbers found for the computing burden suggest that a broad range of trajectory flexibility and clutter resilience will be affordable for an HSR. Shorter-range HSRs should also have this capability but require research into these effects under yet more complex urban conditions.

Chapter 10 pursues the ability of HSR, in observing a CVoR, to exploit the EM uniqueness theorem to the full. When that persistence extends from the time domain to pervade airspace within a coherent network, it appears that the available surveillance information grows substantially. It offers sufficient accuracy and dimensionality for signals arising from separate transmitters, operating at the same frequency, to be resolved numerically and identified, permitting unprecedented spectrum efficiency.

Such a network provides the ability to measure target positions by trilateration, and to measure directly the vector of target motion, rather than only the radial component, thereby simplifying the task of track filtering.

The apparent limitations arising from information processing requirements affect the level of detail that can be extracted and stored in classifying, reporting and displaying CVoR content. They will develop as the speed and capacity of computing systems continue to grow. The limitations in real radar performance – sensitivity, resolution and accuracy – depend as always on the radar power and aperture, and are not degraded by a staring configuration.

Staring radar offers a direction of travel for radar technology such that the normal surveillance process of detecting, tracking and reporting may be updated to that of cell discovery, target capture, information assembly, target analysis and reporting, with the aim of assessing all aspects of the CVoR. Issues of target dynamics, surface scattering and target complexity and classification should, as predicted from the electromagnetic uniqueness theorem, be accessible for solution by an HSR.

Investment in improvements in radar technology has increased rapidly in recent years. These investments have tended to focus on radar electronics, yielding enhanced scanning, waveform and signal processing agility, with the aim of maximising attention on targets of interest while dividing a limited scan return time between multiple functions. However, in some respects the acceleration in performance of fielded systems is seen as disappointing.

This book has addressed the possibility that a logjam has arisen in radar development not because the radar is insufficiently agile, but because the resulting short dwell times restrict the scanned information to a small fraction of what can be encoded in continually scattered and observed target returns.

There are aspects of staring radar that need further confirmation in the field. For example,

1. the growth of surveillance data in quantity, dimensionality and precision with the extension of coherent analysis
2. real-time resilience of HSR sensitivity with dynamic target trajectories
3. internal consistency of multi-look target observations and their processing
4. effectiveness of active multi-static radar networks in airspace surveillance
5. resilience of staring radar to external interference
6. the significance of target data continuity for tracking and reporting methods.

Many measurements have been made using HSR-type sensors in connection with the detection and tracking of drones, with wind turbine mitigation and also with work on spectrum efficiency. Such measurements may be available to those researching the future potential for surveillance and remote measurement systems.

In this book, we have aimed to establish expectations of performance under a finite set of conditions ranging from normal to challenging, based on applying the EM uniqueness theorem through a series of propagation models. In summary, the results suggest that substantial advances in radar surveillance depend on the recognition that continuity in measurement is more important than agility and speed of detection, and that staring, persistent and coherent radar can enable otherwise inaccessible surveillance capabilities.

We also recognise that this book serves as only an introduction to the topic of Holographic Staring Radar, and the results presented only scratch the surface of the 'art of the possible'. New radar processing innovations in the area of 'cognitive signal processing' also align perfectly with staring modes of sensing and promise advances in signal analysis, target capture and classification. Hardware improvements abound in radar and will lead to higher levels of sensitivity and processing capacity. In combination, these developments can be expected to make holographic radar the mode of choice for more and new applications. Indeed, with the advent of vehicular radar sensing, it may be that staring holography already provides the physical basis for the most-used radar form currently in operation.

Appendix 1

Measurement of dynamics in aerial manoeuvres

Chapter 8, Section 1.3 described aspects of air surveillance that can benefit from the assembly of extended Coherent Processing Intervals (CPIs). The ATC concept provides, for example, for the concatenation of 2-second intervals into 8-second continuous sequences, such that analysis of, for example, 4 096- or 2 048-point pulse sequences can be performed at smaller intervals of, say, 1 024 (1 second) or 512 points (0.5 seconds), offering a finer-grained inspection of the target's dynamic sequence.

This appendix offers examples showing that the progression of acceleration and radial speed can be determined linearly and directly from the complex amplitude sequences received. Targets are explored at radial speeds of 0, +/−10, +/−20 and +/−40 m/s, and radial accelerations of +/−5, +/−10, +/−15, +/−20 and +/−25 m/s^2, with both stepped and sinusoidal dynamic sequences.

In Figure A.1, an aircraft flies with an initial positive radial speed **Vr** = 20 m/s for 2 seconds, then radially decelerates (for example, by a roll manoeuvre) for 2 seconds at −20 m/s^2 towards the radar, then returning with acceleration +20 m/s^2 to its previous direction and speed but displaced. This sequence occupies four CPIs as defined for the ATC configuration. These examples are based on the ATC concept since its PRI and CPI are significantly longer than for SRC, making it more susceptible in principle to the effects of target dynamics. Four 2 048-point sequences are concatenated to form an 8-second sequence and spectra are derived for 7 successive CPIs, each lasting 2 seconds and overlapping by 50%.

A number of flight profiles have been modelled and are listed in Table A.1, including a first set in which the model simulates a near-tangential trajectory in which the radial acceleration steps from zero to a negative to a positive value, followed by a set in which a more lifelike profile is used with acceleration in the form of a single sinusoidal cycle.

The panes shown in Figure A.1 illustrate the manoeuvre in acceleration, speed and distance, the resulting Doppler profiles, a Vector Histogram embodying the Doppler data and a plot illustrating the dynamic analysis derived in each case.

This geometry is also illustrated as Flight profile 'I' in Figures A.2–A.5.

The first set of figures (based on cases A–J in Table A.1) illustrates flight profiles in which the target accelerates first inward at a constant rate then outward at the same rate, in what might represent an evasion manoeuvre to avoid collision.

Table A.1 *Flight profiles for dynamics analysis in concatenated pulse sequences, with successively overlapping coherent processing intervals*

Flight profile	Acceleration profile	Vr0 set (m/s)	Ac set (pk m/s^2)	VrMin (mean measured)	AcIn (mean measured)	AcOut (mean measured)
A	Stepped	–40	–10	–59	10	10
B		–20	–10	–39	10	11
C		0	–10	–19	10	10
D		20	–10	1	11	11
E		40	–10	21	10	8
F		–40	–20	X	X	X
G		–20	–20	–58	21	21
H		0	–20	–38	21	21
I		20	–20	–18	21	21
J		40	–20	2	21	21
K	Sinusoidal	–40	–10	–52	7	7
L		–20	–10	–32	7	7
M		0	–10	–12	7	6
N		20	–10	8	7	7
O		40	–10	28	6	6
P		–40	–20	–64	14	14
Q		–20	–20	–44	14	14
R		0	–20	–24	13	13
S		20	–20	–4	13	13
T		40	–20	16	13	13
U		–10	–5	–16	3	3
V		–10	–10	–22	6	6
W		–10	–15	–28	10	10
X		–10	–20	–34	13	13
Y		–10	–25	–40	17	17
Z		+20	–5	14	1	1
AA		+20	–10	8	6	6
AB		+20	–15	2	9	9
AC		+20	–20	–4	14	14
AD		+20	–25	–10	17	17

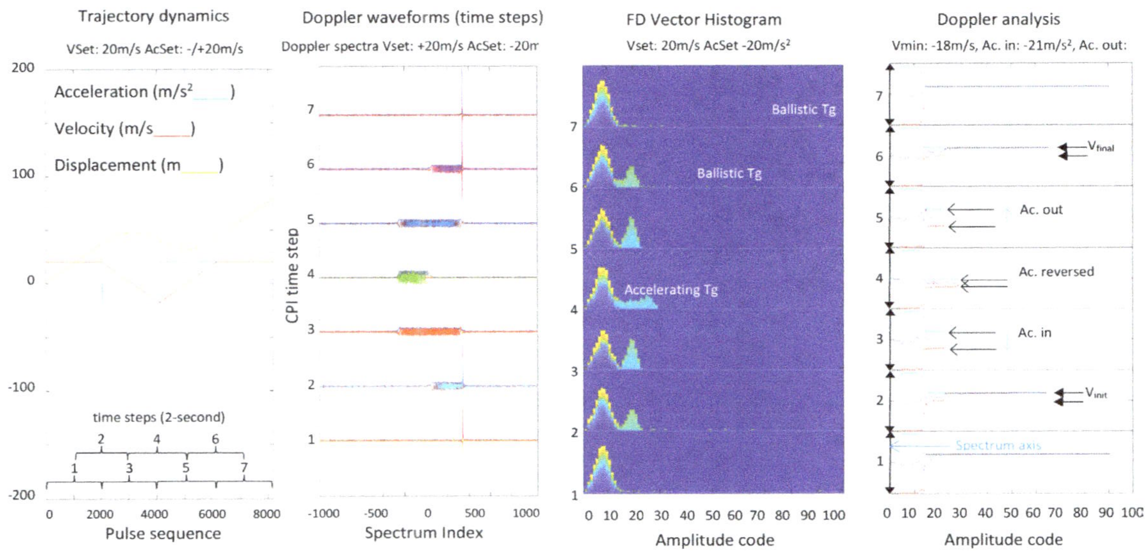

Figure A.1 *Flight profile 'I'; +20 m/s, accelerating and decelerating at −/+20 m/s² steps during the central 4 seconds of the concatenated CPI. Here the SNR at zero acceleration is +40 dB.*

In the second set (Figures A.6– A.9 – Flight profiles K to T), the target moves in a smoother fashion, again in an approach-recede manoeuvre, where the acceleration changes continuously and sinusoidally through negative and positive excursions.

In the third set (Figures A.10-A.13 – Flight profiles U to AD), the target illustrates more finelyspaced accelerations at 10 m/s outward radial speed and 20 m/s inward.

In Figures A.5, A.9 and A.13, analysis of the Doppler range yields estimates of the minimum outward speed at the centre of the manoeuvre and the mean acceleration in its inward and outward segments. These estimates are given at the top of each pane of each of these figures. Note that the sinusoidal segments have the same peak acceleration as the stepped sequences but the mean accelerations are found to be reduced by close to sqrt(2), as should be expected.

Measurements arise from signals with SNR ~30 dB at zero acceleration. No filtering has been applied to recover peak losses due to acceleration.

The results for **Vr**, **AcIn** and **AcOut** indicate that the measurements are independent and linear, subject to minor errors arising from Doppler ambiguities. These can be treated by a suitable non-conditional Doppler Walk process within the Vector Histogram.

Pane 1 sets the form of the manoeuvre in terms of radial acceleration (blue), radial velocity (red) and displacement (orange) in m/s^2, m/s and metres in seven 2-second overlapping time steps.

Pane 2 shows the Doppler spectra within each of the seven time steps. In step 4, the acceleration reverses.

Pane 3 shows the frequency domain vector histograms at each time step. Negative (inward) speeds are blue, low speeds green and positive speeds yellow. There are no threshold conditions.

Pane 4 shows the result of analysis in which the solid line illustrates the mean radial speed corresponding to each amplitude code; red dots are the minimum and blue dots the maximum. At levels above the noise spectrum, the mean indicates the speed **Vr** and the red–blue vertical interval indicates the accelerations **Ac**. Values given above Pane 2 are the set values and those above Pane 4 are the measurements. Mean acceleration is maximum in time steps 3 and 5. Acceleration reverses during time step 4 at **VrMin**. Note that these dynamics cases do not cater for radial speed ambiguities (above 60m/s). Further development is needed in reporting accurate dynamic in those cases.

Newtonian mechanics and persistent observation ensure that targets do not hop across range or Doppler resolution cells. With successive, concatenated CPIs, targets can then be followed by linear analysis through a succession of resolution cells, with increasing detail and precision as time progresses.

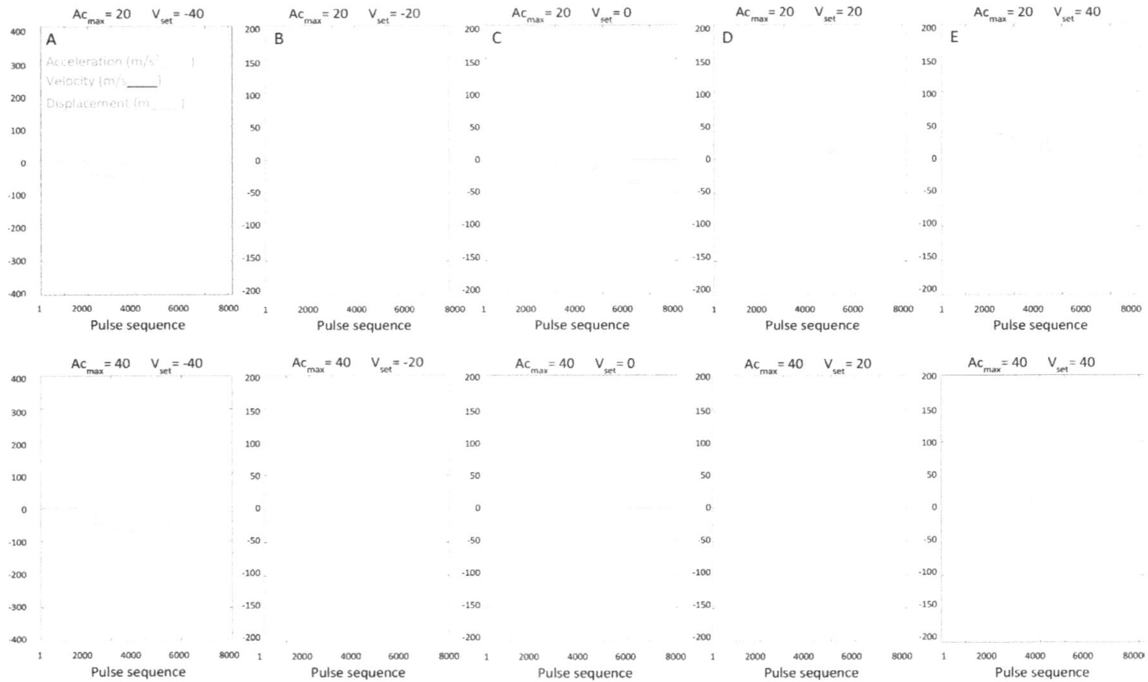

Figure A.2 *Flight profiles A to J showing acceleration (blue), radial speed (red) and relative distance (orange) at stepped inward and outward accelerations AcIn and AcOut of 10 and 20 m/s (~1 and 2 g), and at radial speeds Vr0 of −40, −20, 0, +20 and +40 m/s*

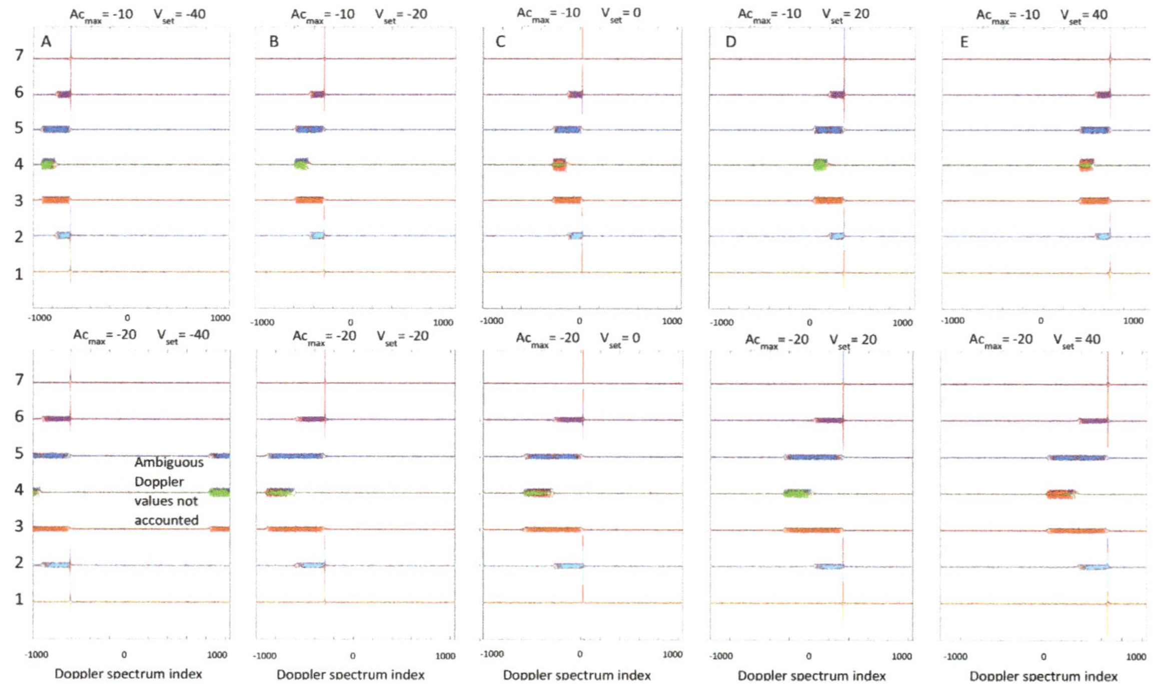

Figure A.3 *Doppler spectra for step-acceleration flight profiles A to J at* $-10\ m/s^2$ *and* $-20\ m/s^2$*. At* $-20\ m/s^2$ *the return is spread over 600 Doppler bins. It is reduced in amplitude by about 25: signal energy is conserved. Doppler ambiguities appear for Vr* $-40\ m/s$*, Ac* $-20\ m/s^2$ *prior to Doppler Walk processing. Flat power spectra result from step-steady acceleration profiles.*

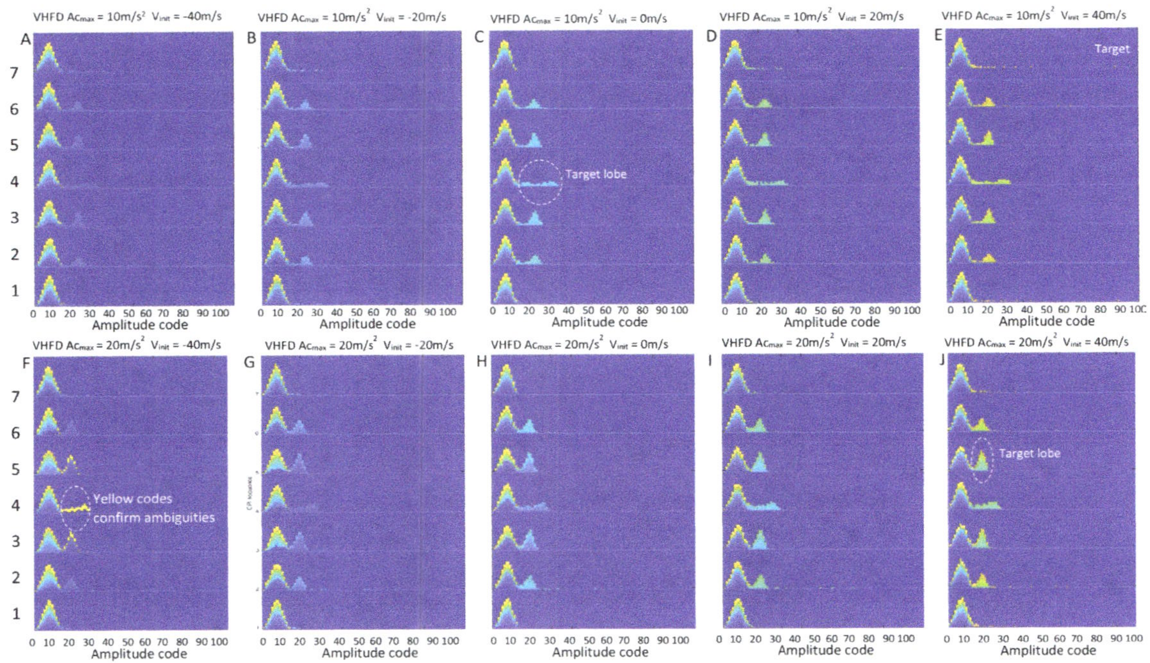

Figure A.4 *VH FDs for flight profiles A to J at −10 and −20 m/s². Colour coding of target signals: inward radial speeds blue, low in/out speeds green, outward speeds yellow. Radial speed derived from mean Doppler indices in the target lobes; acceleration derived from the range of indices. Ambiguities affect the −40 m/s, −20 m/s² VHs before the Doppler Walk process.*

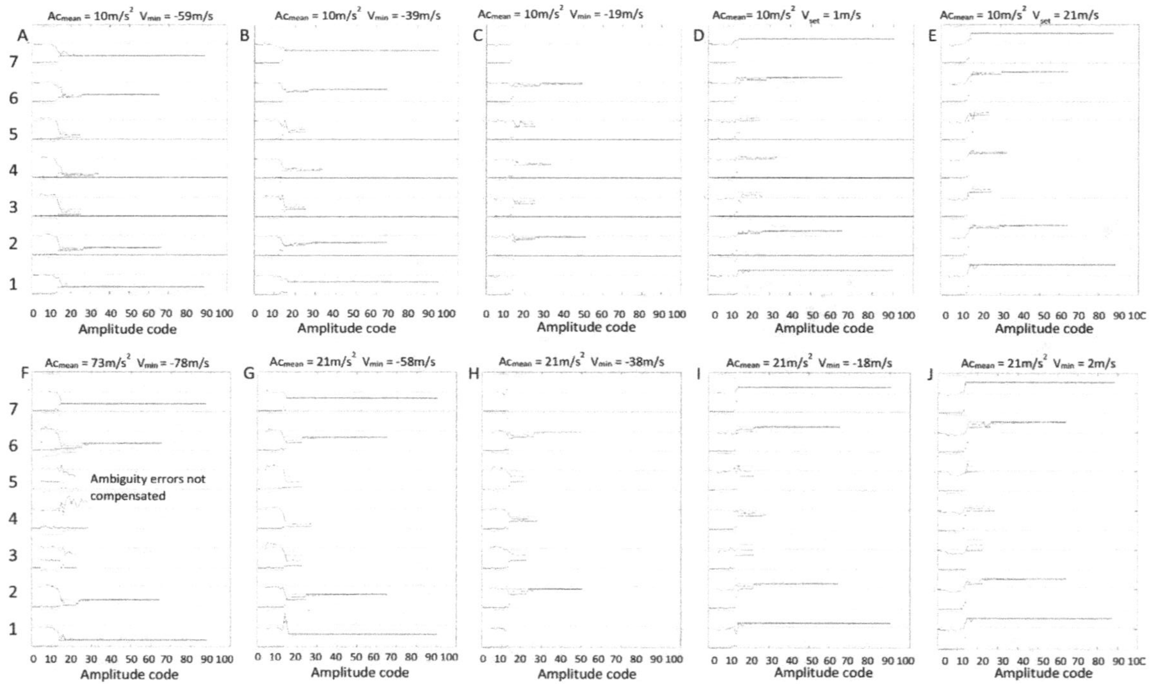

Figure A.5 *Dynamics traces from vector histograms A to J. Blue traces are mean Doppler shifts at each amplitude. Red and blue dots are minimum and maximum Doppler indices at each amplitude code. Minimum manoeuvre speeds and acceleration measurements are given at the upper pane edges. Noise spectrum extends to ~code 16. Acceleration in and out is derived at time steps 3 and 5. Zero Doppler is suppressed.*

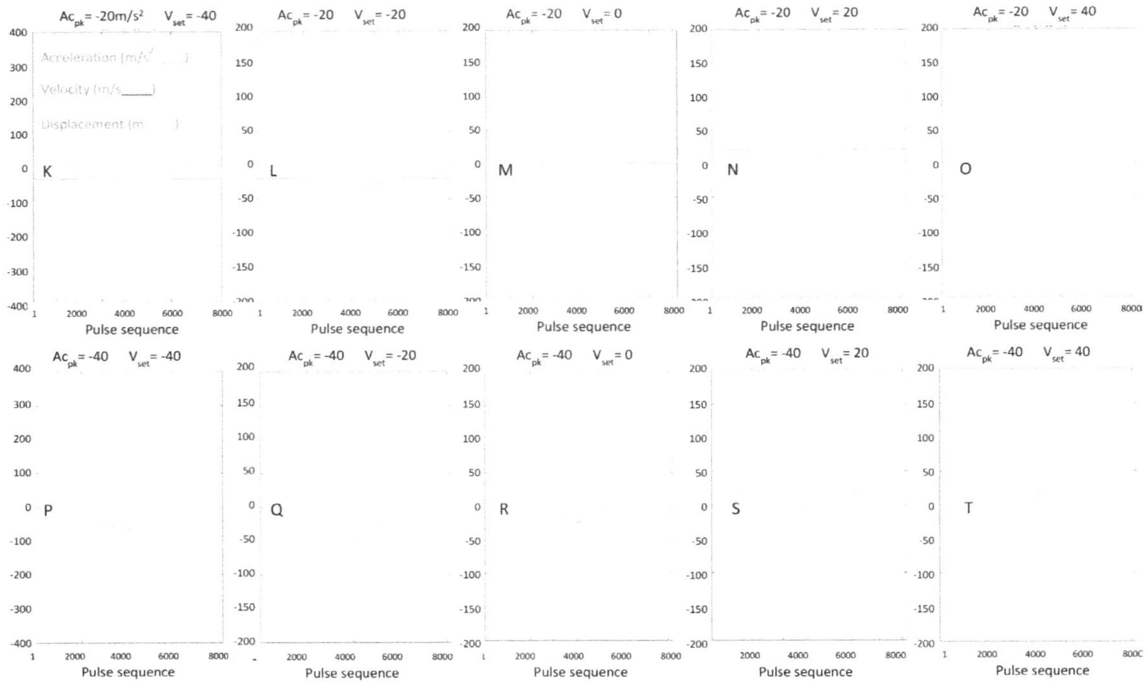

Figure A.6 Flight profiles K to T: Ac0 = −10, −20 m/s²; Vr0 = −40, −20, 0, +20, +40 m/s

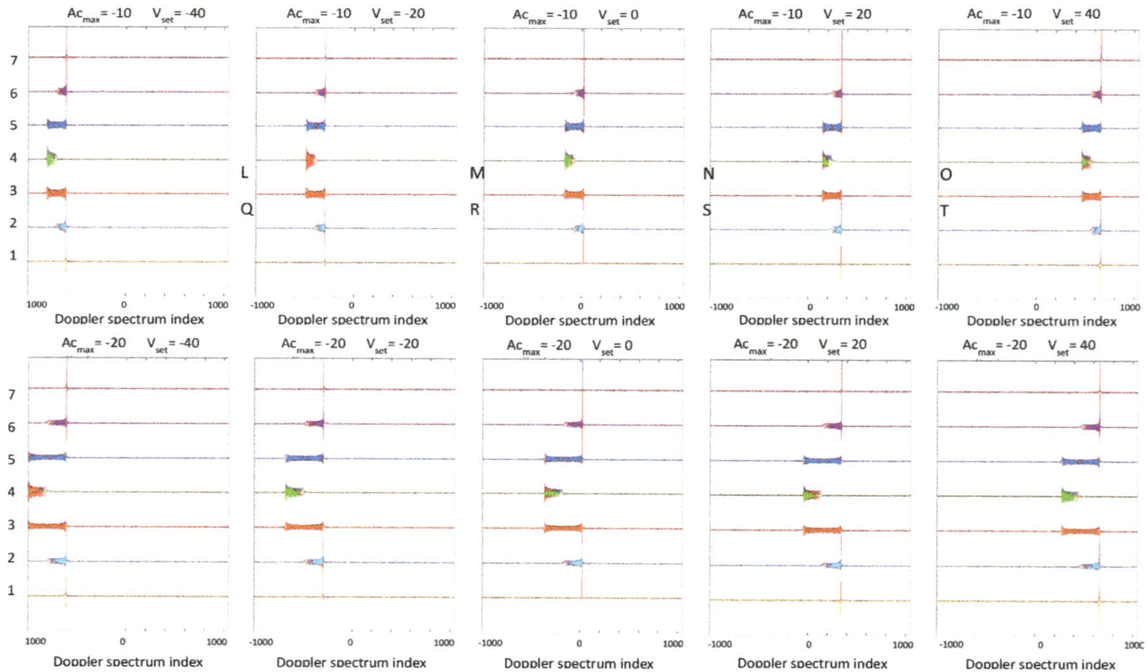

Figure A.7 *Doppler spectra for sinusoid-acceleration trajectories K to T at: Ac0 = −10, −20 m/s²; Vr0 = −40, −20, 0, +20, +40 m/s. At −10 m/s² the return is spread over ~150 Doppler bins; at −20 m/s². it is spread over 300 Doppler bins. It is then reduced in amplitude by about 1/17 and signal energy is conserved. Peaked spectra result from sinusoidal acceleration profiles.*

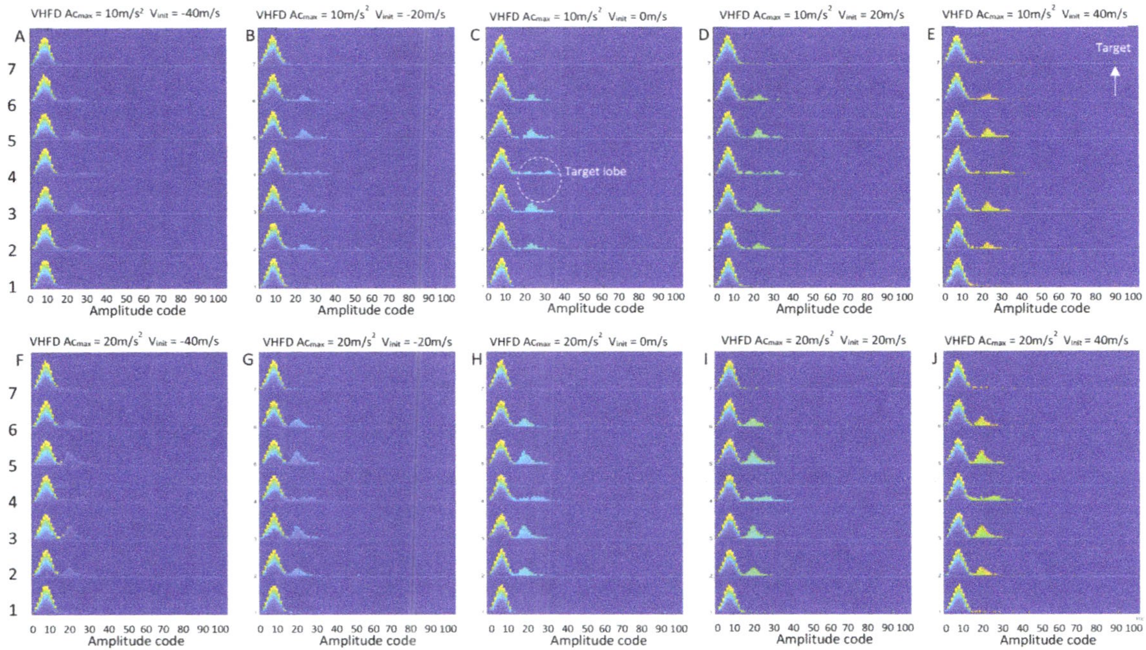

Figure A.8 *VH FDs for flight profiles K to T with accelerations at −10 and −20 m/s*^{2}*, radial speeds −40, −20, 0, +20 m/s and +40 m/s*

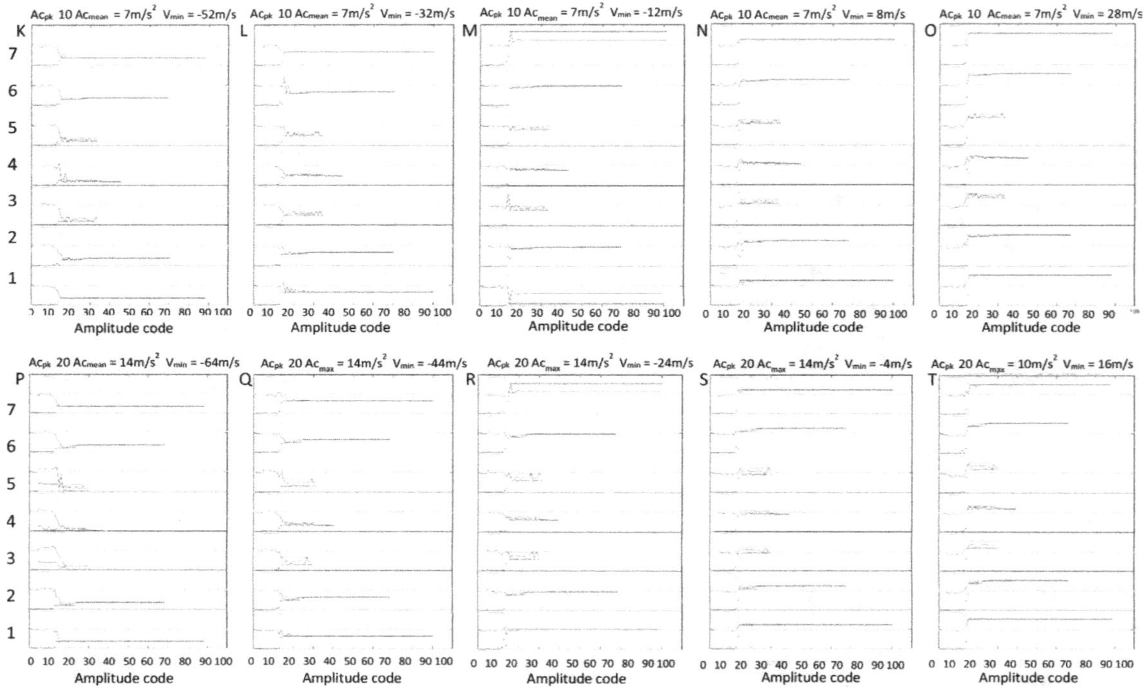

Figure A.9 *Dynamics plots and measurements for sinusoidal manoeuvre flight profiles K–T; mean speeds −40 to +40 m/s and accelerations −10 and −20m/s²*

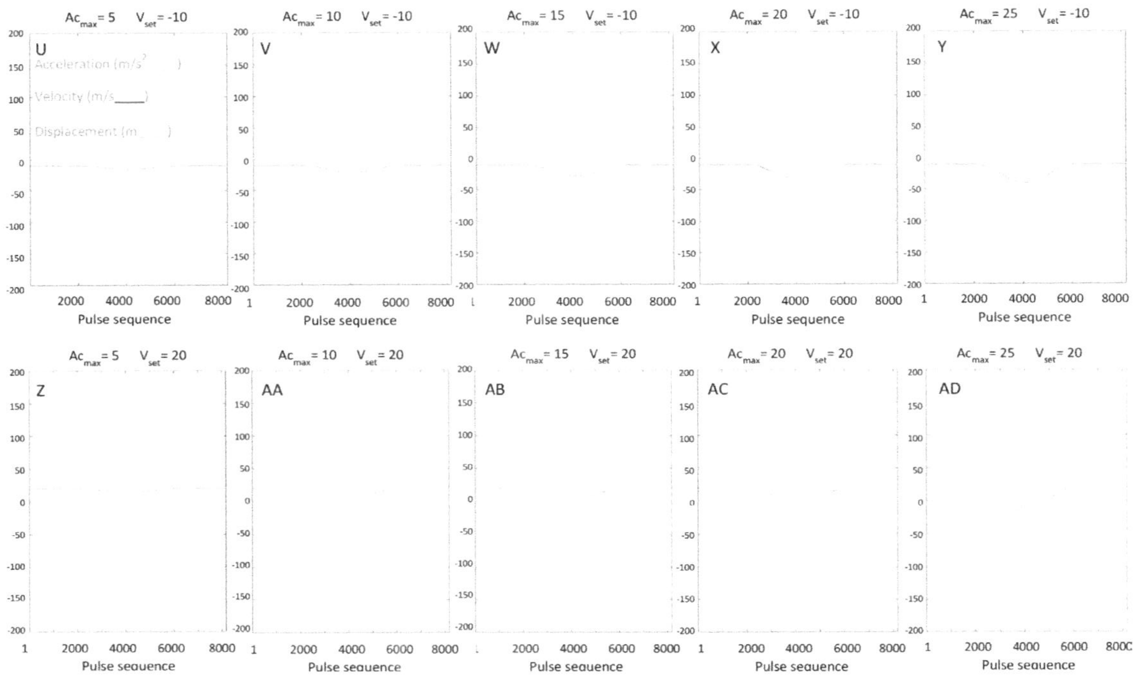

Figure A.10 *Trajectory plots for profiles U to AD: starting speeds of −10 and +20 m/s and a sequence of 5 acceleration rates −5, −10 ^{2}, −15, −20 and −25 m/s^2*

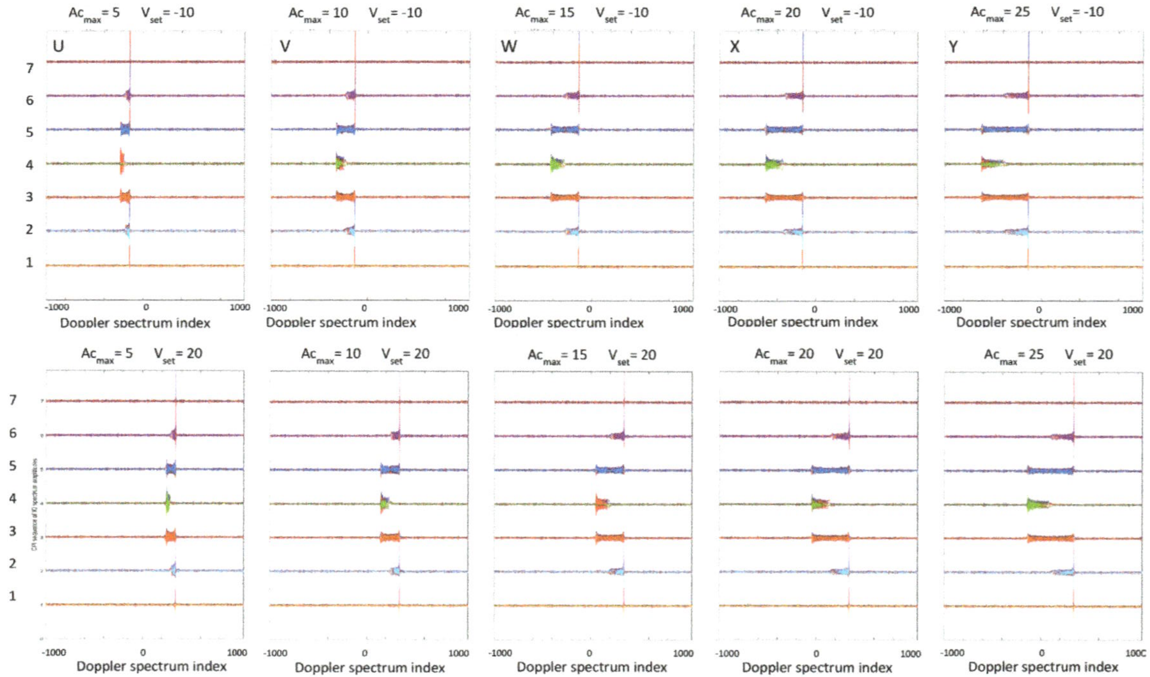

Figure A.11 *Doppler profiles for flight profiles U to AD as Figure A.10; speeds of −10 and +20 m/s and peak accelerations −5, −10, −15, −20 and −25 m/s²*

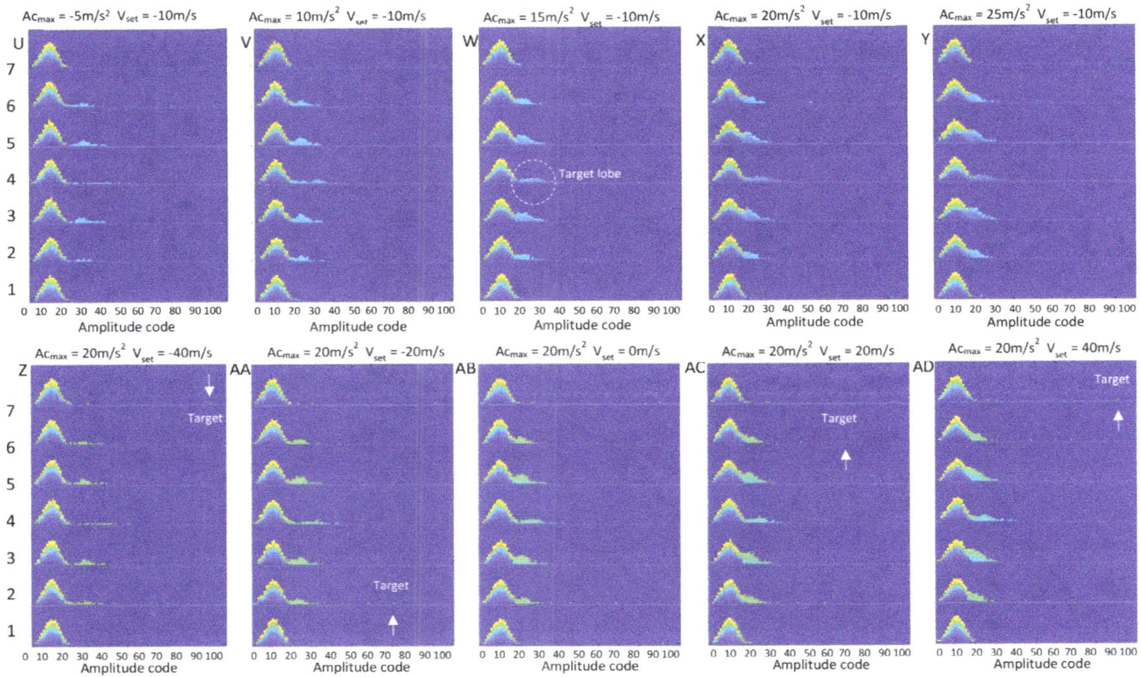

Figure A.12 *VH FDs for sine manoeuvres U to AD as Figure A.10: speeds of −10 and +20 m/s, with peak accelerations −5, −10, −15, −20 and −25 m/s²*

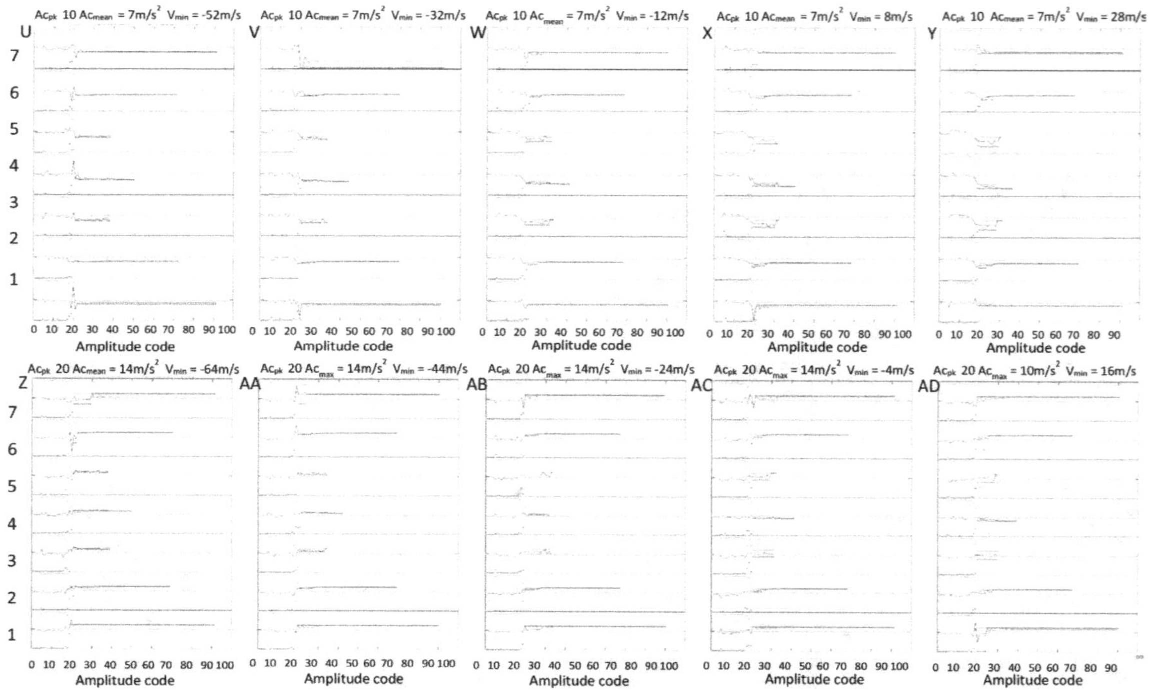

Figure A.13 *Minimum speed and mean acceleration measurements for sine manoeuvres, for each profile U to AD. Radial speeds −10 and +20 m/s, peak accelerations −5, −10, −15, −20 and −25 m/s*^{2}*. Mean accelerations are near peak acceleration divided by sqrt(2); results are shown in Table A.1.*

Index

www.ingramcontent.com/pod-product-compliance
Lightning Source LLC
LaVergne TN
LVHW011843200926
848484LV00002B/393